P. Lanzer A.P. Yoganathan (Eds.)

Vascular Imaging by Color Doppler and Magnetic Resonance

With 222 Figures, Some in Color and 7 Tables

Springer-Verlag
Berlin Heidelberg New York
London Paris Tokyo
Hong Kong Barcelona
Budapest

Dr. med. Peter Lanzer
Abteilung für Radiologie
Universitätsklinikum Rudolf Virchow
Standort Charlottenburg
Spandauer Damm 130
W-1000 Berlin 19, FRG

Prof. Dr. Ajit P. Yoganathan
Cardiovascular Fluid Dynamics Lab.,
School of Chemical Engineering
Georgia Institute of Technology
Atlanta, GA 30322-0100, USA

ISBN-13: 978-3-642-76196-6 e-ISBN-13: 978-3-642-76194-2
DOI: 10.1007/ 978-3-642-76194-2

Library of Congress Cataloging-in-Publication Data.
Vascular imaging by color doppler and magnetic resonance / P. Lanzer, A.P. Yoganathan, (eds). p. cm. ISBN-13: 978-3-642-76196-6
1. Blood-vessels–Ultrasonic imaging. 2. Blood-vessels–Magnetic resonance imaging. 3. Blood flow–Measurement. I. Lanzer, P. (Peter), 1950– . II. Yoganathan, A. P. (Ajit P.), 1951– . [DNLM: 1. Magnetic Resonance Imaging–methods. 2. Vascular Diseases–diagnosis. 3. Echocardiography, Doppler–methods. WG500 V3318] RC691.6.N65V35 1991
616.1'307543–dc20 DNLM/DLC for Library of Congress 91-5153 CIP

27/3140/543210 – Printed on acid-free paper

Foreword

Vascular diagnostics traditionally rely on x-ray angiography. This approach remains even today essential in the clinical work-up of patients with vascular pathology. Recently, however, newer imaging modalities have been introduced to assess vascular disease. Among these the color Doppler flow and magnetic resonance imaging appear the most promising. Due to their noninvasive character both methods are ideally suited for screening as well as serial follow-up of vascular patients.

The emergence of color Doppler flow and magnetic resonance vascular imaging coincides with new concepts in vascular medicine. Today, vascular prevention and percutaneous interventions are becoming the leading components in modern vascular care. It is in this new and exciting environment of novel vascular concepts where the demand for reliable noninvasive vascular imaging is becoming a high priority. Color Doppler flow and magnetic resonance vascular imaging are well on the way to satisfy this demand.

This textbook provides the long awaited information on vascular imaging by color Doppler flow and magnetic resonance. The text covers the essentials of vascular anatomy, physiology and noninvasive imaging technology before providing a state-of-the-art review of their current clinical applications. All chapters are written by competent scientists and clinicians in a clear, conscise and yet thorough and exhaustive manner. The coherent and didactic composition of the textbook allows the reader an easy access to the elementary as well as the advanced principals and clinical applications of the modern noninvasive vascular imaging.

The textbook signals an exciting era in noninvasive vascular diagnostics and I hope that it will become a valuable companion to all of us concerned with better vascular care.

Professor Dr. med. Roland Felix
Berlin

Introduction

Doppler ultrasound, in particular color Doppler flow imaging (CDFI), and magnetic resonance vascular imaging (MRVI) have opened powerful windows into viewing blood flow in the cardiovascular system. These non-invasive techniques provide not only an excellent definition of the vascular morphology but also a fundamental diagnostic hemodynamic information. CDFI and MRVI are already today reducing the need for invasive techniques, such as noncoronary ateriography. Furthermore, they provide for the first time non-invasive tools for serial monitoring of vascular disease and related therapeutic interventions.

However, despite of the impressive clinical results both CDFI and MRVI are still very much in their infancy with a full clinical potential not yet realized. Both modalities suffer at present from the inability to deal accurately with complex and turbulent blood flow fields. Improved hardware and better pre- and postprocessing techniques of the Doppler and MRVI signals will be required to allow a consistent visualization and quantitation of regional hemodynamics. With reaching maturity during the next decade, the clinical applications of both imaging modalities will grow and the emphasis will shift from qualitative data analysis to quantitative clinical decision making.

This textbook, the first of its kind, has been written for physicians and other professionals with primary interest in vascular disease and its diagnostics. To them it provides a comprehensive and yet easily accessible information on color Doppler flow and magnetic resonance imaging of the vascular system. The text is divided into five parts and thirteen chapters providing a quick reference to the specialists as well as a systematic lecture to the beginner. Part I reviews the basics of vascular anatomy and blood flow physiology applicable to non-invasive vascular imaging. Parts II and III are devoted to the principles of color Doppler flow and magnetic resonance imaging technology, respectively. Since the primary objective of both non-invasive methods is to image and quantitate blood flow in the vascular system, reading, rereading and understanding these basic chapters is essential if the reader is to utilize the full potential of CDFI and MRVI either in the clinical and/or research setting. Parts IV and V review the current state-of-the-art clinical applications of CDFI and MRVI methods.

We hope that the textbook will prove a helpful guide to all vascular professionals establishing a common ground and a future tradition to the modern noninvasive imaging vascular diagnostics. We wish to thank to the staff of Springer Verlag in Heidelberg, in particular to Dr. Wieczorek and Ms. Wilbertz, for making this project possible.

P. LANZER
Berlin

A. YOGANATHAN
Atlanta

Contents

List of Contributors

BOESIGER, PETER
Institut für Biomedizinische Technik und Medizinische Informatik, MR-Zentrum, Universitätsspital, Rämistraße 100, CH-8091 Zürich

BOHNING, DARYL E.
Philips Medical Systems North America, 710 Bridgeport Ave., Shelton, CT 06484, USA

CAPE, EDWARD G.
Cardiovascular Fluid Mechanics Laboratory, School of Chemical Engineering, Georgia Institute of Technology, Atlanta, GA 30332-0100, USA

CASTER, JOHN
Department of Surgery, Division of Vascular Surgery, Oregon Health Sciences University and the Portland Veterans Affairs Medical Center, Portland, OR 97201, USA

CROOKS, LAWRENCE E.
Radiologic Imaging Laboratory, University of California, 400 Grandview Drive, South San Francisco, CA 94080, USA

CUMMINGS, C.
Department of Surgery, Division of Vascular Surgery, Oregon Health Sciences University and the Portland Veterans Affairs Medical Center, Portland, OR 97201, USA

DUMOULIN, C. L.
General Electric Corporate Research and Development, Magnetic Resonance Unit, P.O. Box 8, Schenectady, NY 12301, USA

FREUND, JOHN G.
Acuson Corporation, Mountain View, CA 94039, USA

HAAG, K.
Gastroenterologische Abteilung, Medizinische Klinik, Klinikum der Albert-Ludwigs-Universität, Hugstetter Straße 55, 7800 Freiburg i. Br., FRG

HENNERICI, MICHAEL
Neurologische Klinik der Universität Heidelberg, Klinikum Mannheim, Theodor-Kutzer-Ufer, 6800 Mannheim, FRG

Hjortdal, Jesper Ø.
Department of Cardiovascular Surgery, Aarhus University Hospital, Skejby Sygehus, 8200 Aarhus N, Denmark and Institute of Experimental Clinic Research, Aarhus University, 8000 Aarhus C, Denmark

Hylton, Nola M.
Radiologic Imaging Laboratory, University of California, 400 Grandview Drive, South San Francisco, CA 94080, USA

Lanzer, Peter
Abteilung für Radiologie, Universitätsklinikum Rudolf Virchow, Standort Charlottenburg, Spandauer Damm 130, W-1000 Berlin 19, FRG

Laub, Gerhard
Siemens AG, Henkestraße 127, 8520 Erlangen, FRG

Lefebvre, Xavier P.
Cardiovascular Fluid Mechanics Laboratory, School of Chemical Engineering, Georgia Institute of Technology, Atlanta, GA 30332-0100, USA

Maier, Stephan E.
Institut für Biomedizinische Technik und Medizinische Informatik, MR-Zentrum, Universitätsspital, Rämistraße 100, CH-8091 Zürich

Maslak, Samuel H.
Acuson Corporation, Mountain View, CA 94039, USA

Moneta, Gregory L.
Department of Surgery, Division of Vascular Surgery, Oregon Health Sciences University and the Portland Veterans Affairs Medical Center, Portland, OR 97201, USA

Pedersen, Erik M.
Department of Cardiovascular Surgery, Aarhus University Hospital, Skejby Sygehus, 8200 Aarhus N, Denmark and Institute of Experimental Clinic Research, Aarhus University, 8000 Aarhus C, Denmark

Pinheiro, Luiz
2932-A Columbiana Ct, Birmingham, AL 35216, U.S.A.

Porter, John M.
Department of Surgery, Division of Vascular Surgery, Oregon Health Sciences University and the Portland Veterans Affairs Medical Center, Portland, OR 97201, USA

Sousa, S. P.
General Electric Corporate Research and Development, Magnetic Resonance Unit, P.O. Box 8, Schenectady, NY 12301, USA

Steinke, Wolfgang
Neurological Institute New York, Columbia Presbyterian Hospital, 710 West 168th Street, New York, NY 10032, USA

Sung, Hsing-Wen
Cardiovascular Fluid Mechanics Laboratory, School of Chemical Engineering, Georgia Institute of Technology, Atlanta, GA 30332-0100, USA

Yoganathan, Ajit P.
Cardiovascular Fluid Mechanics Laboratory, School of Chemical Engineering, Georgia Institute of Technology, Atlanta, GA 30332-0100, USA

Part I

Chapter 1

Vascular Anatomy

L. PINHEIRO

No matter what imaging method is utilized to assess the vascular bed, the first requirement for a correct interpretation of the results is a thorough knowledge of vascular anatomy. Although the different territories have their own peculiarities and anatomic variants, these are usually functional. Generally, anatomic arrangements such as patterns of ramification and termination in arteries or patterns of confluence in veins remain preserved.

The functional impairment resulting from an arterial stenosis or obstruction essentially dependens on three factors: the site and severity of the lesion, the metabolic demands of the tissue and the magnitude of the collateral flow. The collateral arteries are preexisting pathways, originating from the medium and large-sized arteries. Following an arterial occlusion, they enlarge, increasing the amount of blood supply and therefore acting as natural bypassing conduits. In the presence of an acute occlusion, the resistance offered by these vessels will be the most important factor in determining the survival of ischemic tissues.

The purpose of this chapter is to provide the vascular physician with a concise description of clinical vascular anatomy as a basis for the correct interpretation of the findings obtained with the ultrasonic and magnetic resonance imaging. For further study of vascular anatomy, the reader is referred to standard textbooks on the subject (1–3).

Cerebrovascular Anatomy and Collateral Pathways

Blood supply for the brain is primarily provided by four major vessels: the paired internal carotid and vertebral arteries. They originate from the arterial trunks which arise from the aortic arch in the superior mediastinum, the innominate artery, the left common carotid artery, and the left subclavian artery.

Although most of the collateral vessels providing blood supply to the brain in the presence of impaired cerebral circulation are derived from the carotid and vertebral systems, the subclavian arteries may become involved, and the description of their anatomy will be included in this section.

Innominate Artery

The innominate artery is the first and largest branch of the aortic arch. It arises anterior to the left common carotid artery, at the level of the right second costal cartilage. It is usually 4–5 cm in length. After its origin, the innominate artery passes upward, posteriorly, and to the right. The division into the right common carotid artery and right subclavian artery typically occurs at the level of the upper border of the right sternoclavicular joint. In some instances, the vessel has a higher division due to elongation which is mainly age-related. In some of these cases, the artery is easily palpable at the right base of the neck. The initial segment of the innominate artery is crossed by the left innominate vein. The trachea is located behind and to the left of the trunk. The innominate artery typically does not have any branches. However, the *thyroidima artery* arises from it in 10% of the cases.

The most frequent variants are

1. a common origin with the left common carotid artery,
2. the left common carotid artery originating from the innominate artery, and
3. the presence of bilateral innominate arteries.

Other variants (such as a common innominate artery from which both right and left carotid and subclavian arteries arise) are extremely rare, representing altogether less than 0.6% of cases [1].

Common Carotid Arteries

The right common carotid artery arises from the innominate artery. The left common carotid arises directly from the aortic arch as its second branch (Fig. 1). The right common carotid artery typically originates at the level of the right sternoclavicular joint. The left common carotid artery arises between the innominate artery and the left subclavian artery, at the level of the sixth thoracic vertebra. This location can vary depending on the configuration of the thoracic aorta. The trachea lies medial to the origin of the left common carotid artery.

After the sternoclavicular joint, both arteries run in parallel on each side of the neck, medial to the internal jugular vein, with an upward course within the carotid sheath. The sheath also encloses the vagus nerve, which lies between the artery and the vein.

The common carotid arteries typically divide into their two terminal branches, the external and internal carotid arteries, at the level of the fourth to fifth cervical vertebra. Bifurcation at a lower level is more frequent than at a higher level, especially on the left side. Division at a higher level is more frequent in short-necked than in long-necked individuals, and usually there is no significant difference between either side. The portion immediately prior to the bifurcation, the *carotid bulb,* is frequently dilated with some extension of this dilatation into the internal carotid artery. This tends to be more pronounced

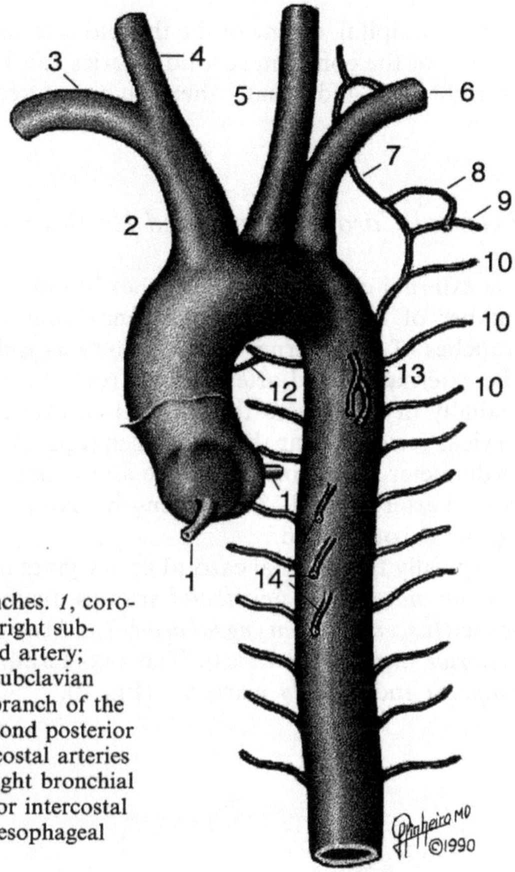

Fig. 1. Thoracic aorta and major branches. *1*, coronary arteries; *2*, innominate artery; *3*, right subclavian artery; *4*, right common carotid artery; *5*, left common carotid artery, *6*, left subclavian artery; *7*, supreme intercostal artery (branch of the costocervical trunk); *8*, *9*, first and second posterior intercostal arteries; *10*, posterior intercostal arteries (3rd–11th); *11*, subcostal artery; *12*, right bronchial artery (branch of the 3rd right posterior intercostal artery); *13*, left bronchial arteries; *14*, esophageal arteries

with advanced age. Also, a characteristic division of the common carotid artery with a wide angle between both internal and external carotid arteries is frequently seen in the elderly. This contrasts with younger individuals, whose arteries divide at an acute angle with an almost parallel course after that. However, similar features can be related to body habitus. An acute angle is more often found in asthenic subjects, while a large angle of division is more commonly seen in pyknic individuals.

The average diameter of the common carotid artery is 0.7 cm, although a difference is commonly seen between the two sides. The length varies considerably with regard to individual morphologic characteristics. The left-sided carotid artery is usually about 3 cm longer than the right-sided one, because of its normal origin at the aortic arch.

Anatomic variations are infrequent and usually have no clinical significance. The most commonly observed are early division of the artery, origin of the left common carotid artery in the brachiocephalic trunk, and the presence of supernumerary arteries arising from the common carotid artery, such as the lingual, occipital, or one of the thyroid arteries. Occasionally, tortuosities and kinking of the common carotid arteries can be observed, more frequently on the right side, and usually they are associated with an elongated innominate artery.

External Carotid Artery and Its Branches

The external carotid artery supplies blood to the neck and face. However, a number of clinically important anastomotic connections with intracranial branches of the internal carotid artery as well as the subclavian artery exist. The external carotid artery arises from the carotid bifurcation frontally and medially in relation to the internal carotid artery at the height of the third cervical vertebra near the anterior margin of the sternocleidomastoid muscle. Its diameter is extremely variable and it can rarely be larger than the internal carotid artery. It is 7–8 cm in length, giving off terminal branches after crossing the parotid gland.

Typically the external carotid artery gives off three anterior branches (*superior thyroid, lingual,* and *facial* arteries, from distal to proximal), one medial branch (*ascending pharyngeal artery*) and two posterior branches (*occipital* and *posterior auricular* arteries). The two terminal branches are the *superficial temporal* and *maxillary* arteries (Fig. 2).

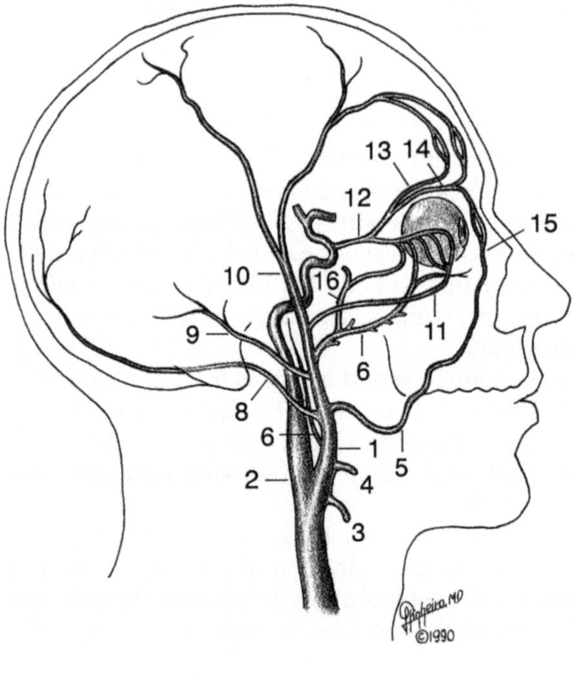

Fig. 2. Schematic representation of the external carotid artery and branches. Anastomotic connections with the ophthalmic artery are also shown. *1*, external carotid artery; *2*, internal carotid artery; *3*, superior thyroid artery; *4*, lingual artery; *5*, facial artery; *6*, maxillary artery; *7*, ascending pharyngeal artery; *8*, occipital artery; *9*, posterior auricular artery; *10*, superficial temporal artery; *11*, transverse facial artery; *12*, ophthalmic artery; *13*, supraorbital artery; *14*, supratrochlear artery; *15*, angular artery; *16*, middle meningeal artery

The anatomic variants can be related either to the course of the external carotid artery itself, which can run lateral and/or posterior to the internal carotid artery, or to the origin of its branches, which can arise separately or can be grouped in common trunks such as linguofacial or thyrolinguofacial trunks.

The clinically important collateral pathways are formed by the branches communicating with the ophthalmic artery and the vertebral arteries.

The frontal and parietal branches of the superficial temporal artery communicate with the supraorbital arteries. The descending branch of the occipital artery communicates with the vertebral artery. The angular artery, a terminal branch of the facial artery, anastomoses with the ophthalmic artery. However, the most important anastomosis with the external carotid artery in the presence of internal carotid artery stenosis or occlusion is usually provided by the branches of the maxillary artery. Less significant are the extracranial anastomoses between the anterior thyroid artery and the inferior thyroid artery, which arise from the thyrocervical trunk, a branch of the subclavian artery.

The noninvasive imaging methods presently available are capable of visualizing only the proximal segments of the external carotid arterial branches. Smaller branches and collaterals are inconsistently visualized.

Internal Carotid Artery and Its Branches

The internal carotid artery is typically, the larger branch of the common carotid artery. After its origin, the vessel ascends in the groove between the pharynx and the deep muscles of the neck. Normally, there are no branches arising from the internal carotid artery in its cervical course. However, anomalous branches replacing the branches of the external carotid artery have been described [2]. Other anatomic variations include aplasia or hypoplasia of one or both internal carotid arteries, independent origin from the aorta or brachiocephalic trunk, and deep origin from the common carotid artery.

The artery can be tortuous, often at the level of the second or third cervical vertebra. Commonly, this tortuosity is S- or C-shaped. In some cases, there is a sharp angulation in one or more segments with or without hemodynamically significant stenosis. Such angulation, referred to as *kinking,* occurs at a distance of 4 cm or more from the bifurcation and is thought to be acquired, mainly due to hypertension and arteriosclerotic degeneration in the elderly. Another kind of tortuosity is represented by a complete curve or loop of the internal carotid artery (sometimes more than one), usually not accompanied by significant blood flow disturbances. This form of tortuosity, known as *coiling,* is less frequently encountered than kinking, and is believed to be congenital.

The segment of the internal carotid artery between its entry into the carotid canal of the petrous portion of the temporal bone, and its intracerebral course is called the *carotid siphon.* An S- or U-shaped course is the rule at this level (Fig. 3). The first branches are the *caroticotympanic branches,* originating

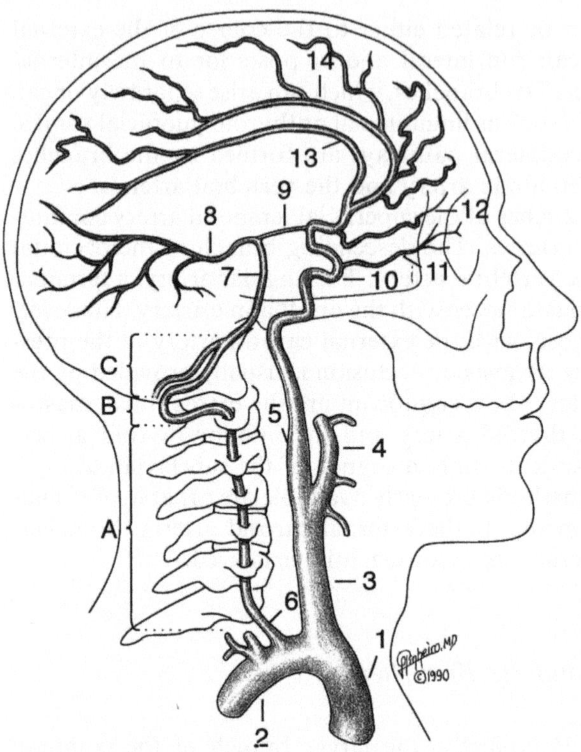

Fig. 3. Schematic representation of the major branches of both the internal carotid and vertebral arteries. *1*, innominate artery; *2*, subclavian artery; *3*, common carotid artery; *4*, External carotid artery; *5*, internal carotid artery; *6*, vertebral artery (*A*, Cervical segment; *B*, Suboccipital segment; *C*, intracranial segment); *7*, basilar artery; *8*, posterior cerebral artery; *9*, posterior communicating artery; *10*, carotid siphon; *11*, ophthalmic artery; *12*, frontoorbital artery; *13*, pericallosal artery; *14*, callosomarginal artery

within the petrous bone. The *meningohypophyseal branches* arise in the cavernous sinus region. Immediately distal to the cavernous sinus, the internal carotid artery gives off the *ophthalmic artery* from the medial side of the anterior convexity of the carotid siphon. Less frequently it may arise together with the middle meningeal artery. The ophthalmic artery lies on the medial side of the anterior clinoid process and enters the orbital cavity inferolateral to the optic neve [3].

Immediately before the bifurcation into the two terminal branches, the anterior and middle cerebral arteries, the internal carotid artery gives rise to the *posterior communicating artery*. The typical origin and course are present in 50% of cases. The vessel connects the internal carotid artery with the posterior cerebral artery, a branch of the basilar artery. There are many normal anatomic variants. Among the most frequent are hypoplasia of the posterior communicating artery on one side, its unilateral or bilateral absence (replaced in the first case by the anterior choroidal artery), and hypoplasia on one side with associated hypoplasia of the posterior cerebral artery on the other side. According to these variations, the circle of Willis (described below) may be divided into variants with adequate connection between the internal carotid and basilar arteries and those with inadequate connection.

The *anterior cerebral artery* is a terminal branch of the internal carotid artery which originates at the lateral side of the optic chiasma. It passes anteriorly and medially above the optic nerve toward the beginning of the longitudinal cerebral fissure. It is connected to its contralateral counterpart by the anterior communicating artery. After this point, both arteries run parallel in the longitudinal cerebral fissura, curve round the genu of the corpus callosum, and turn posteriorly where they end by anastomosing with the posterior cerebral arteries. Here, variations are also common, with hypoplasia and aplasia of the vessel occurring more frequently. The duplication of the vessel is very rare, and in a few cases both vessels may unite to form a common trunk, the *anterior cerebral azygos artery*. The branches of the anterior cerebral artery are: the *anterior communicating artery,* which connects both anterior cerebral arteries, forming the anterior anastomosis of the *circle of Willis;* the *pericallosal artery,* which is the continuation of the anterior cerebral artery itself; the *orbital branches;* the *frontal branches* (*frontopolar* and *callosomarginal* arteries) and the terminal *anterior* and *posterior parietal* branches (Fig. 3).

The *middle cerebral artery* (Fig. 4) is the second terminal and typically the largest branch of the internal carotid artery. It has a lateral course at the base of the brain, curving upward and backward along the cerebral lateral sulcus, where it ramifies. Its diameter and course are usually very regular, but duplication of the artery has been described [2].

The *anterior choroidal artery* arises from the posterior aspect of the internal carotid artery immediately above the origin of the anterior communicating artery. Occasionally, it originates directly from the middle cerebral, the posterior communicating, or the posterior cerebral artery. The vessel passes backward along the optic tract and supplies the choroid plexus and certain groups of basal ganglia. Usually a slender vessel, it becomes larger when the posterior communicating artery is absent. Anastomoses with the posterior choroidal, posterior cerebral, or middle cerebral artery are sometimes present.

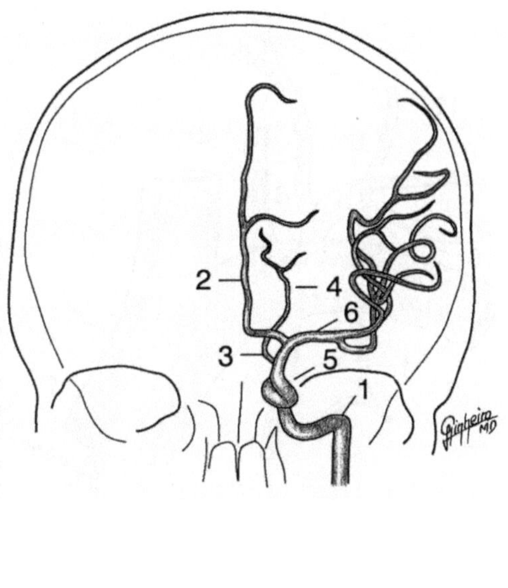

Fig. 4. Branches of the internal carotid artery (frontal view). The posterior cerebral artery (branch of the vertebral artery) is also represented. *1*, internal carotid artery; *2*, anterior cerebral artery; *3*, posterior communication artery; *4*, posterior cerebral artery; *5*, carotid siphon; *6*, middle cerebral artery

Vascular ultrasound visualize consistently the proximal extracranial segments of the internal carotid artery. Due to the high acoustic impedance of the skull, at present only Doppler ultrasound can be used to obtain flow velocities. Magnetic resonance imaging shows considerable promise to visualize the extra- and intracerebral vessels consistently.

The Subclavian Artery and Its Branches

The right and left subclavian artery have different origins (Fig. 1). The right subclavian artery originates from the innominate artery behind the upper level of the right sternoclavicular articulation at the height of the second or third thoracic vertebra, and passes upward and laterally to the medial portion of the scalenus anterior muscle. The left subclavian artery arises from the aortic arch posterior to the left common carotid artery at the level of the fourth thoracic vertebra, ascending to the base of the neck. At this level, both arteries arch over the cervical pleura and continue their inferolateral course until they reach the outer border of the first rib, to continue as the axillary arteries. The diameter of the vessel ranges from 0.6 to 1.1 cm, the left artery usually being larger. In the presence of a cervical rib, the subclavian artery always runs above the uppermost rib.

Anatomic variations can be found, such as the separate origin of the right subclavian artery from the aorta, occuring in any position, but usually originating as the first or last aortic branch. When it is the last branch, the artery arises from the left extremity of the arch, coursing obliquely to the right side.

The *vertebral artery, internal mammary artery, thyrocervical trunk, suprascapular artery* (frequently originating in the thyrocervical trunk), and *costocervical trunk* are the normal subclavian branches (Fig. 5).

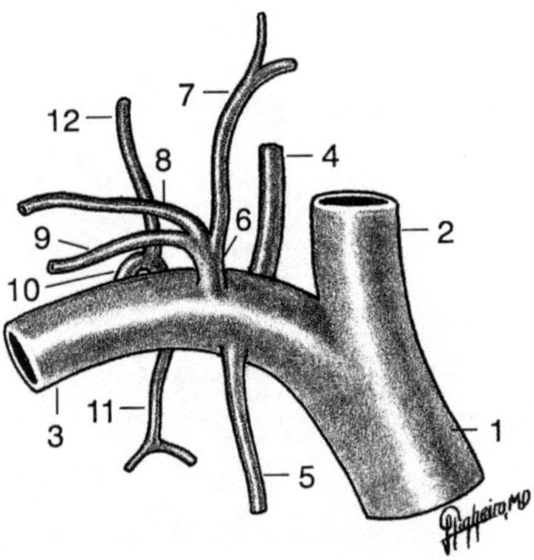

Fig. 5. Right subclavian artery and branches. The most frequent pattern is represented. *1*, innominate artery; *2*, common carotid artery; *3*, subclavian artery; *4*, vertebral artery; *5*, internal mammary artery; *6*, thyrocervical trunk; *7*, inferior thyroid artery; *8*, transverse cervical artery; *9*, suprascapular artery; *10*, costocervical trunk; *11*, supreme intercostal artery; *12*, deep cervical artery

Current noninvasive imaging methods cannot reliably define the entire course of these vessels, especially the left subclavian artery. The right subclavian artery can usually be followed from its origin to the level of the axillary artery by ultrasound, but the branches in both sides are not consistenly seen.

Vertebral Arteries

The vertebral artery, usually the first and largest branch of the subclavian artery, arises from the posterosuperior surface of that vessel approximately 2.5 cm from the subclavian origin on the right side and 3.5 cm on the left. The common variants are the unilateral duplication, origin from the common or internal carotid artery, and unilateral hypoplasia or aplasia. Three different segments of the artery can be recognized (Fig. 3). The first (cervical) is situated between its origin and the entrance into the foramen transversarium of the atlas. The second (suboccipital) extends up to the atlanto-occipital membrane, and the third portion (intracranial) is inside the skull.

In its cervical portion, the vertebral artery ascends behind the anterior scalenus muscle, proceeding cephalad and slightly posteriorly, entering the transverse process of the sixth (90%), fifth or seventh (10%) cervical vertebra. It then runs inside the canal formed by the foramina in the transverse process of the other vertebrae. The accompanying vein courses slightly laterally and anteriorly to the artery, forming within the transverse foramina the vertebral venous plexus surrounding the artery. After passing the foramen of the axis, the vessel curves laterally, enters the transverse foramen of the atlas with an anterior convexity, then curves abruptly backward and medially to reach the groove of the posterior arch of the atlas. Without these curves of the artery free movement of the head would not be possible. After leaving the arterial groove of the atlas, the vessel turns upward and forward, crosses the atlanto-occipital membrane, and reaches the cranial cavity through the foramen magnum. The establishment of a regular pattern is difficult due to the wide variation in the loops formed by the vertebral artery in its suboccipital portion [2].

The intracranial portion of each vessel has an initial course lateral to and then in front of the medulla oblongata. At the lower aspect of the pons, both vertebral arteries unite to form the basilar artery in the midline. In 30% of the cases, the basilar artery is asymmetrically situated. The vertebral arteries are frequently asymmetric, the left vessel commonly being larger than the right. Intracranial variations are rare. The vertebral artery gives off two groups of branches in its cervical portion (*spinal* and *muscular*) and five branches in its intracranial portion (*meningeal, posterior spinal, anterior spinal, posterior inferior cerebellar*, and *medullary*).

The *basilar artery* is the continuation of both vertebral arteries after their confluence at the level of the junction between the medulla oblongata and the pons. It is a single vascular trunk, although islet formation (division of the vessel with subsequent rejunction) and partial or complete duplication has been described. In the last case, one of the vessels is larger than the other. Dilatation with an S-shaped deformity may be seen. This is not related to age, since young patients often show this variant. The artery ends approximately at the level of the posterior clinoid process, giving off two terminal branches, the *posterior cerebral arteries*. The basilar artery can be shorter if the site of junction of both vertebral arteries is higher than usual. The following arteries are found as its branches on both sides: *pontine, artery of the labyrinth, anterior inferior cerebellar, middle cerebellar* (a rare branch occurring in the absence of the inferior cerebellar artery), *superior cerebellar,* and *posterior cerebral artery.*

The *posterior cerebral arteries* are the paired terminal branches of the basilar artery and form the posterior part of the circle of Willis. They are connected to the internal carotid arteries (from which they originate embryologically) through the posterior communicating arteries. The most frequent variation is the unilateral origin from the internal carotid artery (an embryologic remnant). It can also originate from the posterior communicating artery, anterior choroidal artery, contralateral posterior cerebral artery, or may be unilaterally doubled.

The vertebrobasilar circulation is not at present fully and reliably accessible to noninvasive imaging. The presence of such bony structures as the tranverse processes of the vertebrae prevents uninterrupted visualization of the vertebral arteries by ultrasound, while the typical tortuosity encountered in the suboccipital segment may interfere with a reliable definition by magnetic resonance imaging.

Collateral Circulation to the Brain

The importance of collateral circulation in the presence of insufficient blood supply to the brain has been known since the first description of the arterial circle at the base of the brain by Willis in 1664.

The term "collateral circulation" has been long used to describe the anastomotic connections occurring distal to a vascular occlusion. This use will be maintained in this chapter, although by definition collaterals are vessels running parallel to one another which carry blood in the same direction [4]. The anastomotic channels for which we will use the term "collaterals", on the other hand, occur between different systems with essentially no pressure gradient across the communications. In some circumstances, the direction of the blood flow can be reversed. All these channels are usually present in every individual, but their efficacy as an alternative source of blood supply depends on their caliber, which is variable in the population, and on the effectiveness of the remaining anastomotic connections supplying the same territory [4, 5]. For example, if the flow from the circle of Willis into the internal carotid artery distal to an occlusion is high enough to maintain normal pressure in the ophthalmic artery, the normal blood flow direction in this artery will be preserved [4].

The different possibilities of anastomotic connections which can supply a region distal to a vascular occlusion in the cerebral circulation can be grouped into two main categories [4]: 1. arterial bypass without participation of other arterial systems and 2. arterial bypass involving neighboring vascular systems.

In the first group, we can find the following possibilities:

1. Collaterals between branches of the external carotid artery and the ophthalmic artery occurring with occlusion of the internal carotid artery.
2. Collateral circulation to the posterior portion of the brain by way of the vertebral artery – posterior inferior cerebellar artery – superior cerebellar artery – basilar artery in occlusion of the middle segment of the basilar artery.
3. Collaterals involving the anterior spinal artery in the occlusion of the intracranial vertebral artery.
4. Collateral circulation through the callosomarginal artery in the case of occlusion of the pericallosal artery.
5. Collaterals from the lenticulostriate arteries bypassing an occluded middle cerebral artery.

Other potential sources of anastomosis include channels between the dural branches of the external carotid artery and dural branches of the meningohypophyseal trunk, and the so-called rete mirabile (dural pathways between extracranial and intracranial portions of the carotid artery).

The major representative of the second group is the circle of Willis, which connects anteriorly both carotid systems and posteriorly the carotid and basilar systems. The circle of Willis, when completely developed, provides an extensive intercommunication among the four major vessels which participate in the cerebral blood supply. However, many variations can be found in at least 50% of cases, and they assume clinical importance in pathological conditions. These variations are mainly related to the anterior and posterior communicating arteries, or in other words, the anterior and posterior components of the arterial circle (Fig. 6).

In rare situations, collateral circulation is established through the *primitive trigeminal artery* or the *hypoglossal artery*.

The development of one specific type of collateral circulation is mainly dependent on the site of narrowing or occlusion in one of the extracranial major arteries. The most frequent patterns are summarized in Fig. 7.

The precise knowledge of cerebral collateral circulation is not only important in evaluating its role in the presence of an arterial occlusion, it is also important in providing the surgeon with the knowledge to optimally treat the intracranial aneurysms or arteriovenous fistulae. At present, the collateral circulation is more consistently visualized with X-ray arteriography, although magnetic resonance and intracranial color Doppler show some promise.

Thoracic Aorta and Its Branches

The thoracic aorta arises from the left ventricle within the pericardium at the aortic annulus and ends at the diaphragm, where it passes into the abdominal

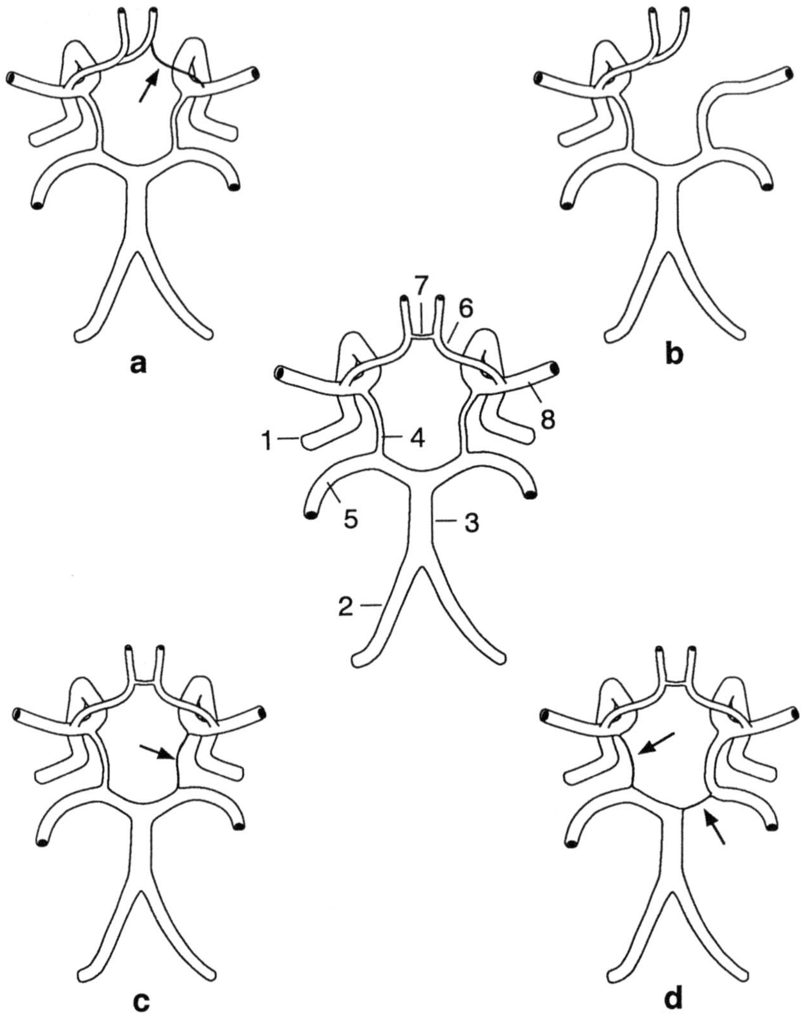

Fig. 6a–d. Diagram showing the "normal" anatomy of the arterial circle of Willis (*center*) and some of the possible variations. *1*, internal carotid artery; *2*, vertebral artery; *3*, basilar artery; *4*, posterior communicating artery; *5*, posterior cerebral artery; *6*, anterior cerebral artery; *7*, anterior communicating artery; *8*, middle cerebral artery. **a** Absence or hypoplasia (*arrow*) of the proximal segment of the anterior cerebral artery on one side; **b**, absence of one internal carotid artery with the middle cerebral artery originating from the ipsilateral posterior cerebral artery; **c** absence or hypoplasia (*arrow*) of the posterior communicating artery on one side; **d**, absence or hypoplasia of the posterior communicating artery on one side and the proximal portion of the posterior cerebral artery on the other side (*arrows*). The situation shown in **b** is rare, occurring in about 0.1% of the population

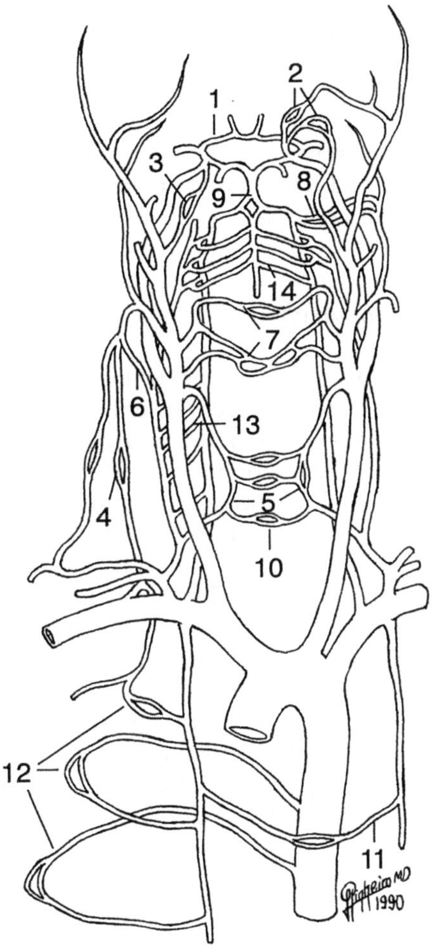

Fig. 7. Diagram representing the major potentially available collateral pathways in occlusion of the internal carotid, common carotid, vertebral, and subclavian arteries. *1*, anastomoses at the circle of Willis; *2*, communication between branches of the external carotid and ophthalmic artery; *3*, communication between the maxillary artery and internal carotid through the caroticotympanic branches; *4*, communication between the deep cervical artery and occipital artery; *5*, communication between the inferior and superior thyroid arteries; *6*, communication between the ascending cervical and occipital arteries; *7*, anastomoses between branches of the external carotid artery; *8*, anastomoses between the occipital and vertebral arteries; *9*, vertebrovertebral anastomosis; *10*, anastomoses between both inferior thyroid arteries; *11*, anastomoses between both internal mammary arteries; *12*, communications between the posterior intercostal arteries and the internal mammary arteries; *13*, anastomoses between the ascending cervical and vertebral arteries; *14*, vertebrovertebral and vertebro-ascending cervical communications via spinal radicular branches

cavity. According to its topography, the artery is divided into three segments: ascending aorta, aortic arch (or transverse aorta), and descending thoracic aorta.

Ascending Aorta

In its proximal half, the ascending aorta is covered by the visceral pericardium, and shares a common sheath with the pulmonary trunk. It is dilated at its origin (the aortic bulb) in the form of three pouchlike formations, the *sinuses of Valsalva*. From its origin, the ascending aorta passes obliquely superiorly, ventrally, and to the right, reaching the height of the second right intercostal joint ending by definition just before the origin of the innominate artery. The pulmonary trunk is located on its left side and the superior vena cava runs along its right side, slightly posterior to the proximal portion. The right pulmonary artery is behind it.

The total length of the ascending aorta varies from 5 to 6 cm and the diameter from 3 to 3.5 cm. The *right* and *left coronary arteries,* arising from the right and left sinuses of Valsalva, respectively, are the only branches of the ascending aorta.

Aortic Arch

The aortic arch begins behind the second sternocostal joint, and courses in an upward convex arch, oriented anteroposteriorly and to the left. Rarely, the highest point of the aortic arch reaches above the upper border of the sternum. The trachea, esophagus, and thoracic duct lie behind the posterior surface of the arch, and the right pulmonary artery runs below it toward the right pulmonary hilum. The left major bronchus passes below the arch. The segment between the origin of the left subclavian artery and the ligamentum arteriosum is slightly narrowed after birth (aortic isthmus). The length of the aortic arch is 5–6 cm and the diameter ranges between 1.7 and 3.0 cm.

The aortic arch normally gives off three branches: the *innominate artery,* the *left common carotid artery,* and the *left subclavian artery* (Fig. 1). The carotid artery takes off from the middle of the arch nearer to the innominate artery than to the subclavian artery [2]. The left subclavian artery is 1–1.5 cm from the left common carotid artery. This pattern of branching occurs in approximately 65% of cases [6]. Variations can be found in relation to either the aortic arch itself (less frequently) or the origin of its branches, either in association or isolated.

Variations of the aortic arch are (Fig. 8)

1. Right-sided aortic arch. If no additional variations occur such as a different pattern of branching, the result is a mirror image of the "normal" arch.
2. Double aortic arch. Most of the cases are better classified as abnormalities, since they can cause symptomatic compression of the trachea and/or esophagus and are frequently associated with other cardiac malformations. However, in rare instances, the double arch can occur as an isolated variant causing no symptoms. Frequently one of the arches is hypoplastic.
3. Circumflex aortic arch. The aortic arch is normal, but the descending thoracic aorta runs on the left side (left circumflex aortic arch), or there is a right-sided aortic arch with the descending aorta on the right side (right circumflex aortic arch). In both situations the origin of the branches varies considerably.

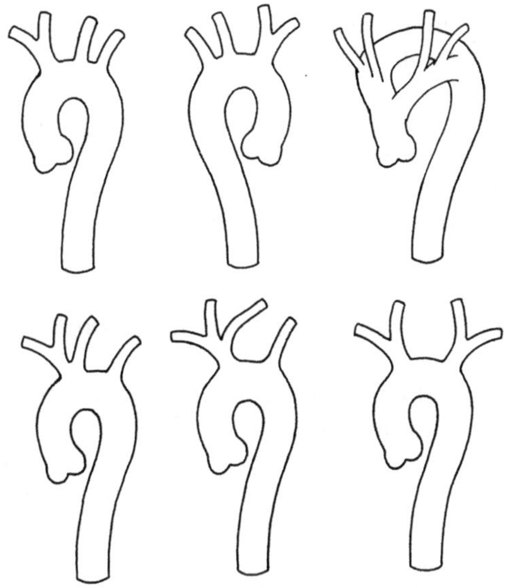

Fig. 8a–f. Examples of anatomic variations of the aortic arch and its branching pattern. **a**, "normal" situation; **b**, right-sided aortic arch; **c**, double aortic arch; **d**, common origin of the innominate artery and left common carotid artery; **e**, origin of the left common carotid artery from the innominate artery; **f**, double innominate artery

Variations of the pattern of branching are (Fig. 8):

1. Common origin of the innominate artery and left common carotid artery (≈13%).
2. Left common carotid artery originating from the innominate artery (≈9%).
3. Right and left innominate artery (≈1%) [1].

Less frequent are the occurrence of a common trunk for both common carotid arteries, the presence of common innominate artery, and the absence of an innominate artery.

Descending Thoracic Aorta

The descending thoracic aorta lies in the posterior mediastinum. It begins at the left side of the fourth thoracic vertebra, running downward and obliquely toward the midline, and ends at the lower border of the 12th thoracic vertebra, where it enters the aortic hiatus on the diaphragm.

The left pulmonary artery crosses the vessel at its proximal portion, above the left bronchus. At the midportion, the aorta is separated from the left ventricle and the left atrium by the pericardium. On the right side are the azygos vein and the thoracic ductus. The average length of the descending aorta is 30 cm and the average diameter is 2.5 cm, with no significant decrease along its course due to the small size of its branches.

The most important branches of the descending thoracic aorta are the *bronchial*, the *esophageal*, and the paired *intercostal* arteries (Fig. 1). The remaining branches (*pericardial, mediastinal, subcostal, and superior phrenic*) are very small and do not represent an important source of collateral blood supply.

The ability to visualize the branches of the thoracic aorta noninvasively has not yet been systematically investigated. The vessel itself is routinely interrogated with ultrasound and magnetic resonance imaging for example in patients with dissecting and nondissecting aneurysms.

Abdominal Aorta and Its Branches

The abdominal aorta enters the abdominal cavity through the aortic hiatus of the diaphragm in front of the body of the 12th thoracic vertebra. In 70% of the cases it runs to the left side of the spinal column, although in a few cases the vessel has a right-sided course (5%). This segment of the aorta describes a curve that has an anterior convexity which follows the lordosis of the lumbar spine, its peak corresponding to the third lumbar vertebra. When there is a curvature in the frontal plane, it is usually to the left. The bifurcation in its two terminal branches, the *common iliac arteries,* usually occurs at the junction between the fourth and fifth lumbar vertebrae, projecting approximately 2.5 cm below the umbilicus. The division is usually lower in the elderly. The *middle sacral artery* is the rudimentary continuation of the aorta. It passes in front of the sacrum, giving off the fifth *lumbar arteries* and the *sacral branches*. The average aortic diameter is 2.3 cm and the total length about 20 cm. In the lower half of the abdomen, the aorta is in contact with the inferior vena cava, while in the upper half the inferior vena cava deviates to the right.

The abdominal aorta gives off branches to the abdominal wall and the viscera. Figure 9 enumerates the aortic branches, giving their level of origin in relation to the vertebral bodies. The abdominal aorta is usually well-depicted using noninvasive methods.

Inferior Phrenic Artery

The inferior phrenic artery is a small paired vessel which arises from the anterior surface of the aorta immediately below the diaphragm. It may also originate from the celiac trunk. In one-third of cases, the two inferior phrenics take off from a common stem, either from the aorta or from the celiac trunk. The artery supplies the diaphragm and sends several branches to the suprarenal gland on each side.

Lumbar Arteries

There are four pairs of lumbar arteries arising from the posterolateral aspect of the aorta, opposite the bodies of the first four lumbar vertebrae. A small

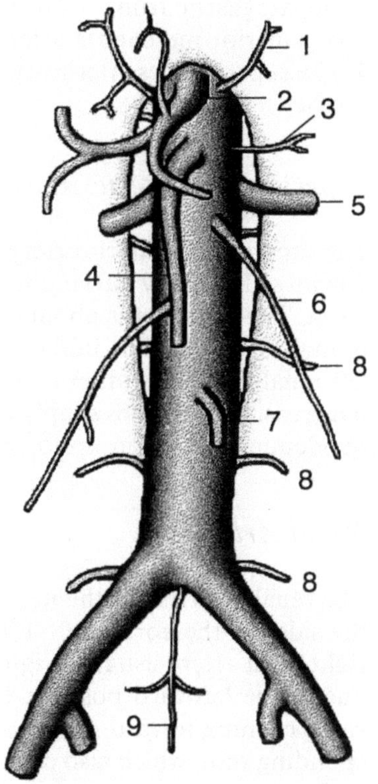

Fig. 9. Branches of the abdominal aorta. The level of origin according ot the vertebral bodies is given in parenthesis. *1*, inferior phrenic artery (12th thoracic); *2*, celiac trunk (12th thoracic – 1st lumbar); *3*, middle suprarenal artery (1st–2nd lumbar); *4*, superior mesenteric artery (1st–2nd lumbar); *5*, renal artery (1st–2nd lumbar); *6*, testicular (ovarian) artery (2nd–3rd lumbar); *7*, inferior mesenteric artery (3rd–4th lumbar); *8*, lumbar arteries (1st–4th lumbar); *9*, middle sacral artery (4th–5th lumbar)

fifth pair sometimes arises from the middle sacral artery. The right lumbar arteries pass posterior to the inferior vena cava and the first two on each side are covered by the crus of the diaphragm. They supply the muscles and skin of the back, the bodies of the lumbar vertebrae and send small branches to the capsule of the kidney.

Celiac Trunk

The celiac trunk arises just below the aortic hiatus of the diaphragm at the level of the first lumbar vertebra and after a short course (1–2 cm) divides into three branches: *left gastric, hepatic,* and *splenic* arteries. It is the major source of blood supply to the stomach, liver, gallbladder, and spleen, and part of the duodenum. According to the different arrangements of the branches, the following variations can be found:

1. Gastrohepatosplenic trunk – This is the usual pattern, occurring in 89% of cases.
2. Hepatosplenic trunk – The left gastric artery arises independently from the aorta or hepatic or splenic artery.
3. Gastrosplenic trunk – The hepatic artery arises separately from the aorta or superior mesenteric artery.

4. Hepatogastric trunk – The splenic artery arises individually from the aorta or superior mesenteric artery.
5. The superior mesenteric artery originates from the celiac trunk in 1–3% of cases.

Superior Mesenteric Artery

The superior mesenteric artery arises 1–20 mm from the celiac trunk and runs downward, usually curving to the right and dividing on the right side of the aorta. The diameter is about 0.5 cm, and the length from its origin to the first branch is 4.0 cm. The initial portion is crossed obliquely from behind by the left renal vein. It is anteriorly related to the splenic vein and the head of the pancreas. Its branches supply the small intestine except the superior part of the duodenum, the colon up to the splenic flexure, and part of the pancreas.

Renal Arteries

The renal arteries are the next major aortic branches, arising bilaterally from the sides of the aorta, 1.0–1.5 cm below the superior mesenteric artery. The right renal artery usually originates superior to the left, and the vessel is longer due to the leftward position of the aorta. It passes behind the inferior vena cava, running toward the hilum. The left renal artery is covered by the corresponding vein, which also covers the proximal segment of the right renal artery while running to join the inferior vena cava. Just before reaching the hilum of the kidneys, the renal arteries divide into four or five branches, the *segmental arteries*. Each segmental artery supplies a group of pyramids. Within the renal parenchyma, they divide progressively into *interlobar* (running between the pyramids) and the *arcuate arteries*. The latter curve around the pyramids, running parallel to the cortical surface.

The abdominal aorta decreases in caliber rather abruptly after the emergence of the renal arteries, due to the fact that these arteries receive around 25% of the cardiac output. At the renal hilum, the renal arteries lie between the renal vein (anterior), and the renal pelvis (posterior). The renal arteries also supply the suprarenal glands, the ureter, and surrounding tissue and muscles. In 23% of cases, there are one or two accessory renal arteries. They are more frequently found on the left side and enter the kidney in its upper or lower pole (*polar arteries*). Other variations include the presence of multiple renal arteries (usually up to three) entering the hilum. They may or may not be associated with abnormalities such as duplication of the renal pelvis or horseshoe kidney.

Inferior Mesenteric Artery

The inferior mesenteric artery arises about 3 or 4 cm above the aortic division into the common iliac arteries, at the level of the third lumbar vertebra. In rare instances, it can arise together with the superior mesenteric artery. It is usually much smaller than the superior mesenteric and runs initially anterior to the aorta, coursing immediately to the left. This artery supplies the colon from the midportion of the transverse colon to a greater part of the rectum.

Middle Suprarenal and Testicular (Ovarian) Arteries

In 70% of the population, the middle suprarenal arteries arise below the origin of the celiac trunk from each side of the aorta, running laterally to supply the suprarenal gland.

The testicular (ovarian) arteries originate from the anterolateral side of the aorta. The testicular artery crosses the iliac artery and ureter, entering the inguinal canal and ramifying in the testicle. The ovarian arteries are shorter, and do not pass out of the abdominal cavity. These vessels turn medially at the inlet of the pelvis, dividing between the folds of the broad ligamentum of the uterus to supply the ovary.

The celiac trunk, superior mesenteric artery, and the renal arteries are frequently depicted in their proximal segments by noninvasive imaging methods. Color Doppler flow imaging is capable to assess the intrarenal segments of the renal arteries, however, the precise diagnostic value in clinical settings has not yet been determined. The inferior mesenteric artery is rarely identified.

Arteries of the Pelvis and Lower Extremities

Common Iliac Arteries

The common iliac arteries, the terminal branches of the aorta, supply the pelvic organs and the lower limbs. After their origin, both arteries have a downward and lateral course, with the right common iliac artery usually in a more anterior position than the left-sided vessel. The confluence of both iliac veins to form the inferior vena cava lies behind the right common iliac artery. Both arteries are crossed frontally by the ureters. The angle of the aortic division is around 65° in males and 75° in females. The normal diameters range between 0.5 and 1.2 cm, with the right iliac commonly larger than the left iliac. The arteries are 6.5 cm (3.5–12 cm) long before they bifurcate into the external and internal iliac arteries. This normally occurs between the last lumbar vertebra and the sacrum. In some cases, the common iliac artery is very short, with the iliac bifurcation occurring close to the aorta. Bifurcation beyond the sacrolumbar joint is rare.

In addition to the terminal branches, the common iliac arteries give off some small branches to supply the peritoneum, the ureters, and the psoas major muscle.

The common and external iliac arteries are usually adequately defined using noninvasive vascular imaging methods.

Internal Iliac Artery

Formerly known as the hypogastric artery, the internal iliac artery is responsible for the blood supply to the pelvis and gluteal region. It runs downward and medially from its origin, anterior to the sacroiliac joint, toward the greater sciatic foramen, and gives off branches along its course. The vessel is smaller in adults and larger in fetuses and newborns. The frequent variation in terms of the origin of its branches makes the description of a typical pattern rather difficult. However, an *anterior* and *posterior trunk* can be defined in more than 50% of the cases. The most frequent distribution of the branches is shown in the Fig. 10. In this situation, from the posterior trunk originate:

1. The *iliolumbar artery,* which courses upward and laterally, ascending between the vertebral column and the psoas major muscles, giving off two branches at the level of the iliac fossa, the *lumbar branch* (which supplies the psoas major muscle and the quadratus lumborum, anastomosing with the fourth lumbar artery) and the *iliac branch,* which supplies the pelvic muscles.
2. The *lateral sacral arteries,* which normally arise from the posterior trunk itself but sometimes from the iliolumbar artery, and which supply the sacral canal and part of the pelvic musculature.
3. The *superior gluteal artery,* a continuation of the posterior trunk and the largest branch of the internal iliac artery, which leaves the pelvis via the greater sciatic foramen and supplies the gluteal muscles by means of its superficial and deep branches.

The anterior trunk, on the other hand, gives off:

1. The *umbilical artery,* which is usually the first visceral branch of the internal iliac artery and has little importance in collateral blood supply.
2. The *obturator artery,* the second branch of the anterior trunk, which courses ventrally on the lateral wall of the pelvis toward the obturator canal, and passing through it, divides into an *anterior* and a *posterior* branch, between the adductor muscles of the thigh. The artery gives off the *iliac, vesical,* and *pubic* branches before it leaves the pelvis.
3. The *uterine artery,* which originates from the medial surface of the anterior trunk, and follows a tortuous course toward the cervix of the uterus. It then ascends on each side of the uterus to reach the junction of the uterine tube and the uterus, curves laterally, and ends by anastomosing with the ovarian artery. In the male, the vessel corresponds to the *artery of the ductus deferens.*
4. The *inferior vesical artery,* which is usually a single branch, but sometimes can be doubled or give off the *middle vesical artery,* a vessel not always present. These arteries, plus the middle rectal artery, converge to form an arterial network, supplying the bladder.
5. The *middle rectal artery,* usually an independent branch of the internal iliac artery which may sometimes arise from the internal pudendal artery or inferior vesical artery [2]. The artery forms important anastomoses with the inferior mesenteric artery and the internal pudendal artery.

6. The *internal pudendal artery,* one of the two terminal branches of the anterior trunk. It is much smaller in the female (same course, different distribution of the branches) and supplies the perineum and external genital organs.
7. The *inferior gluteal artery,* the other terminal branch of the anterior trunk, which leaves the pelvis through the greater sciatic foramen below the piriformis muscle. Then, it turns to the gluteal region and supplies the gluteus maximus and the back of the thigh, where it terminates. A thin branch, the *sciatic artery,* supplies the sciatic nerve. In rare instances, this artery may constitute the main artery of the lower limb, continuing distally in the popliteal artery. In such cases, the femoral artery is small, supplying the area usually related to the profunda femoris artery [1, 2].

The proximal segments of the internal iliac artery are frequently visualized by noninvasive methods, however more peripheral branches are not consistently depicted.

External Iliac Artery

The external iliac artery is the larger branch of the common iliac artery. From its origin at the level of the sacroiliac joint, the vessel runs obliquely downward and laterally along the psoas major muscle. After it passes the inguinal ligament, it reaches the thigh as the femoral artery. The length of the vessel ranges from 6 to 12 cm, and the right iliac is frequently larger than the left, its diameter being between 0.4 and 1.0 cm [2]. The origin of the artery is crossed by the ureter and, in females, by the ovarian artery and vein. In males, the testicular artery crosses its distal portion. Proximally, the external iliac vein lies behind the artery but the distal course of the vein is entirely medial to the artery. Besides the possibility of a tortuous or curved course, significant variations have not been described.

The external iliac artery gives off two major branches:

1. The *inferior epigastric artery*, which originates 0.5 cm above the inguinal ligament. It runs medially and then upward on the dorsal surface of the rectus abdominis muscle, anastomosing with the terminal branches of the internal mammary artery.
2. The *deep circumflex iliac artery,* which curves along the crest of the ilium and anastomoses with the iliolumbar artery. Its ascending branch forms important anastomoses with the lumbar arteries and the inferior epigastric artery. No significant variations have been described.

Common Femoral Artery and Its Branches

The common femoral artery is actually the continuation of the external iliac artery (Fig. 10). It begins after passing beneath the inguinal ligament, running downward lateral to the femoral vein. After a course of 3–4 cm, it divides into

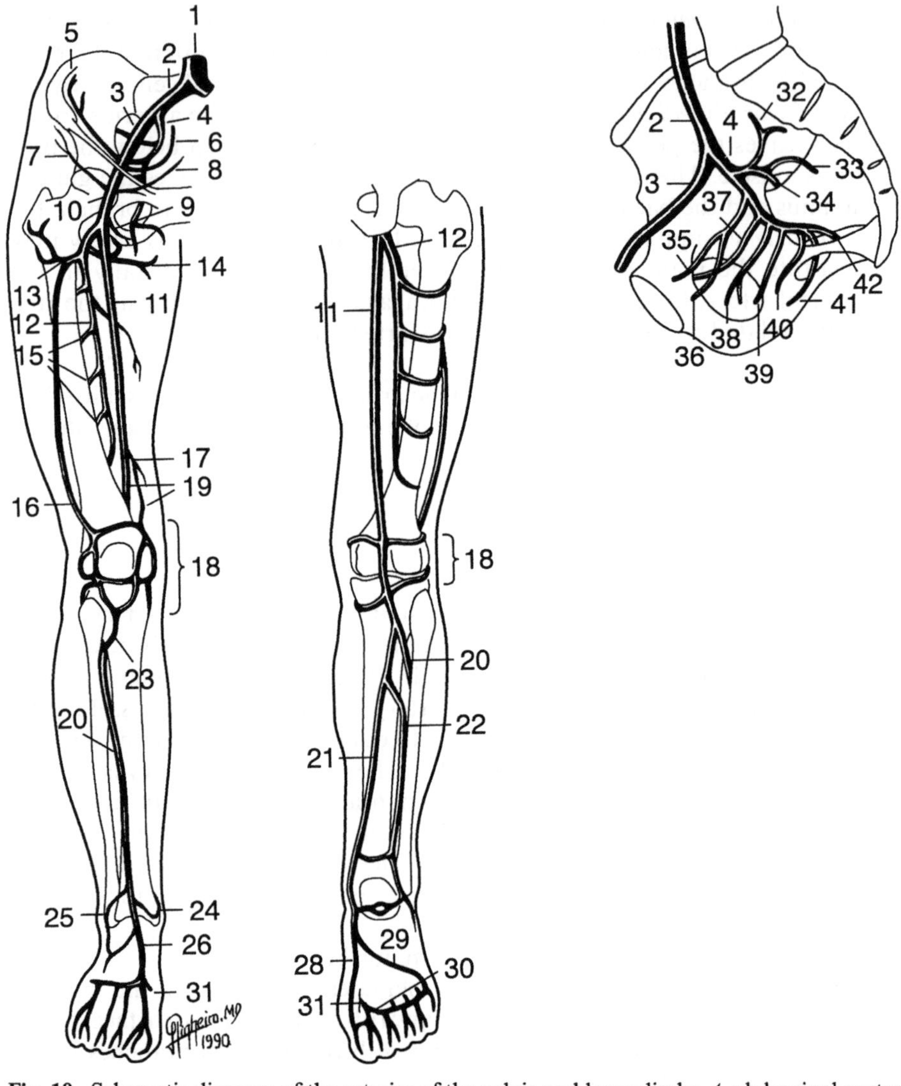

Fig. 10. Schematic diagram of the arteries of the pelvis and lower limbs. *1*, abdominal aorta; *2*, common iliac artery; *3*, external iliac artery; *4*, internal iliac artery; *5*, deep circumflex artery; *6*, inferior epigastric artery; *7*, superficial circumflex artery; *8*, superficial epigastric artery; *9*, external pudendal artery; *10*, common femoral artery; *11*, superficial femoral artery; *12*, profunda femoris artery; *13*, lateral circumflex femoral artery; *14*, medial circumflex femoral artery; *15*, perforating arteries; *16*, descending branch; *17*, descending genicular artery; *18*, collateral network of the knee; *19*, popliteal artery; *20*, anterior tibial artery; *21*, posterior tibial artery; *22*, peroneal artery; *23*, anterior tibial recurrent artery; *24*, anterior medial malleolar artery; *25*, anterior lateral malleolar artery; *26*, dorsalis pedis artery; *27*, communicating artery; *28*, medial plantar artery; *29*, lateral plantar artery; *30*, plantar arch; *31*, deep plantar artery; *32*, iliolumbar artery; *33*, lateral sacral artery; *34*, superior gluteal artery; *35*, umbilical artery; *36*, superior vesical artery; *37*, obturator artery; *38*, uterine artery; *39*, inferior vesical artery; *40*, middle rectal artery; *41*, internal pudendal artery; *42*, inferior gluteal artery

the *superficial femoral artery* and *profunda femoris artery,* its two terminal branches. Prior to this bifurcation, the following branches can be identified: *superficial epigastric artery* (medially), *superficial circumflex artery* (laterally), and *external pudendal artery* (medially). In relation to the inguinal ligament, the site of the bifurcation can vary from 0 to 8 cm, which creates some confusion regarding the nomenclature based on this anatomic reference. Some authors [2, 3] prefer to consider the common and superficial fermoral artery as only one vessel, named the *femoral artery,* with the profunda femoris artery being one of its branches. Thus, the entire vessel below the inguinal ligament is the femoral artery, no matter what the level of origin of the profunda femoris artery. Despite some obvious advantages to this usage, the nomenclature adopted in this chapter will follow the more widespread method, as referred to by Strandness [5], Keen [7], and Muller et al. [8].

Superficial Femoral Artery

From the femoral bifurcation, the superficial femoral artery continues distally in the thigh under the sartorius muscle within a fascial tunnel called the adductor (Hunter's) canal [3, 5]. The canal contains the femoral artery and vein, and the saphenous nerve. At the level of the opening in the tendon of the adductor magnus muscle, the femoral vessels pass through it reaching the popliteal fossa. The *descending genicular artery* arises from the superficial femoral artery just before the artery enters the opening in the tendon of the adductor magnus muscle. This artery is an important source of collateral supply in cases of occlusion of the popliteal artery. Also, during surgical exposure of the femoropopliteal trunk, care must be taken to prevent damage to this vessel, as well as to the saphenous branch of the femoral nerve [5].

Significant variations are infrequent. In the few cases where the artery is absent, it is replaced by the sciatic artery, normally a branch of the internal iliac artery. In such cases, the femoral artery is small and supplies the area of the profunda femoris artery. More rarely, the superficial femoral artery can be replaced by the saphenous artery, which runs along the saphenous vein [1, 2].

Profunda Femoris Artery

After its origin from the common femoral artery the profunda femoris artery usually courses posterolateral to the superficial femoral artery (Fig. 10), but sometimes it can arise dorsal (40%) or medial (10%) to that vessel [1]. It has a downward course, medial to the femur and posterior to the femoral vein. At the middle of the thigh it turns posteriorly and passes beneath the adductor longus muscle, ending as a small branch, the *fourth perforating artery,* which anastomoses with the upper branches of the popliteal artery. Along its course, the profunda femoris artery gives off the *medial* and *lateral circumflex arteries,* the three *perforating arteries,* and the *muscular branches.* These vessels supply most of the musculature of the anterior and medial compartment of the thigh. The medial circumflex artery also participates in the blood supply to the head of the femur and the perforating branches also contribute to the supply of the hamstrings in the posterior compartment of the thigh.

Variations are frequent and mainly related to the circumflex femoral vessels. Both of them can originate from the profunda, one can arise from the common femoral artery and the other from the profunda, or both may originate from the common femoral artery just proximal to the profunda femoris artery origin [5].

Popliteal Artery

The popliteal artery is the direct continuation of the superficial femoral artery. It extends from the adductor hiatus to the inferior border of the popliteus muscle, where it divides into terminal branches. Another landmark for the end of the popliteal artery is the tendinous arch of the upper end of the soleus muscle. The vessel has a total length of about 18 cm, and its average diameter is 0.7 cm. The point of division lies about 6 cm below the knee joint [2]. The popliteal vein has a course initially lateral to the artery, and then crosses it posteriorly to reach the medial side. The tibial nerve occupies the most superficial position.

Between its origin and division, the popliteal artery has the following branches: the *medial* and *lateral superior genicular arteries,* which form the arterial network of the knee, anastomosing with the descending genicular artery; the *middle genicular artery,* which pierces the articular capsule to supply the sinovia and ligamentum of the knee joint; the *medial* and *lateral inferior genicular arteries,* which also participate in the arterial network of the knee; the *sural arteries,* distributed to the gastrocnemius, soleus, and skin; and the *cutaneous branches,* which arise either from the popliteal artery or from some of its branches, and supply the skin on the back of the leg.

The most important anatomic variants are related to the pattern and location of the division. In the normal situation (95%), the popliteal artery divides at the upper border of the popliteal muscle and its first terminal branch is the *anterior tibial artery.* After this, the popliteal artery continues as the tibioperoneal trunk, which divides into *posterior tibial* and *peroneal arteries,* after a distance that ranges from 0 to 6 cm [5]. A trifurcation at the same level occurs in 4% of cases. Rarely (1%), the peroneal artery arises from the anterior tibial artery. If the artery divides above the upper border of the popliteal muscle, it is called a "higher division" (5%). In this case, the usual branching pattern can be found in 3% of cases, in 1% of cases the peroneal artery can arise from the anterior tibial artery, or the anterior tibial can run ventrally to the popliteal muscle with an otherwise normal branching pattern [1].

Anterior Tibial Artery

After its origin, the anterior tibial artery passes anteriorly through the interosseous membrane, running steeply downward between the extensor muscles of the leg. As it descends anterior to the interosseous membrane, the vessel gradually approaches the tibia. In the distal part of the leg it lies on the tibia, becoming more superficial at the ankle joint, where it continues on the dorsum of the foot as the *dorsalis pedis artery*. The diameter of this vessel is 1.2–3.5 mm [2].

The following branches are identified: *posterior tibial recurrent artery*, which curves back to the knee joint; *anterior tibial recurrent artery*, which ascends to contribute to the arterial network of the knee; several *muscular branches* are given off to supply the adjacent muscles; and the *anterior medial* and *lateral malleolar arteries*, which supply structures of the ankle joint.

Variations concern the more superficial or lateral course of the vessel [2, 3]. Occasionally, the anterior tibial artery can be rudimentary or completely absent, being replaced by perforating branches of the peroneal or posterior tibial arteries [1].

Posterior Tibial Artery

The posterior tibial artery begins at the lower border of the popliteal muscle, after the origin of the peroneal artery. It is usually larger than the anterior tibial artery (diameter from 2.2 to 4.1 mm). It courses downward and medially between the superficial and deep muscles of the leg, toward a point between the medial malleolus and the medial tubercle of the calcaneus, dividing into terminal branches on the sole. The branches of the posterior tibial artery are: *circumflex fibular branches*, sometimes arising from the tibioperoneal trunk; the *nutrient artery of the tibia*, which enters the tibia in its proximal third and which is the largest bone nutrient artery in the body; the *medial malleolar artery;* the *communicating artery*, which links the posterior tibial artery and the peroneal artery 4 cm above the malleoli; the *medial calcaneal artery*, which forms a rich network over the heel; and the *medial* and *lateral plantar arteries*, the latter forming the main plantar arch together with the dorsalis pedis by means of a branch known as the *deep plantar artery*.

The anatomic variants consist in the underdevelopment or absence of the posterior tibial artery. In these cases, its territory is supplied by an enlarged peroneal artery, which can either join the rudimentary posterior tibial or continue to the sole of the foot [3].

Peroneal Artery

The peroneal artery arises from the tibioperoneal trunk about 2.7 cm [7] below the origin of the anterior tibial artery, passes obliquely toward the fibula, and runs along it to the lateral malleolus. It supplies vessels to the structures in the lateral aspect of the leg. The peroneal artery gives off: *muscular branches*, supplying the soleus, tibialis posterior, flexor hallucis, longus, and peroneal muscles; the *nutrient artery of the fibula; perforating branches*, piercing the interosseous membrane and anastomosing with the anterior lateral malleolar artery in the anterior part of the leg; the *communicating branch*, which joins the communicating branch of the posterior tibial artery just posterior to the interosseous membrane; *lateral malleolar branches*, which contribute to the malleolar network; and *lateral calcaneal branches,* the terminal vessels of the peroneal artery, which supply the heel and anastomose with the medial calcaneal branches of the posterior tibial and the anterior malleolar branch of the anterior tibial artery.

The peroneal artery is the most consistently encountered artery in the lower leg. Some authors prefer to consider the peroneal artery as the main artery of the leg and the tibial arteries as its branches in order to better explain the variations in the pattern of the leg arteries [1, 5, 7]. Thus, it is usually the peroneal artery that replaces the anterior tibial artery when it is hypoplastic or forms the origin of the dorsalis pedis in its absence. Similarly, when the posterior tibial artery is absent, the peroneal artery forms the plantar arteries [1].

Arteries of the Foot

The foot derives its blood supply from the dorsalis pedis artery, usually the continuation of the anterior tibial artery, and from the lateral and medial plantar arteries, terminal branches of the posterior tibial artery.

After its origin at the ankle joint, the *dorsalis pedis artery* proceeds on its course toward the big toe. The artery gives off branches which run to the dorsum of the foot (*lateral* and *medial tarsal arteries*), a branch which curves along the tarsometatarsal line (the *arcuate artery*). This penetrates to the plantar side as its terminal branch, the *deep plantar branch,* which assists in the formation of the plantar arch. Five *dorsal metatarsal arteries* are derived from the arcuate artery. These continue as the *dorsal digital arteries* after communicating with the plantar vessels via the perforating branches.

The *lateral plantar artery* is usually the dominant branch of the posterior tibial artery. It courses through the deep plantar aspect of the foot to reach the fifth metatarsal bone and then curves medially in a plantar equivalent of the arcuate artery, anastomosing with the deep plantar branches of the dorsalis pedis. After its origin at the medial malleolus, the *medial plantar artery* continues its course along the medial border of the first metatarsal bone in the form of its *superficial branch,* which supplies the first and second toes. The blood supply for the second to fourth toes is derived from the *plantar metatarsal arteries* (branches of the plantar arch), by means of the *digital arteries.* The fifth toe is supplied by a vessel which arises directly from the lateral plantar artery [2]. Numerous variations reflect the dominance of the dorsal or plantar arteries [7]. Variations of the metatarsal and digital arteries are of no clinical importance.

Collateral Circulation Between Thoracoabdominal Aorta and Lower Limbs

Anastomotic connections between the abdominal vessels were first described by Riolan in 1649 (whose name identifies the arch-shaped communicating system of the colon). When circulation is interrupted in any segment of the aorta after the origin of the left subclavian artery, the circulation may be preserved through the communications between the subclavian and external iliac arteries or the abdominal aorta and the external iliac artery.

The available collateral pathways are grouped into two categories: *upper circulation,* when the occlusion occurs just below the origin of the renal artery, and *lower circulation,* when it occurs near the aortic bifurcation.

The upper circulation uses the following routes:

1. Subclavian artery – internal mammary artery – superior and inferior epigastric artery, external iliac artery
2. Abdominal aorta – superior mesenteric artery – Riolan's arcade – inferior mesenteric artery – superior and inferior rectal arteries – internal pudendal artery – internal iliac artery – external iliac artery
3. Abdominal aorta – testicular (ovarian) artery – anterior scrotal (labial) branches – external iliac artery

The lower circulation is maintained through:

1. Abdominal aorta – posterior intercostal artery and lumbar arteries – deep circumflex iliac artery – external iliac artery
2. Abdominal aorta – lumbar arteries – iliolumbar arteries – internal iliac artery – external iliac artery

Collateral circulation involving the three unpaired abdominal aortic branches also plays an important role in the blood supply of the abdominal viscera. Anastomoses between the celiac trunk and the superior mesenteric artery are formed by the *anterior* and *posterior pancreaticoduodenal arches;* communications along the greater and lesser curvature of the stomach connect the left gastric and hepatic arteries, and the splenic and gastroduodenal arteries; Riolan's arch connects the superior to inferior mesenteric artery via the *middle* and *left colic arteries.* Communications between the inferior mesenteric artery and splenic arteries may be found even in normal anatomic conditions [2]. Additional anastomoses exist between the esophageal arteries and the left inferior phrenic artery, among the various vessels at the hilum of the liver, and between the 12th posterior intercostal artery and the ileocolic artery.

The collateral circulation of the iliac system is not as good as that in the upper limbs, particularly at the scapular region. If occlusion occurs immediately above the femoral artery bifurcation, for instance, the exclusion of the profunda femoris artery prevents the developments of adequate circulation of the lower extremity [9].

Three groups of anastomotic connections are recognized: (1) connections between the aorta, internal iliac artery, external iliac artery, and femoral artery; (2) genicular anastomosis between the common femoral artery, profunda femoris artery, popliteal artery, and anterior tibial artery; (3) anastomoses between the anterior tibial, posterior tibial, and peroneal arteries at the level of the ankle and heels through the malleolar and calcaneal arteries, and the communicating and perforating branches of the peroneal artery, and vessels which form the plantar arch.

Arteries of the Upper Limbs

Anatomic description of the vessels of the upper limbs has less clinical significance than in the legs, since significant vascular disease is relatively infrequent in this region, and when present it is often clinically silent due to the presence of abundant collateral communications.

Axillary Artery

The axillary artery is the continuation of the subclavian artery. It begins at the inferior margin of the clavicle and runs up to the inferior border of the teres major muscle, where it becomes the *brachial artery*. Its entire course is within the axillary fossa, lateral to and deeper than the axillary vein. The vessel is covered anteriorly by the pectoralis major muscle and it is surrounded by the upper third of the brachial plexus. The length of the artery is 8.2–12 cm, and the diameter 0.6–0.8 cm [2].

Six branches are usually derived from the axillary artery:

1. The *supreme thoracic artery* which supplies branches to the pectoralis minor and major muscles and to the chest wall.
2. The *thoracoacromial artery,* which divides into four branches, the *acromial branch* (contributing to the acromial network), the *deltoid branch,* the *pectoral branch,* and the *clavicular branch.*
3. The *lateral thoracic artery,* which runs along the lateral side of the chest, supplying the muscles of the chest wall and the breast in the female (via its *lateral mammary branches*).
4. the *subscapular artery,* which is usually the largest branch of the axillary artery. It divides into the *circumflex scapular artery* and the *thoracodorsal artery.*
5. the *anterior and posterior humeral circumflex arteries,* both of which encircle the proximal neck of the humerus.

There is wide variability in the origin of these branches. These most frequently pertain to the origin of some branches of the axillary artery and certain branches of the brachial artery from a common trunk. The "normal" pattern, with all the branches arising directly from the axillary artery, accounts for only 10% of cases [1].

Brachial Artery

The origin of the brachial artery is considered at the lower border of the teres major muscle and it runs deep to the median nerve up to the cubital fossa, dividing into terminal branches approximately 1 cm below the elbow joint [2]. The vessel lies medial to the humerus in its initial portion, curving gradually to the front of the arm as it descends, accompanied by the brachia veins and the nerves of the brachial plexus.

The following branches are usually identified (Fig. 11):

1. The *deep brachial artery* arises from the medial side of the main stem, and goes around the humerus in its posterior aspect in company with the radial nerve. It is the largest branch of the brachial artery and gives off four named

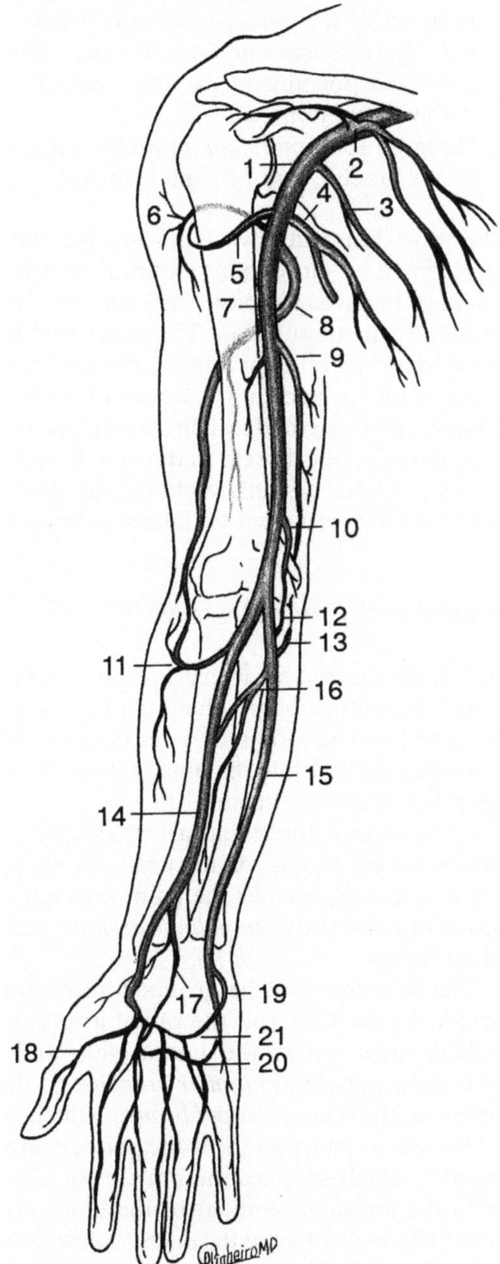

Fig. 11. Diagram showing the major arteries of the upper limb. *1*, axillary artery; *2*, thoracoacromial artery; *3*, lateral thoracic artery; *4*, subscapular artery; *5*, anterior humeral circumflex artery; *6*, posterior humeral circumflex artery; *7*, brachial artery; *8*, deep brachial artery; *9*, superior ulnar collateral artery; *10*, inferior ulnar collateral artery; *11*, radial recurrent artery; *12*, anterior ulnar recurrent artery; *13*, posterior ulnar recurrent artery; *14*, radial artery; *15*, ulnar artery; *16*, common interosseous artery; *17*, superficial palmar branch; *18*, principal artery of the thumb; *19*, deep palmar branch; *20*, superficial palmar arch; *21*, deep palmar arch

branches, in addition to the muscular vessels: the *deltoid branch*, the *nutrient branches of the humerus*, the *middle collateral artery*, and the *radial collateral artery*.

2. The principal nutrient artery of the humerus originates from the midportion of the brachial artery and enters the nutrient canal of the humerus near the insertion of the coracobrachialis muscle.
3. The *superior ulnar* and *inferior ulnar collateral arteries* arise from the middle and lower portions of the arm, respectively, and contribute to the network of the elbow joint.
4. Three or four *muscular branches* are distributed to the coracobrachialis, biceps brachii, and brachialis muscle.

The most important variants involve the deep brachial artery. It is often represented by more than one trunk which follow the brachial nerve [7]. High origin of the artery (above the lower border of the teres major muscle) occurs in 26% of the population. The superficial brachial artery (running anterior to the median nerve) is present as the only main stem in 9%. Two main arterial stems (with variance in the height of division) are present in 13% of cases [1]. Despite the rich collateral circulation at the elbow joint and the low incidence of atherosclerotic disease in the arm, external compression such as that caused by tourniquets or burn eschars may occlude not only the superficial main stems but the recurrent collateral arteries as well.

Radial and Ulnar Arteries

Although the radial artery appears to be the continuation of the brachial artery due to its course, the ulnar artery is usually the dominant vessel of the forearm [7]. The radial artery is the most superficial, lying under the brachioradialis muscle, while the ulnar artery runs deeper between the superficial and deep flexor muscle groups.

In its course, the radial artery describes a straight line toward the styloid process of the radius. At the wrist, the vessel curves round the lateral aspect of the carpus and reaches the palm through the muscle of the first interosseous space to form the *deep palmar arch* by joining the deep palmar branch of the uluar artery.

The branches can be grouped according to the different segments of the artery. In the forearm, the radial artery gives off: the *radial recurrent artery* (which arises from the lateral side and supplies the brachioradialis and brachialis muscle); *muscular branches* to the muscles on the lateral side of the forearm; the *palmar carpal branch*, which anastomoses with the carpal branch of the ulnar artery to form the *palmar carpal arch;* and the *superficial palmar branch*, which supplies the muscles of the thenar eminence and anastomoses with the terminal segment of the uluar artery to form the *superficial palmar arch*. At the wrist arise the *dorsal carpal branch* (it anastomoses with the dorsal carpal branch of the ulnar artery to form the *dorsal carpal arch*) and the *first dorsal metacarpal* (which supplies part of the thumb and index finger). At the hand arise the *principal artery of the thumb*, the *radial artery of the index*, and the *deep palmar arch* (the terminal branch of the radial artery). The latter artery gives off four *palmar metacarpal arteries*, which anastomose with branches of the superficial palmar arch.

The ulnar artery passes downward and medially from the cubital fossa to the medial side of the forearm, running beneath the pronator teres muscle. At first running parallel to the median nerve, the vessel crosses it posteriorly at the midportion of the forearm to reach the medial side. In the distal two-thirds of the forearm, the ulnar artery lies lateral to the ulnar nerve. At the wrist, the artery crosses the flexor retinaculum and, beyond the pisiform bone, becomes the *superficial palmar arch*. The following branches are identified in the forearm: the *anterior ulnar artery;* the *posterior ulnar recurrent artery;* the *common interosseous artery* (which divides in an anterior and posterior branch); and *muscular branches*. The branches at the wrist are the *palmar carpal branch* and the *dorsal carpal branch*. Finally, on the hand, the branches are the *deep palmar branch*, which contributes to the deep palmar arch, and the *superficial palmar arch*, from which the three *common palmar digital arteries* originate.

Absence of the radial or ulnar artery rarely occurs and the anterior interosseous artery is usually the branch which replaces their function in this eventuality. Persistence of the median artery, and embryonic remnant, is found in 10% of cases. Both radial and ulnar artery may have a more superficial course, representing the *superficial arteries* of the forearm (*antebrachialis superficialis*) [1]. Numerous variations of the arterial patterns of the hand have been described, however, their clinical significance is limited.

Noninvasive imaging methods, in particular the color Doppler flow imaging, are increasingly recognized as the primary clinical means to assess the arteries of the extremities.

Venous Anatomy

The veins are channels which carry the blood returning from the capillary network of every tissue in the body to the heart. They start as small channels that progressively increase in size as they receive tributaries in their course toward the heart. The veins are larger than their concomittant arteries and their capacity usually surpasses that of the systemic arteries. The veins are cylindrical channels with a high degree of deformability depending on the transmural pressure. In contrast to arteries the veins communicate more readily, in particular in the cranial cavity, spinal canal, the venous plexuses of the abdomen and pelvis.

Of the two venous systems in the body, the *pulmonary* and the *systemic*, only the latter will be described in this chapter.

Veins of the Head and Neck

The head and neck veins have as their principal function the drainage of the brain. The course of the cerebral veins is independent of the arteries and variation may occur among individuals or between the hemispheres of the same brain [4]. However, anastomoes between the intracerebral and extracerebral veins generally have a uniform pattern, differing from the highly variable connections found between the veins of the neck and thorax [2]. The head and neck veins may be divided into three categories:

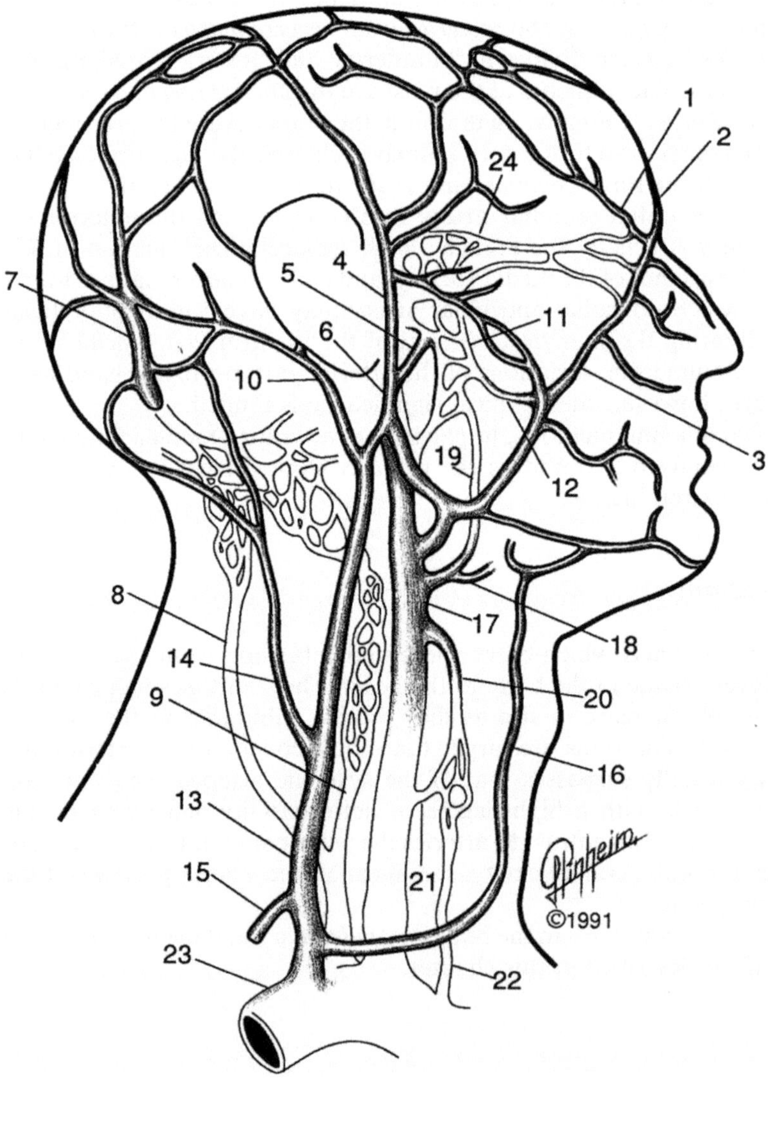

1. extracranial veins,
2. intracranial extracerebral veins, and
3. cerebral veins (Figs. 12, 13).

Two groups of cerebral veins may be identified: the superficial (external) and the deep (internal) veins.

Extracranial Veins

The *facial vein* is formed by the junction, near the medial angle of the eye, of the *supratrochlear* and the *supraorbital* veins. It has an initial downward course

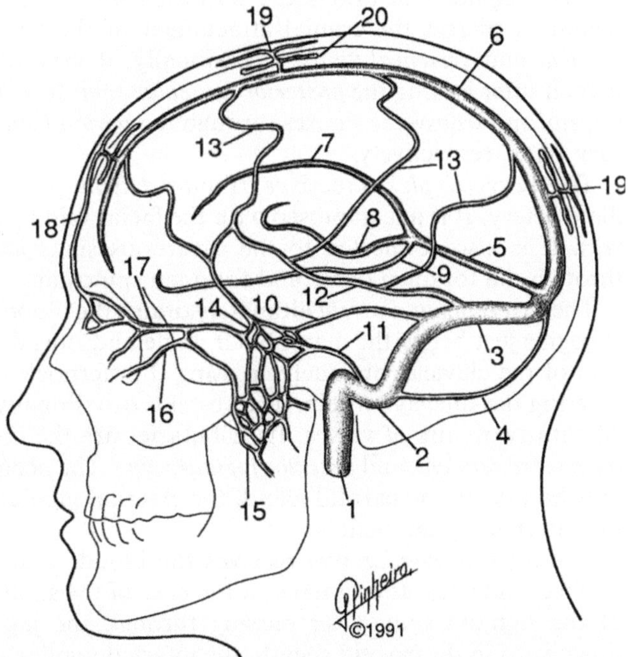

Fig. 12. Diagram representing the veins of the head and neck. *1*, supraorbital vein; *2*, supratrochlear vein; *3*, facial vein; *4*, superficial temporal vein; *5*, maxillary vein; *6*, retromandibular vein; *7*, occipital vein; *8*, deep cervical vein; *9*, vertebral vein; *10*, posterior auricular vein; *11*, pterygoid plexus; *12*, deep facial vein; *13*, external jugular vein; *14*, posterior external jugular vein; *15*, transverse cervical vein; *16*, anterior jugular vein; *17*, internal jugular vein; *18*, lingual vein; *19*, pharyngeal vein; *20*, superior thyroid vein; *21*, middle thyroid vein; *22*, inferior thyroid vein; *23*, subclavian vein

Fig. 13. Schematic representation of the major intracranial, extracerebral and cerebral veins. Some of the external veins are shown. *1*, internal jugular vein; *2*, sigmoid sinus; *3*, transverse sinus; *4*, occipital sinus; *5*, straight sinus; *6*, superior sagittal sinus; *7*, inferior sagittal sinus; *8*, great cerebral vein; *9*, basal vein; *10*, superior petrous sinus; *11*, inferior petrous sinus; *12*, inferior anastomotic vein; *13*, superior cerebral veins; *14*, cavernous sinus; *15*, pterygoid plexus; *16*, inferior ophthalmic vein; *17*, superior ophthalmic vein; *18*, supratrochlear vein; *19*, emissary veins; *20*, diploic veins

on the side of the root of the nose and then runs downward and posteriorly, approximately following the course of the facial artery, and, after receiving the anterior division of the *retromandibular vein,* finally enters the internal jugular vein below the angle of the mandible. The facial vein drains the scalp and soft tissues of the face.

The *superficial temporal vein* begins above the zygomatic arch as a result of the confluence of anterior and posterior tributaries originating with the venous network extended over the scalp. It then descends anterior to the auricle to penetrate the parotid gland, where it joins the *maxillary vein* to form the *retromandibular vein.* The latter divides into the anterior and posterior divisions, the posterior entering the external jugular vein.

The *occipital vein* drains the posterior scalp and the superficial occipital region. It pierces the cranial attachment of the trapezium to join the *deep cervical* and *vertebral* veins. Occasionally, it may terminate in the external jugular after joining the *posterior auricular vein.* It is connected to the *superior sagittal* and *transverse* sinuses through the *parietal emissary* and *mastoid emissary* veins, respectively.

The *pterygoid plexus* receives tributaries from the area supplied by the maxillary artery. It is anastomosed with the facial vein by means of the *deep facial vein.* It is also connected to the cavernous sinus and internal jugular vein through the foramen ovale and foramen spinosum.

The *external jugular vein* ultimately drains the blood from the scalp and face. It begins just below the mandibular angle and runs downward to the midportion of the clavicle, obliquely crossing the sternocleidomastoid in its course, entering the subclavian vein. The vessel is occasionally doubled and is provided with two pairs of valves. Its tributaries are the *posterior external jugular, transverse cervical,* and *anterior jugular veins.* The occipital vein is an occasional tributary. In the parotid gland, the external jugular receives a branch from the internal jugular vein.

The *internal jugular vein* receives the blood return from the brain, facial surface, and neck. It originates at the base of the skull as a direct continuation of the sigmoid sinus after passing through the jugular foramen. Running downward in the carotid sheath, the internal jugular joins the subclavian vein behind the sternoclavicular joint, where it dilates slightly (the *inferior bulb*). A pair of valves is found just before its termination. The thyrocervical trunk vertebral vein, and the first part of the subclavian artery are located posteriorly while the common carotid, internal carotid, and vagus nerve run medially. Its tributaries are the *inferior petrosal sinus* and the *facial, lingual, pharyngeal, superior thyroid,* and *middle thyroid* veins. The thoracic duct and the right lymphatic duct open into the respective subclavian vein at its junction with the internal jugular vein.

The *vertebral veins* begin at the suboccipital triangle with the confluence of several small tributaries originating in the internal vertebral plexuses. They form a dense plexus which descends surrounding the vertebral artery on its course through the foramina in the transverse processes of the cervical vertebra, and become a single vessel at the level of the sixth cervical vertebral. The vein then courses lateral to the vertebral artery to drain into the upper and posterior part of the innominate vein. The vertebral vein is posteriorly related to the internal jugular vein and runs in front of the initial segment of the subclavian artery. The anterior vertebral and deep cervical veins drain in the lower part of the vertebral vein.

Intracranial Extracerebral Veins

The *diploic veins* run within the channels in the diploë of the cranial bones. They are large, pouch-like sinuses, fully developed only at about 2 years of age [3]. The more constant channels are the *frontal diploic, anterior temporal diploic, posterior temporal diploic,* and *occipital diploic* veins.

The *meningeal veins* originate from the plexiform vessels of the dura mater and drain into the efferent vessels in the outer layer of the dura. These communicate with the cranial sinuses and the diploic veins.

Cerebral Veins

Several different venous pathways drain the blood from the intracranial structures and, in fact, each is capable of transporting the entire venous return in the presence of venous occlusion elsewhere [4]. In normal conditions, however, most of the cerebral veins drain into the internal jugular vein. The *superficial veins* drain the cortical and subcortical regions of the cerebrum and cerebellum, emptying directly into the sinuses. The *deep veins* drain the basal ganglia, corpus callosum, and part of the limbic system. They also terminate in the sinuses.

The *cerebral* and *cerebellar veins* are devoid of valves and have extremely thin walls. They drain into the cranial sinuses after piercing the arachnoid and dura mater.

The *cranial dural venous sinuses* drain the blood from the brain and bones. They lie between the two layers of the dura mater and have no valves. Two groups can be identified: the *posterosuperior* and the *anteroinferior.*

The *emissary veins* establish communications between the intracranial venous sinuses and the extracranial veins.

Although the intracranial veins are not currently accessible to ultrasound investigation, the larger channels of the neck are usually easily imaged. The clinical experience with magnetic resonance intra- and extracranial venography is limited at present.

Veins of the Thorax

The *innominate veins* are formed by the junction of the internal jugular and subclavian veins (Fig. 14). The *right innominate* vein is from 2.6 to 5.5 cm long [2] and begins at the right sternoclavicular joint. From there it runs downward almost vertically to join its counterpart on the right side of the sternum, behind the lower border of the first costal cartilage. It courses anterolateral to the right innominate artery and the right vagus nerve. Its tributaries are the *right verte-*

bral, right internal mammary, and *right inferior thyroid* veins. Occasionally it receives the first *right posterior intercostal* vein. The *left innominate vein* is 6–8 cm long [2] and begins behind the left sternoclavicular joint in front of the cervical pleura. It runs to the right and downward behind the upper half of the manubrium sternum, joining the contralateral vessel to form the superior vena cava. In its course, it passes in front of the left internal mammary, left subclavian, and left common carotid arteries, the left phrenic and vagus nerves, the trachea, and the innominate artery. The regular tributaries are the *left vertebral, left internal mammary, left inferior thyroid,* and *left superior intercostal* veins. The most common variation is represented by its separate opening into the right atrium via a left-sided superior vena cava, which drains into the coronary sinus. This situation has also been referred to as replacement of the coronary sinus by the left innominate vein [3]. No valves are present in either innominate vein.

The *internal mammary veins* are doubled in the lower half of the thorax, running on each side of the internal mammary arteries. They are provided with valves. At the level of the third or fourth costal cartilage, they converge into a common trunk which runs upward medial to the artery to drain into the respective innominate veins. Tributaries are the *anterior intercostal* and *pericardiophrenic veins.* They are the continuation of the *superior epigastric veins* in the abdomen, through which communication with the portal system is established.

The *inferior thyroid veins* begin in the thyroid gland in a venous network. The left vein then runs downward to join the left innominate vein and the right takes on oblique rightward course to join the right innominate vein. Frequently they form a single trunk which open into the superior vena cava or left innominate vein. Several branches may drain into the internal jugular vein.

The *left supreme intercostal vein* is formed by the confluence of the second and third (and sometimes the forth) left posterior intercostal veins. It runs obliquely upward and forward on the left side of the aortic arch to open into the posteroinferior aspect of the left innominate vein. The left bronchial veins drain into it and a communication with the accessory hemiazygos vein is normally present. The *right supreme intercostal vein* is most often tributary to the azygos vein.

Superior Vena Cava

The superior vena cava drains the blood from the upper half of the body (Fig. 14). It is about 7 cm long and 2 cm wide [2, 10] and is devoid of valves. It begins at the inferior border of the first right intercostal cartilage and descends vertically to enter the posterosuperior portion of the right atrium. Its lower half is covered by the fibrous pericardium. The superior vena cava is related to the ascending aorta and initial portion of the aortic arch on the left side, to the right upper lobe of the lung on the right side, and to the right pulmonary artery and right upper pulmonary vein posteriorly. Its tributaries are the *innominate* veins, the *azygos* vein and the small *pericardial* and *mediasti-*

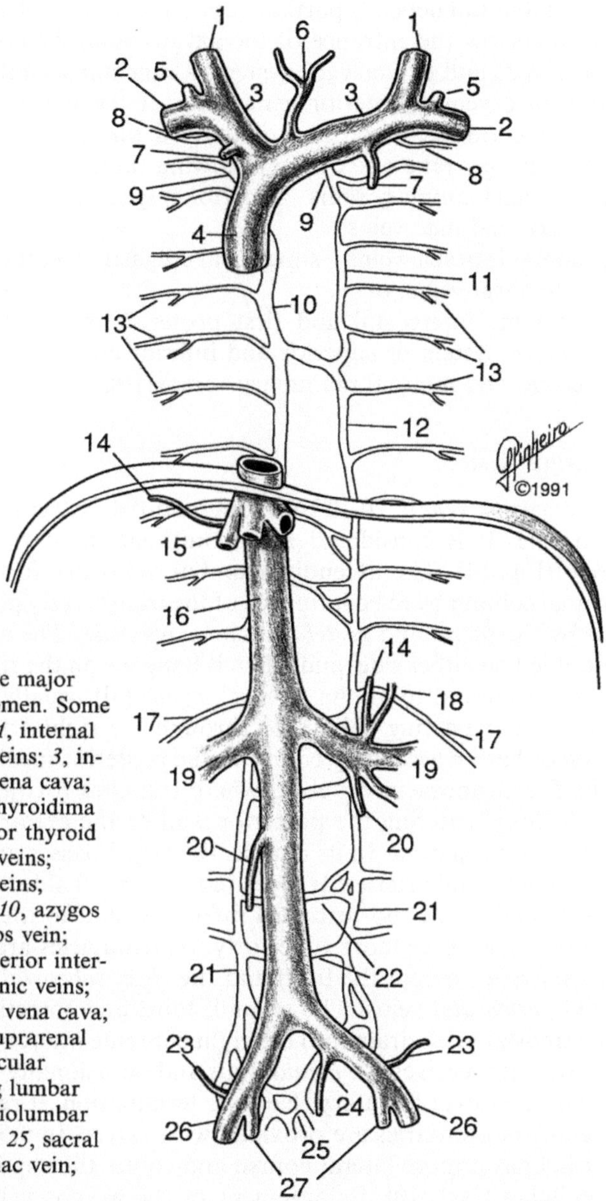

Fig. 14. Diagram showing the major veins of the thorax and abdomen. Some pelvic veins are also shown. *1*, internal jugular veins; *2*, subclavian veins; *3*, innominate veins; *4*, superior vena cava; *5*, external jugular veins; *6*, thyroidima vein (confluence of the inferior thyroid veins); *7*, internal mammary veins; *8*, first posterior intercostal veins; *9*, supreme intercostal veins; *10*, azygos vein; *11*, accessory hemiazygos vein; *12*, hemiazygos vein; *13*, posterior intercostal veins; *14*, inferior phrenic veins; *15*, hepatic veins; *16*, inferior vena cava; *17*, subcostal veins; *18*, left suprarenal vein; *19*, renal veins; *20*, testicular (ovarian) veins; *21*, ascending lumbar veins; *22*, lumbar veins; *23*, iliolumbar veins; *24*, median sacral vein; *25*, sacral venous plexus; *26*, external iliac vein; *27*, internal iliac vein

nal veins. Persistence of the left superior vena cava is the most frequent variation, occurring in 0.3% of the general population, although this number rises to 2%–4.4% in patients with congenital heart disease [11]. A communication may be present between the two venae cavae (occurring in 50% of cases), or the right superior vena cava may be hypoplastic or absent [2, 12]. The distribution of the collateral circulation in cases of occlusion of the superior vena cava is variable and depends partially upon the level of obstruction. If the occlusion occurs below the entrance of the azygos vein, the collateral pathway follows the azygos and hemiazygos systems – ascending lumbar vines – renal or common iliac veins – inferior vena cava. If the occlusion is located above the azygos entrance, blood flow is derived through the innominate veins, from where it may take one of the following three pathways:

1. internal mammary veins – superior epigastric veins – inferior epigastric veins – external iliac veins;
2. lateral thoracic veins – superficial epigastric veins – greater saphenous veins – femoral veins; or
3. supreme intercostal and first posterior intercostal veins – azygos/hemiazygos system or superior and inferior epigastric veins (via anterior intercostal and internal mammary veins) [10].

Azygos System

The *azygos vein* is 20–25 cm long and 0.4–1.0 cm wide. Its origin is widely variable. It is considered as a continuation of the *right ascending lumbar vein* (Fig. 14). The ascending lumbar veins are formed on each side of the spinal column by the confluence of the transversely placed *lumbar veins,* as well as by the *sacral* and *vertebral venous plexuses*. The ascending lumbar may be doubled on either side, and when it happens on the right side, the more medial vessel (referred to as the *lumbar azygos* [3]) usually continues as the azygos vein. After passing behind or through the right crus of the diaphragm, the azygos has an ascending course in the posterior mediastinum until the level of the fourth thoracic vertebra, where it arches forward above the root of the right lung, entering the posterior wall of the extrapericardic segment of the superior vena cava. In its course, the vessel passes anterior to the right posterior intercostal arteries. It is related on the left side to the descending thoracic aorta and to the thoracic duct inferiorly, and to the esophagus, trachea, and right vagus nerve more superiorly. Its tributaries are the *right posterior intercostal veins* (except the first) and the *right subcostal, esophageal, mediastinal,* and *pericardial* veins. The second, third and sometimes the fourth posterior intercostal veins drain into the right supreme intercostal vein. The azygos vein frequently receives the *hemiazygos* and occasionally the *hemiazygos accessory* vein as a direct tributary. Near its termination, it receives the *right bronchial* veins. Its tributaries are provided with valves. In 0.5% of the population, the vessel has a more lateral course and enters the superior vena cava at a more cephalad level [10]. Enlargement of the azygos vein (more than 5–7 cm in diameter) may be present in congestive heart failure or hepatic cirrhosis, or when it is participating in the collateral circulation in cases of occlusion or absence of the inferior vena cava.

The *hemiazygos vein* begins on the left side of the spinal column similarly to the azygos and ascends up to the level of the eighth or ninth thoracic vertebra, then crosses behind the aorta (or, less frequently, anterior to it [2]), the esophagus, and the thoracic duct to enter the azygos vein (Fig. 14). Its tributaries are: the lower three *posterior intercostal veins,* the *left subcostal vein,* and some of the *esophageal* and *mediastinal* veins. At its lower end, the hemiazygos frequently communicates with the left renal vein.

The *accessory hemiazygos vein* descends on the left side of the spinal column. It receives the *posterior intercostal veins* (4th–8th) on the left side and sometimes the *left bronchial veins.* The mode of termination is extremely variable [3, 6]. It may cross the body of the seventh thoracic vertebra to join the azygos vein or it may proceed downward to join the hemiazygos vein. A communication with the left innominate vein by means of the supreme intercostal vein is frequently present (Fig. 14).

Considering the entire azygos system, right side predominance with a variable configuration of less prominent veins on the left side is the most frequent presentation, occurring in 98% of cases; two separate parallel azygos veins of approximately equal sizes occurs in 1%; a single azygos vein running in the midline anterior to the thoracic vertebrae is also present in 1% of the population [6]. The number of retroaortic transvertebral connections between the right and left sided veins may vary from one to five, and they occur most frequently at the level of the eighth thoracic vertebra [6, 10].

Presently, the major veins of the azygos system cannot be consistently imaged by ultrasound in adult patients. In children, the terminal azygos vein may be visualized, especially in those conditions which cause its dilatation.

Veins of the Abdomen

Although both the inferior vena cava and the portal vein and their respective tributaries are considered part of the same group of systemic veins, they are usually described separately due to their anatomic and functional peculiarities.

Inferior Vena Cava

The inferior vena cava is responsible for the drainage of blood from the abdomen and lower extremities. After its origin at the junction of the two common iliac veins, the inferior vena cava runs upward in front of the vertebral column, initially to the right and slightly posterior to the aorta, gradually adopting a more anterior position as it ascends. Reaching the liver, it runs in a groove in the posterior aspect of that organ, passes through the tendinous part of the diaphragm and then the fibrous pericardium, and enters the inferoposterior aspect of the right atrium. Its channel is valveless except for the *valve of the inferior vena cava* or *Eustachian valve,* located at its entrance into the right atrium. The thoracic segment of the vein is short, and is usually divided into an extrapericardial and an intrapericardial portion. Variations are rare. Two caval veins have been reported to be present at a frequency which varies from 0.2% to 3.0% [10]. The duplication occurs below the renal veins with no cross-communication between the common iliacs. The right vein is usually larger and the left vessel is usually continuous with the left renal vein. A left-sided infrarenal inferior vena cava (in the presence of situs solitus) occurs in 0.2% to 0.5% of the population; the suprarenal segment courses normally on the right side of the column. The main tributaries of the inferior vena cava are the *lumbar, right testicular or ovarian, renal, right suprarenal, inferior phrenic* and *hepatic* veins (Fig. 14).

A total of four *lumbar veins* is found on each side, with the first and second often draining into the ascending lumbar vein. The latter connects the common iliac, iliolumbar, and lumbar veins.

The *right testicular* (*ovarian*) vein joins the inferior vena cava below the entrance of the renal veins, while the left vessel is a tributary of the left renal vein.

The *left renal vein* is three times the length of the right and it courses to the right, posterior to the splenic vein and the body of the pancreas, crosses between the abdominal aorta and the superior mesenteric artery, and usually enters the inferior vena cava at a higher level than the right vein. The left renal vein may be doubled, and in that case, one runs anterior and the other posterior to the aorta.

The *right suprarenal vein* joins the posterior aspect of the inferior vena cava, while the left-sided vein drains into the left renal vein.

The *inferior phrenic* veins follow the course of the corresponding arteries, with the right vein draining into the inferior vena cava and the left, usually paired, draining into the inferior vena cava and left renal or suprarenal vein.

The *hepatic veins* drain the liver, opening into the inferior vena cava at the groove in the posterior surface of the liver. An upper and lower group can be identified. The upper group is subdivided into right, middle, and left hepatic veins, draining the right, caudate, and left lobes of the liver, respectively. The lower group is variable, composed of small veins, and returns blood from the right and caudate lobes of the liver. The hepatic veins are devoid of valves. A large *accessory hepatic vein,* related to the lower group, is present in 15% of cases, draining part of the right lobe [3].

Portal System

The *portal vein* drains all the veins from the abdominal part of the digestive tract (except the terminal colon and anal canal), including the spleen, pancreas, and gallbladder. It is about 6.5 cm [6] long and begins at the level of the second lumbar vertebra, anterior to the inferior vena cava and posterior to the neck of the pancreas, by the junction of the splenic and superior mesenteric veins (Fig. 15). Numerous variations can be found in its formation. From its origin, it runs upward and to the right towards the porta hepatis, where it divides into left and right stems, following the branches of the hepatic artery. Inside the liver, the portal vein divides like an artery until the level of capillary-like vessels (sinusoids), which ultimately converge to drain into the hepatic veins. In an adult individual, no valves are present in the portal vein and tributaries. The

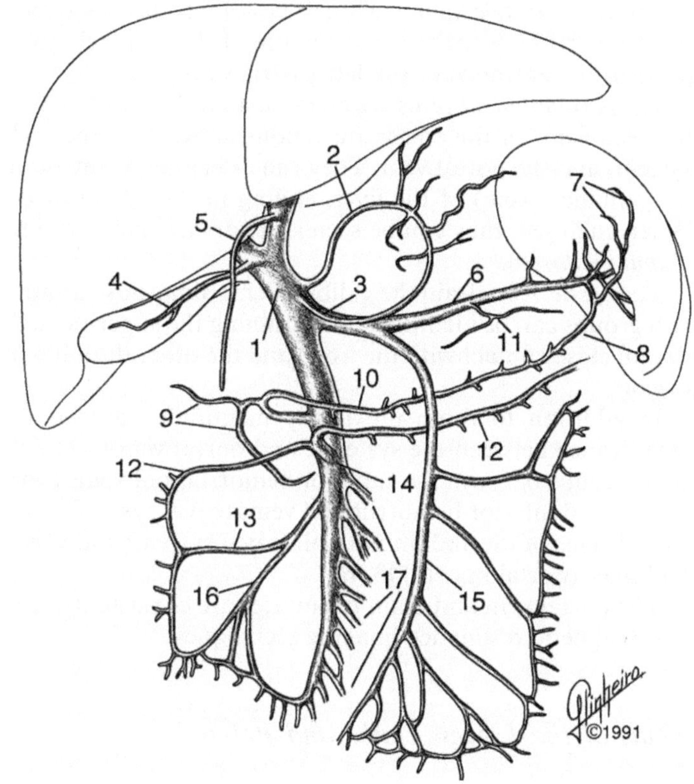

Fig. 15. Schematic representation of the portal vein and its tributaries. *1*, portal vein; *2*, left gastric vein; *3*, right gastric vein; *4*, cystic vein; *5*, paraumbilical vein; *6*, splenic vein; *7*, short gastric veins; *8*, left gastroepiploic vein; *9*, pancreaticoduodenal veins; *10*, right gastroepiploic vein; *11*, pancreatic veins; *12*, middle colic vein; *13*, right colic vein; *14*, superior mesenteric vein; *15*, inferior mesenteric vein; *16*, ileocolic vein; *17*, jejunal and ileal veins

following veins are tributaries of the portal vein: *splenic, superior mesenteric, left gastric, right gastric, paraumbilical,* and *cystic* veins.

The *splenic vein* is a large vessel formed by the confluence of two to six venous trunks which return the blood from the spleen [6]. It runs downward and to the right, posterior to the body of the pancreas and anterior to the left kidney. It is separated from the aorta by the superior mesenteric artery and left renal vein. Its tributaries are the *short gastric, left gastroepiploic, pancreatic,* and *inferior mesenteric* veins.

The *superior mesenteric* vein drains the small intestine, cecum, and ascending and transverse colon. It begins at the right iliac fossa and ascends on the right of the superior mesenteric artery. Its tributaries are the *jejunal, ileal, ileocolic, right colic, middle colic, right gastroepiploic,* and *pancreaticoduodenal* veins.

The *left gastric* vein drains both surfaces of the stomach, and ascends along the lesser curvature towards the distal esophagus. It then curves downward and to the right to enter the initial segment of the portal vein. At the distal end of the esophagus, it communicates with the esophageal veins (caval system). In cases of portal obstruction, the anastomotic channels are enlarged, resulting in large submucosal esophageal varices.

The *right gastric* vein is a small vessel which runs from left to right at the pyloric portion of the lesser curvature of the stomach, joining the portal vein prior to the entrance of the left gastric vein.

The *paraumbilical* veins are an important source of anastomotic connections between veins of the epigastric venous network in the abdominal wall (caval system) and the portal vein. They run along the ligamentum teres (remnant of the umbilical vein) of the liver, ending in the left branch of the portal vein. When enlarged, they can be seen as a varicose network around the umbilicus (*caput medusae*).

The *cystic veins* drain the gallbladder. Numerous variations exist but at least two groups can be identified, one draining the superior surface of the gallbladder (in close contact with the liver) and the other draining the remainder of the organ.

In addition to those previously mentioned, anastomoses of less clinical significance between the systemic and portal venous circulation include communications between the superior hemorrhoidal veins (portal system) and the middle and inferior hemorrhoidal veins (caval system), and between retroperitoneal veins of the abdominal wall (caval system) and venous radicles of colon and liver (portal system) [3, 6].

The portal vein and main tributaries are consistently imaged by ultrasound and magnetic resonance imaging techniques.

Veins of the Lower Limbs and Pelvis

Two venous systems can be identified in the lower limb: the *superficial* and the *deep venous systems,* the first running between the two layers of the superficial fascia and the second accompanying the arteries with which it shares common sheaths. Distal to the popliteal vein, the deep veins are doubled, forming plexuses surrounding the arteries. Valves are encountered more frequently in the deep veins. The *communicating* and *perforating* veins have a transverse course. The communicating veins connect the deep and superficial venous systems and the perforating veins link the deep system to the muscle veins. In these veins, the valves are oriented in such a way that the blood flow is directed from the surface to the deeper layers. Figure 16 shows diagrammatically both the superficial and deep venous systems.

Superficial Venous System

The major vessels which form the superficial venous system are the *greater* and *lesser saphenous veins*. The greater saphenous vein originates from the medial side of the dorsal venous arch in the foot. It then ascends in front of the medial malleolus of the tibia, running upward in the medial aspect of the leg and thigh to finally drain into the femoral vein through the saphenous opening in the deep fascia, about 2.5 cm below the inguinal ligament. Along its course in the leg and thigh, the greater saphenous vein receives several cutaneous tribu-

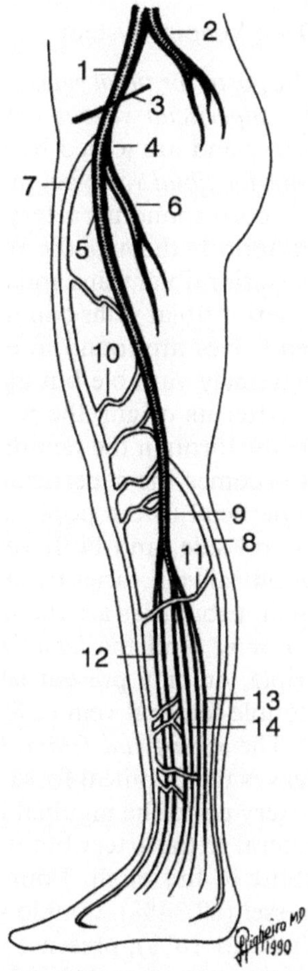

Fig. 16. Schematic representation of the superficial and deep venous systems. *1*, external iliac vein; *2*, internal iliac vein; *3*, inguinal ligament; *4*, common femoral vein; *5*, superficial femoral vein; *6*, profunda femoris vein; *7*, greater saphenous vein; *8*, lesser saphenous vein; *9*, popliteal vein; *10*, perforating veins; *11*, superficial crosscommunication; *12*, anterior tibial veins; *13*, posterior tibial veins; *14*, peroneal veins

taries, and at the level of the saphenous opening, the *superficial epigastric, superficial circumflex iliac,* and *external pudendal veins.* The valves are more numerous in the thigh than in the leg, and the total number varies from two to six.

The lesser saphenous vein commences at the lateral side of the dorsal venous arch in the foot and ascends obliquely, posterior to the lateral malleolus of the fibula, to reach the posterior aspect of the leg. At the lower part of the popliteal fossa, this vein perforates the deep fascia and passes between the two heads of the gastrocnemius muscle to drain into the popliteal vein. Numerous cutaneous tributaries join the lesser saphenous along its course. Just before perforating the deep fascia, the lesser saphenous is seen to communicate with the greater saphenous through the *femoropoplital vein.* The number of valves varies from two to twelve.

Deep Venous System

The *posterior tibial veins* are formed by the union of the *plantar venous arch* and the *superficial venous network* of the foot. They follow the posterior tibial artery and are joined by the *peroneal veins* in the upper third of the leg. The *anterior tibial veins* begin at the *dorsal venous network* and follow the path of the corresponding artery. They pass between the tibia and fibula, curving posteriorly through the aperture above the interosseous membrane to unite the posterior tibial veins and form the *popliteal vein*. At the dorsum of the foot, the anterior tibial veins communicate with the superficial veins of the leg. At least ten valves are found in each deep vein of the leg. The level of confluence is extremely variable but agenesis of any of the leg veins is rare.

After its origin, the *popliteal vein* runs upward to the medial aspect of the femur through the tendinous aperture in the adductor magnus muscle, where it becomes the superficial femoral vein. The popliteal vein lies posterior and superficial to the popliteal artery, running obliquely from the medial to the lateral side, and both vessels share a common sheath. The diameter of the popliteal vein varies from 8 to 15 mm, and it usually contains four valves. The main tributaries are the *sural veins* from the gastrocnemius muscle, the *articular veins,* and the *lesser saphenous vein.* The vein may be single, double, or triple, or may present islet formation. Duplication may be associated with double femoral vein (2.8% of the cases).

The *superficial femoral vein* is the continuation of the popliteal vein as it leaves the popliteal fossa. It courses upward following the superficial femoral artery up to the inguinal ligament. In the lower part of its course, the vein lies lateral to the artery but it becomes posterior to it when approaching the upper third of the thigh. Four variations have been described [13]: normal single vessel (62.34%), double vessel in the lower segment with subsequent confluence (21.16%), presence of multiple femoral veins (13.72%), and duplication of the entire vessel (2.78%). In case of agenesis, the femoral vein is replaced by the accompanying vein of the sciatic nerve or a superficial collateral [2]. The diameter of the superficial femoral vein ranges from 0.9 to 1.0 cm. Two to five valves may be present.

The *profunda femoris vein* is formed by the confluence of the *venae comitantes* which follow the perforating branches of the profunda femoris artery. It drains into the lateral aspect of the superficial femoral vein about 3.5–4 cm below the inguinal ligament, at which confluence the common femoral vein begins.

External Iliac Veins

The *external iliac veins* begin on each side at the level of the inguinal ligament and run medially to join the internal iliac veins at the sacroiliac joint. The vein on the right side courses medial and then posterior to the corresponding artery; the left-sided vein has a course medial to the artery. Usually no valves are present. Its tributaries are: the *inferior epigastric, deep circumflex iliac,* and *pubic* veins. The inferior epigastric communicates with the internal thoracic, and the deep circumflex with the iliolumbar vein.

Internal Iliac Veins

The paired *internal iliac veins* drain the blood flow from the pelvic veins. They run posterior and slightly medial to the corresponding artery and join the common iliac vein at the sacroliliac joint. They have no valves. The vessel on each side may be single, double, or plexiform. Its branches follow closely the arteries in pairs. The parietal tributaries are the *superior* and *inferior gluteal veins,* which connect to the *obturator veins.* The latter form anastomoses with the external iliac vein. The *lateral sacral veins* form the sacral venous plexus. The visceral tributaries form numerous anastomoses at the floor of the pelvis, except for the dorsal vein of the penis or the clitoris. The internal pudendal vein receives the veins from the scrotum and labium majus and the deep vein of the penis and clitoris. The internal pudendal vein anastomoses with the rectal venous plexus. The vesical veins form the vesical venous plexus around the bladder, which communicates with the prostatic and rectal veins. The prostatic venous plexus receives the flow from the dorsal vein of the penis. The rectal venous plexus forms an important anastomosis with the inferior mesenteric vein. The largest pelvic plexus in the woman is the uterine and vaginal venous plexus, a network surrounding the body and neck of the uterus and the upper part of the vagina. It is drained by the uterine veins into the internal iliac vein.

Common Iliac Veins

Each common iliac vein starts in front of the sacroiliac joint and runs parallel to the common iliac arteries to join the inferior vena cava. They receive the venous return from the pelvis and lower extremities. The average length is 7.5 cm on the left and 5.5 cm on the right common iliac vein. The left vein runs medial to its corresponding artery and the right one posteromedial. Each common iliac vein receives the *iliolumbar* and sometimes the *lateral sacral* veins. The *median sacral* vein drains into the left common iliac. Variations at this level are rare. Agenesis is associated with simultaneous absence of the inferior vena cava. In some instances, the left common iliac vein does not join the contralateral vessel and runs upward on the left side of the aorta, joining the inferior vena cava more superiorly.

Communicating (Perforating) Veins

The communicating veins connect the superficial venous system with the deep veins in the leg, knee, and thigh. At each of these levels, three groups are identified: the *medial, lateral,* and *posterior* veins [14].

At the leg, the medial veins consist of three groups which connect the great saphenous vein and the posterior tibial veins. The lateral veins form four groups, two of them connecting the greater saphenous vein and the anterior tibial veins, another the greater saphenous with the peroneal veins, and the fourth the greater saphenous with the veins from the gastrocnemius muscle. The posterior veins connect the veins from the gastrocnemius muscle with the lesser saphenous vein.

At the knee joint, the medial veins allow communication between the greater saphenous and popliteal veins. The lateral veins run from the subcutaneous veins to the popliteal veins. The posterior veins connect the lesser saphenous vein to the popliteal vein.

At the thigh, the medial veins are divided in three groups which connect the accessory saphenous vein and the muscular veins and which also run from the greater saphenous vein to the superficial femoral vein. Two groups of lateral veins connect the subcutaneous and the muscular veins. The posterior veins run from the femoropopliteal vein to the superficial femoral vein.

The major veins of the pelvis and lower limbs are usually adequately visualized by magnetic resonance imaging. Ultrasound examination allows the visualization of the major trunks but more distal evaluation is usually based on Doppler (either pulsed or color-coded) information.

Veins of the Upper Limbs

As in the lower limbs, the veins of the upper limbs can be divided into two groups, the *superficial* and *deep* venous systems. In this case, however, in contrast to venous function in the lower extremities, the superficial system plays a more important role in returning the blood from the limbs. The deep veins are usually paired and accompany the corresponding arteries. They have a larger number of valves than the superficial veins, with which they freely communicate, particularly in the vicinity of the joints. The muscular veins drain into the superficial veins and, in contrast with the lower limbs, the muscles of the arms do not have significant pumping function to help the venous return, as the major channels lie superficial to the muscular layers, with resultant low venous tension.

The *radial, ulnar, brachial,* and *axillary* veins constitute the deep venous system. The radial and ulnar are companion veins of the radial and ulnar arteries. Each pair unites at the level of the elbow joint to continue as the paired brachial veins. Near the lower margin of the subscapularis muscle, the brachial veins join the axillary vein. The vein medially located, however, frequently joins the basilic vein. The *axillary vein* at the outer border of the first rib. The vessel courses medial to the corresponding artery and is provided with one valve. A double axillary vein is occasionally found. Its major tributaries are the *lateral thoracic* and the *thoracoepigastric* veins, the latter being an important potential source of collateral communication between the inferior and superior vena cava.

The *subclavian vein* is a continuation of the axillary vein up to the medial border of the scalenus anterior muscle, where it joins the jugular vein to form the innominate vein. It is provided with a valve approximately 2 cm from its termination. Duplication of the subclavian vein occurs in 1% of the population, and an accessory subclavian vein as well as islet formation may be present [2]. The major tributaries are the *external jugular*, the *dorsal scapular*, and sometimes the *anterior jugular* veins. The left vein receives the *thoracic duct* and the right receives the *right lymphatic duct*.

The major superficial veins are the *cephalic*, *basilic*, and *median cubital* veins and the median vein of forearm (Fig. 17). The *cephalic* vein runs upward from the dorsal venous network of the hand, then turns round the radial border of the arm to reach the anterior surface. At the cubital fossa it communicates with the basilic vein via the *median cubital vein*. Subsequently, it ascends subcutaneously lateral to the biceps, then between the deltoid and the pectoralis major, entering the infraclavicular fossa to end, after crossing the axillary artery, in the axillary vein 2–3 cm below the clavicle. Sometimes it communicates with

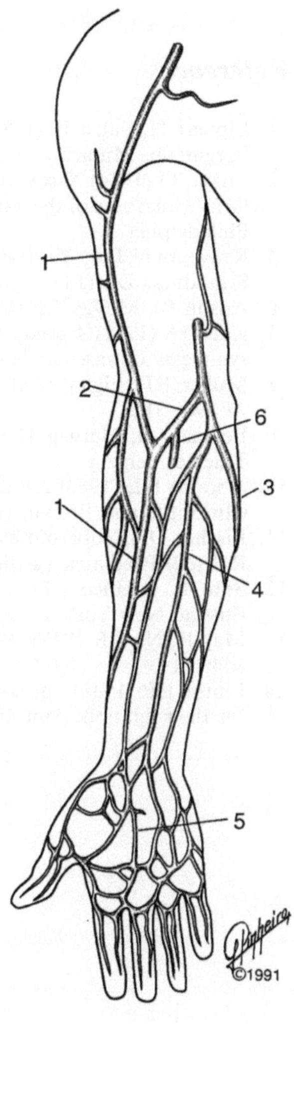

Fig. 17. Superficial veins of the upper limb (ventral aspect). *1*, cephalic vein; *2*, median cubital vein; *3*, basilic vein; *4*, median vein of forearm; *5*, palmar venous plexus; *6*, anastomotic branch from the deep veins

the external jugular, subclavian, or internal jugular vein. When the proximal segment of the cephalic vein is hypoplastic or absent, the median cubital vein may carry all the blood from the distal cephalic into the basilic vein. The cephalic vein is frequently doubled in the forearm.

The *basilic vein* begins at the ulnar part of the dorsal network. It ascends on the ulnar side, turns round anteriorly below the elbow joint, receiving the median cubital vein in the groove between the biceps and the pronator teres, and then runs upward in the medial bicipital groove. After perforating the deep fascia at the midlevel of the arm, it ascends medial to the brachial artery to continue as the axillary vein.

The *median vein of the forearm* originates in the superficial palmar venous plexuses to run upward in front of the forearm. It may end in either the basilic or median cubital veins.

References

1. Lippert H, Pabst R (1985) Arterial variations in man – classification and frequency. Bergmann, Munich
2. Luzsa, G (1974) X-ray anatomy of the vascular system. Lippincott, Philadelphia
3. Gray's anatomy of the human body, 30th edn (1985) Clemente CD (ed) Lea and Febiger, Philadelphia
4. Krayenbuhl HA, Yasargil MG (1968) Cerebral angiography. Butterworth, London
5. Strandness DE (1969) Collateral circulation in clinical surgery. Saunders, Philadelphia
6. Anson BJ, McVay CB (1984) Surgical anatomy, 6th edn. Saunders, Philadelphia
7. Keen JA (1961) A study of the arterial variations in the limbs, with special reference to symmetry of vascular patterns. Am J Anat 108:245–261
8. Muller RF, Figley MM (1957) The arteries of the abdomen, pelvis and thigh. AJR 77:296–311
9. Bellman G, Herwig H (1964) Die Aortographie in der angiologischen Diagnostik. Thieme, Leipzig
10. Abrams HL (1983) Abrams's angiography: vascular and interventional radiology, 3rd edn. Little and Brown, Boston
11. Fischer DR, Zuberbuhler JR (1987) Anomalous systemic venous return. In: Anderson RH (ed) Paediatric cardiology. Churchill Livingstone, New York
12. Soto B, Pacifico AD (1990) Angiocardiography in congenital heart malformations. Futura, New York
13. May R, Nissl R (1966) Phlebographische Studien zur Anatomie der Beinvenen. ROFO 104:171
14. Linton RR (1938) The communicating veins of the lower leg and the operative technique for their ligation. Ann Surg 107:582

Chapter 2

Hemodynamics

X. P. LEFEBVRE, E. M. PEDERSEN, J. Ø. HJORTDAL, and A. P. Yoganathan

Introduction

This chapter focuses on the basic hemodynamic principles as applied to blood flow in the large arteries of the body and across arterial stenoses. The pulsatile nature of blood flow, the elasticity of the arteries and their complex geometry, the nature of peripheral resistance and the non-Newtonian character of blood are just a few important parameters that have to be taken into consideration when describing blood flow in the human body. Due to these complexities, a theoretical description of the hemodynamic principles of the human arterial tree is at best incomplete.

Arterial hemodynamics is defined as the study of pressure, flow rate, and instantaneous velocities. Traditionally, flow rates and pressure can be measured in clinical settings. More recently, the introduction of color Doppler imaging and nuclear magnetic resonance has enabled clinical measurements of the local distribution of instantaneous velocities in the human arterial system under both normal and pathologic conditions. These imaging techniques have been used to enhance our knowledge of the arterial flow field through experiments performed either in in vitro models or in in vivo studies of animals or humans. In addition, in recent years numerical simulations using computer-assisted mathematical models have greatly expanded the present knowledge.

To improve the understanding of regional hemodynamics as characterized by the new imaging techniques, this chapter reviews the basic hemodynamic principles governing blood flow in the human arterial circulation.

Basic Principles of Fluid Mechanics in a Straight Tube

Steady Flow

Although blood flow in the human body is by nature pulsatile, in order to fully understand the basic hemodynamic principles it is necessary to describe first flow concepts applicable to the steady flow of a Newtonian fluid inside a straight, rigid, circular tube.

Viscous Forces, Inertial Forces, and Reynolds Number

The nature of the flow is primarily determined by two forces that act on the fluid particles, the viscous and the inertial forces [1]. When viscous forces are predominant, the flow is laminar and is characterized by smooth motion of the fluid particles. When inertial forces are predominant, the flow is turbulent and the particles display an irregular, or essentially random motion superimposed on the main motion of the flow field. The character of the flow in a tube or in a blood vessel can be approximately described by the ratio of inertial to viscous forces. This dimensionless number, known as the Reynolds number (*Re*), is small when the flow is laminar and large when it is tubulent. *Re* is defined as

$$Re = \frac{\text{inertial force}}{\text{viscous force}} = \frac{\varrho \langle V \rangle d}{\mu} \tag{1}$$

where ϱ [kg/m^3] is the fluid density, $\langle V \rangle$ [m/s] is the average fluid velocity across the tube, d [m] is the tube diameter, and μ [Ns/m^2] is the fluid viscosity. Although *Re* does provide insight into the basic nature of the flow field, the transition from laminar to turbulent flow can also be influenced by other parameters. For example, the presence of upstream obstructions or wall vibrations results in turbulence at lower *Re*, while a very smooth wall surface will delay the appearance of turbulence.

Velocity profiles

The velocity profile can also characterize the nature of the flow field. In a straight tube, as *Re* increases, the velocity profile across the diameter changes from a parabolic (laminar flow, $Re < 1200$) to a flat shape (turbulent flow, $Re > 2300$) [2]. Figure 1 illustrates these changes. When $1200 < Re < 2300$, the flow is neither laminar nor turbulent. Such a flow is called transitional, and little is known about its precise nature. It is generally accepted that the flow in the human body is predominantly laminar with the exception of the ascending aorta, downstream of stenoses or artificial valve, and upstream of insufficient valves.

For laminar flow ($Re < 1200$), the velocity profile and the mean velocity are given by Eqs. 2 and 3, respectively. Applying the Haagen-Poiseuille law, it is also possible to calculate the pressure drop as a function of the flow rate in a tube (Eq. 4) [3].

$$V = V_{\max}\left(1 - \left(\frac{r}{R}\right)^2\right) \quad \text{for} \quad Re < 1200 \tag{2}$$

$$\langle V \rangle = \frac{V_{\max}}{2} \quad \text{for} \quad Re < 1200 \tag{3}$$

$$\Delta P = 128\,\frac{\mu L Q}{\pi d^4} \quad \text{for} \quad Re < 1200 \tag{4}$$

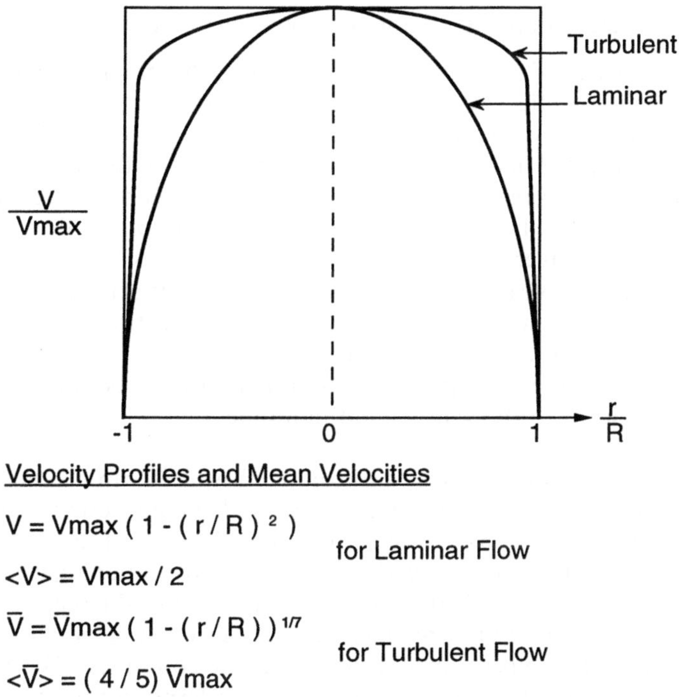

Fig. 1. Velocity profiles in a straight tube for fully developed laminar and turbulent flow

where V [m/s] is the fluid velocity at a distance r [m] from the center of the tube, V_{max} [m/s] the maximum (centerline) velocity, and R [m] the radius of the tube. ΔP [Pa] is the pressure drop along a length L [m] of tube and Q [m^3/s] is the flow rate.

When $Re > 2300$, the flow is turbulent and the instantaneous velocity V at a given location is the sum of a time-averaged velocity $\bar{V}$ [m/s] and a fluctuating velocity V' [m/s].

$$V = \bar{V} + V' \tag{5}$$

The time-averaged velocity characterizes the coherent, or deterministic, structures (eddies) of the flow while V' represents the random structures (pure turbulence). For steady turbulent flow, the velocity profile and the mean velocity are described by Eqs. 6 and 7 respectively:

$$\bar{V} = \bar{V}_{max}\left(1 - \left(\frac{r}{R}\right)\right)^{1/7} \quad \text{for} \quad Re < 2300 \tag{6}$$

$$\langle \bar{V} \rangle = \frac{4}{5}\,\bar{V}_{max} \quad \text{for} \quad Re < 2300 \tag{7}$$

where $\bar{V}$ [m/s] is the time-averaged velocity at a distance r from the center of the tube and $\langle \bar{V} \rangle$ [m/s] is the average of $\bar{V}$ at a given cross section of the pipe. $\bar{V}_{max}$ [m/s] is the maximum (centerline) time-averaged velocity.

Flow Development Region and Boundary Layer Concept

The discussed velocity profiles are valid only for a fully developed flow region, where the effect of any upstream disturbance such as bifurcation, sudden contraction, or curvature can be neglected. The effects of flow disturbances are concentrated in a region located downstream of the disturbance, known as the development region. Downstream of the development region, the flow profile regains its fully developed shape. The nature of the flow in the development region is governed by the following constraints:

1. Due to conservation of mass or continuity principle, the average velocity across the tube remains constant at any location downstream of the disturbance for a constant diameter tube.[1]
2. Due to the no-slip condition at a solid-fluid boundary, the fluid particles adjacent to the wall remain motionless.

The effect of these constraints on the flow field can be understood by examining the flow field downstream of a sudden contraction of a conduit. Immediately after the contraction the flow profile is flat (Fig. 2). Then the motionless particles close to the wall decelerate and slow other adjacent particles through viscous forces, while the fluid near the center of the pipe accelerates to comply with the continuity principle. As a result of these accelerating and retarding forces, the velocity profile development continues. The region where these shearing deceleration and reciprocal acceleration forces have an effect on the velocity profile is called the boundary layer (Fig. 2). In a boundary layer, viscous effects are predominant and inertial forces can be neglected [4].

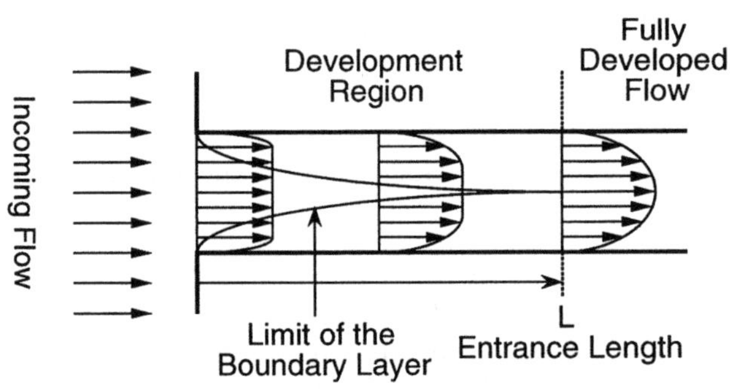

Entrance Length

$L = 0.03\ D\ Re$ for Laminar Flow

$L = 0.693\ D\ Re^{1/4}$ for Turbulent Flow

Fig. 2. Development of velocity profiles in a cylindrical tube after a sudden contraction for laminar flow

[1] The continuity principle states that in the absence of chemical reaction and when there is no accumulation of fluid, the amount of fluid entering any cylindrical element of the pipe is equal to the amount of fluid exiting it.

Entrance Length Concept

The thickness of the boundary layer increases with increasing distance downstream from the disturbance and eventually reaches its maximum when the flow becomes fully developed. For laminar flow, the viscous forces are predominant everywhere in the tube, and the boundary layer eventually occupies the entire cross section of the pipe. For turbulent flow, viscous forces can be neglected everywhere except very close to the wall, and the boundary layer therefore remains very thin. The entrance length, i.e., the length of the development region, can be estimated from Eq. 8 for laminar flow and from Eq. 9 for turbulent flow.

$$L = 0.03\ d\ Re \qquad \text{for} \qquad Re < 1200 \tag{8}$$

$$L = 0.693\ d\ Re^{1/4} \qquad \text{for} \qquad Re > 2300 \tag{9}$$

Shear Stress and Shear Rate

The shear stress, τ [N/m^2], is another important hemodynamic parameter relevant for blood flow in large arteries. It characterizes the drag force caused by viscous effects that the fluid exerts on the vessel walls or on nearby fluid particles. An excessive shear stress can cause mechanical damage or alteration of biological properties of red blood cells, platelets, or endothelial cells. In normal arteries, the shear stress is low and the blood cells, as well as the endothelium, remain intact. However, atherosclerotic lesions or artificial valves increase the shear stress level, which can cause cellular damage. In laminar flow, the shear stress is proportional to the shear rate (dV/dr) and is defined as (Newton's law of viscosity)

$$\tau = -\mu \frac{dV}{dr} \qquad \text{for} \qquad Re < 1200 \tag{10}$$

where V is the velocity at radial position r. The shear rate characterizes the rate of change of the blood velocity with respect to position. In turbulent flow, the boundary layer is thin, and the shear stress is high. Therefore, the sharp increase in wall shear stress that can be seen as the flow becomes turbulent is due to steeper velocity gradients at the walls. Using the pressure drop measurements along a length L of a pipe, it is possible to estimate the shear stress at the wall of the vessel for both laminar and turbulent flow (Eq. 11) [5].

$$\tau_{\text{at the wall}} = \frac{d\,\Delta P}{4L} \tag{11}$$

Turbulence Intensity

It is possible to estimate the level of turbulence in a flow field with a quantity called turbulence intensity, denoted I. This number is defined as:

$$I = \frac{V_{rms}}{\langle \bar{V} \rangle} 100\% \qquad \text{for} \qquad Re > 2300 \tag{12}$$

where $V_{rms} = \sqrt{\overline{V'^2}}$ = root mean square of the fluctuating velocity [m/s].

Unsteady flow

Effect of Pulsatility

Pulsatility (i.e., ratio of maximum to mean axial velocity) and other parameters related to unsteady flow also have important effects on the flow field. Indeed, pulsatility influences not only the velocity and the pressure in the flow field but also the transition from laminar to turbulent flow.

In contrast to a steady flow, which does not vary with time, pulsatile flow changes as a periodic function of time. For pulsatile flow, such as blood flow, the velocity is a function of time within a cardiac cycle. To remove superimposed random velocity components, the time-averaged velocity is obtained by averaging the instantaneous flow velocity at a given location and at a given time during the cardiac cycle over several cardiac cycles (i.e., phase averaging). Pulsatile flow can therefore be either laminar or turbulent, depending on the magnitude of the phase-averaged velocities (i.e., *Re* calculated at a given instant during the cardiac cycle). The temporal, or periodic, changes in a pulsatile flow are associated with flow acceleration and deceleration. Under physiologic conditions, the pulsatile flow remains laminar although the flow field can become extremely complex, such as in the carotid bulb.

The transition to turbulence is delayed by flow acceleration; conversely, the flow field is destabilized by flow deceleration. Thus, the deceleration phase of a pulsatile flow is characterized by an increase of the level of turbulence. In the cardiovascular system, it is possible to observe a transitional or even turbulent flow during a deceleration phase, followed closely by a relamination of the flow field during the subsequent acceleration phase. Since turbulence does not appear instantaneously and requires time to develop, it is possible to have a decelerating flow field in which turbulence does not have enough time to develop between two acceleration phases. Therefore, the pulsatility of the flow field determines the transition from laminar to turbulent flow as well as the propensity to develop turbulence.

The Womersley Number

It is possible to characterize the periodicity of a flow field with a dimensionless parameter, the Womersley number (α), which can be regarded as a supplemental Reynolds number for unsteady flow describing the relative importance of inertial to viscous forces. The Womersley number is given by Eq. 13.

$$\alpha = \frac{d}{2}\sqrt{\frac{2\pi\varrho f}{\mu}} \tag{13}$$

where f [Hz] is the heart rate. When α is less than unity, the viscous forces are predominant everywhere across the tube diameter and the flow is called quasi steady. As α increases, the influence of the inertial forces on the flow field begins in the center of the tube, resulting in a lag of the bulk flow behind the driving pressure gradient, i.e. the velocity profile flattens. In the human body, α varies over a wide range: from 10^{-3} in a capillary to 17.5 in the ascending aorta of a resting subject with a heart rate of 60 beats/min.

The Bernoulli Equation

The principle of conservation of energy states that in the absence of nuclear reaction, the total energy of a system is composed of kinetic energy (velocity), of potential energy (gravity), and of thermal energy (heat). At the same time, when no external forces are applied to the system, the total energy is constant. Using this principle it is possible to derive the Bernoulli equation, which states that the pressure drop between two points located along a streamline[2] is a function of the velocities at these locations and the viscous losses between them (Eq. 14). Thus, according to Bernoulli's equation, the total energy content of the system remains constant; only the form of the energy changes.

$$P_1 - P_2 = 0.5\,\varrho\,(V_2^2 - V_1^2) + \varrho\int_1^2 \frac{dV}{dt}\,ds + VL(V) \tag{14}$$

where P_1 [Pa] and V_1 [m/s] are the pressure and velocity at the upstream location, and P_2 and V_2 are the pressure and velocity at the downstream location. The integral term accounts for the flow acceleration or deceleration between the two locations. If this equation is applied to a flow at peak systole or peak diastole, this term becomes zero. The viscous losses, $VL(V)$, the third term in Bernoulli's equation, can also be neglected in large arteries such as the aorta, where the viscous losses are known to be negligible [5]. Therefore, Eq. 14 can be rewritten as Eq. 15:

$$P_1 - P_2 = 0.5\,\varrho\,(V_2^2 - V_1^2) \tag{15}$$

---

[2] A streamline is defined as a line everywhere tangent to the flow velocity vector at a given time. In steady flow, a streamline is also the path followed by a fluid particle [6].

Furthermore, when the downstream velocity is much higher than that at the upstream location ($V_2 \gg V_1$), it is possible to neglect V_1^2. This condition frequently applies downstream to an arterial stenosis. When the pressure is expressed in mmHg, the velocity in m/s, and when the blood density is assumed to be 1.07 g/cm^3, Eq. 15 can be simplified to Eq. 16:

$$P_1 - P_2 = 4\ V_2^2 \tag{16}$$

The simplified Bernoulli Eq. 16 has been found to correlate well with the measured pressure drop across severe stenoses in the aortoiliac arterial segment. However, its validity in smaller peripheral arteries has not been determined. The simplified Bernoulli equation permits the use of a noninvasive velocimeter (Doppler ultrasound) to characterize stenoses in place of direct and invasive measurements (catheter).

Equation 16 becomes inaccurate in flows where viscous losses are significant, such as in long tunnel-like constrictions with a small diameter, or in regions of flow separation. Here, the viscous forces begin to dominate over the inertial forces and another factor must be considered, namely the energy dissipation due to turbulent stresses. Both viscous and turbulent energy losses should then be taken into consideration, as they reduce the energy content of the fluid.

Effect of Vessel Geometry on the Flow Field

Geometry of a conduit (tube) is a major determinant of regional hemodynamics. Clearly, the geometry of human arteries is complex. Both the intraindividual variations throughout the arterial tree and the interindividual differences between corresponding vascular segments alter the regional hemodynamics significantly [7, 8]. Furthermore, the arterial geometry changes with age [9] and with pathology (e.g., atherosclerosis). Therefore, an accurate description of local pulsatile flow phenomena which takes into account biologic variability is difficult. Because high-fidelity measurements of vascular function and geometry are not readily obtainable with current in vivo measurement techniques, the majority of knowledge was gained from in vitro experiments performed in models of the human vascular system. From these experiments it became evident that an exact simulation of the geometry is not enough if one wants to assess local hemodynamics, and that at least three other parameters are important:

1. Flow rate (*Re* number) [10]
2. Pulsatility and the shape of the pulsatile flow waveform (e.g., Womersley number) [11–13]
3. Flow ratio between branches [14–16]

More recently, it has been argued that even the non-Newtonian character of the blood [17] and the elasticity of the vessel wall [18] must be considered to accurately model the local velocity fields in the human arterial tree. The importance of detailed knowledge of the local hemodynamics has increased along with the increasing evidence that the local velocity fields play a role in atherogenesis [19, 20].

In the following, the local velocity fields at common geometric configurations which characteristically create disturbed flow patterns in the circulation will be discussed, namely:

- Branchings (e.g., carotid bifurcation, renal arteries, abdominal aortic bifurcation)
- Curvatures (e.g., aortic arch, coronary arteries)
- Constrictions/expansions (e.g., stenoses due to atherosclerosis, coarctation of the aorta, carotid bulb)

Since under physiologic conditions turbulent flow exists only in the ascending aorta, the disturbances at curvatures and branches appear to be of a nonturbulent (coherent) nature [21]; i.e., they are not random in time, but are reproduced with each heart beat in a predictable manner. Three main types of coherent flow patterns are created by curvatures and branches: secondary flow, flow separation, and nonparabolic (skewed) velocity profiles.

Secondary Flow, Flow Separation, and Skewed Velocity Profiles

Secondary flow develops in curvatures and branches because the fluid is exposed to a centrifugal force, namely to the inertial effects which tend to move the fluid lamina toward the outer wall (Fig. 3). The centrifugal force causes a radial pressure gradient from the inner to the outer wall, resulting in a secondary radial flow on top of the primary forward motion. The consequence of these forward and radial fluid motions is the formation with each cardiac cycle of two counter-rotating helices, each moving along the tube as two corkscrews (Fig. 3b) [22, 24]. This implies that the flow fields in a curved tube or a branch are three-dimensional in nature, as opposed to the two-dimensional flow fields in a straight tube. When estimating the mean velocity near a branch or in a curved tube, it is thus necessary to obtain knowledge about the velocity distribution not only across one diameter, but across the entire vessel area.

Separation zones develop at vessel expansions, at vessels' branching points, and at curvatures. They are defined as regions near the vessel wall where the blood recirculates, i.e., the blood near the vessel wall flows in an opposite direction to the blood in the center of the vessel [25]. Separation zones appear where there is a retrograde hydrostatic pressure gradient that exceeds the fluid momentum in the forward direction. A retrograde hydrostatic pressure gradient will be created at the expansion of a vessel (Fig. 5) where, according to Bernoulli's theorem (Eq. 14), the high velocity and low pressure in the constriction will change to lower velocity and higher pressure as the vessel area increases. In the center, where the forward fluid momentum is high, no reversal of flow will occur, but at some point near the vessel wall, where the velocity and hence the forward fluid momentum is low, the retrograde pressure gradient may cause the flow to change direction.

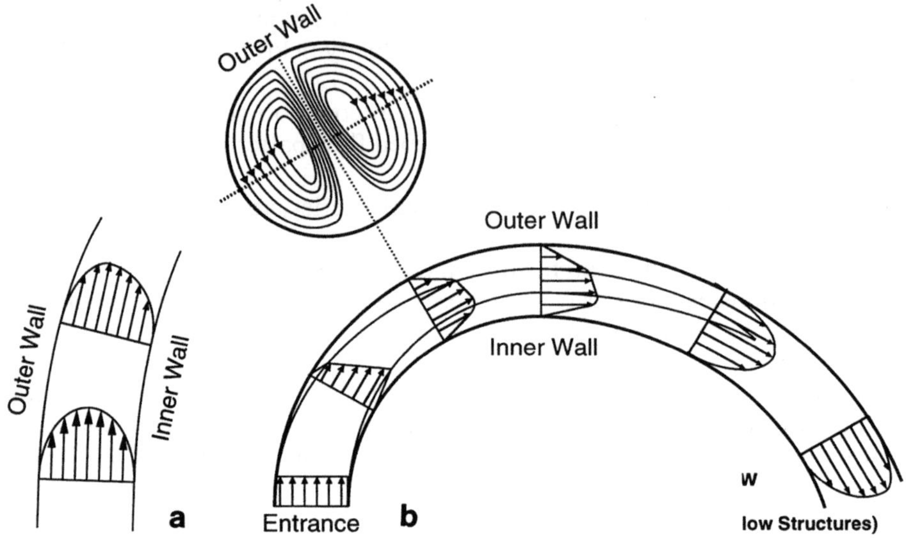

Fig. 3 a, b. Peak velocity profiles in a curvature. **a** Developed flow at entrance; **b** undeveloped flow at entrance (cross section illustrating secondary flow structures)

It follows that the velocity at the boundary between the main flow and such a separation zone will be zero. The "separation point" and the "reattachment point" are defined as the upstream and downstream positions, respectively, where this "zero velocity line" meets the vessel wall (Fig. 4b). In pulsatile flow, the separation and reattachment points move back and forth along the vessel during each cardiac cycle. For some locations a flow separation zone will be present only during diastole.

Effects of Branching

When discussing flow patterns at branches there is a need for defining the geometric terms used (see Fig. 4 for definition). The "lateral angle" (θ), the "branch-trunk" area ratio (β), the shape of the flow divider, and the form of the corners (rounded or sharp) are likely to be the most important geometric parameters that influence the flow field [18].

In general, secondary flow structures will develop in the daughter branches, and separation zones will appear at the outer walls opposite the flow divider (Fig. 4b). The velocity profiles will be skewed, with the highest velocities located at the inner walls near the flow divider (Fig. 4a). The wall shear stress will thus tend to be highest at the inner walls near the flow divider and lowest at the outer walls, where the separation zones are typically formed. The effect of θ on the flow field is reflected by the difference between "Y" (Fig. 4a) and "T" (Fig. 4b) bifurcations. A decrease in θ leads to increased formation of

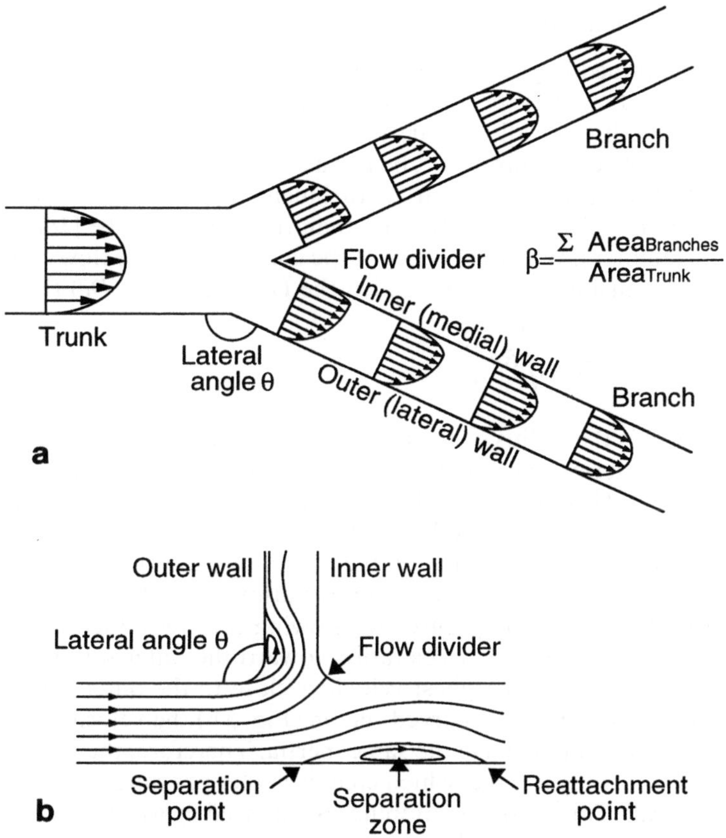

Fig. 4a, b. **a** Peak systolic velocity profiles at a Y bifurcation. **b** Streamlines and separation zones at a T bifurcation

separation zones at the outer wall and at the vessel wall opposite the branch (Fig. 4b), and to more pronounced secondary flow structures. At the same time, the velocity profiles in the branches will become more skewed.

The branch-to-trunk area ratio is important for the characterization of flow stability. When $\beta > 1$, the branching is equivalent to an expansion of the vessel area, while $\beta < 1$ is equivalent to a constriction. It follows from the continuity principle that if the flow ratio between daughter vessels reflects their area ratio, $\beta > 1$ reduces the mean velocity in the daughter vessels while $\beta < 1$ increases it. Therefore, a constriction which is an accelerating effect stabilizes the flow field while an expansion which is a decelerating effect destabilizes it (an increase in β is known to decrease the critical *Re* number; i.e., turbulence develops at lower flow rates [26]). Finally, it is not unexpected that a decrease in θ leads to an increased formation of secondary flow structures and separation zones. It is generally agreed that changes in β influence the flow pattern more than changes in θ [27].

Effects of Curvatures

As mentioned in previous sections, the flow in a curved tube is characterized by skewed velocity profiles, secondary flow patterns, and separation zones along the inner wall, all due to an imbalance between inertial and viscous effects. The curvature of a tube has a stabilizing effect on the flow field. Thus, the transition from laminar to turbulent flow occurs in a curved pipe at a higher critical Reynolds number than in a straight tube [28].

In laminar flow the influence of the curvature on the flow can be determined by a dimensionless number called the Dean number (*De*) (Eq. 17) [29]:

$$De = \frac{1}{2}\sqrt{Re\,\frac{R}{r}} \tag{17}$$

where R is the radius of the cross section and r is the radius of the curvature. *De* characterizes the combined effects of curvature and *Re* on the shape of the flow field. However, the velocity profile also depends on the development of the entrance flow. In developed flow entering the curved region (Fig. 3 a), the fluid lamina in the center of the vessel has the highest flow velocity (and inertia) and becomes the lamina most affected by the centrifugal forces in the entire flow field. The progressive slowing of flow velocities toward the inner wall and flow velocity acceleration toward the outer wall produces a skewed velocity profile with highest velocities close to the outer wall. In an undeveloped flow entering the curved region (Fig. 3 b) the velocity profile is flat, and all the fluid lamina have the same inertia. Therefore, they are equally affected by the centrifugal force, which creates higher pressure at the outer than at the inner wall. This pressure difference is balanced by a difference in kinetic energy (as expressed in Bernoulli's equation, Eq. 14). Thus, at the peak of the curvature higher flow velocities are observed at the inner wall of the vessel. Past the peak the rearrangement of forces produces flow fields with higher velocities close to the outer wall of the conduit. This can be seen, for example, in the thoracic aorta.

In the human vessels the curvatures and branchings are frequently situated close to each other. It is therefore difficult to separate their respective influences on the flow field. The resulting flow fields are complex and not easily defined [30].

Arterial Stenosis

An arterial stenosis can be defined as a sudden reduction of the cross-sectional area of an arterial segment, followed by an area of expansion. The hemodynamic consequences of these two geometric qualities can, to some extent, be estimated using the basic fluid dynamic principles already introduced.

Characteristics of an Arterial Stenosis

Constriction – Stenotic Segment

As the diameter of the stenotic segment decreases, the average velocity across the artery increases, the velocity profile flattens, and the boundary layer becomes thinner. The increase in kinetic energy is accompanied by a drop in hydrodynamic pressure. These changes are due to the law of preservation of energy, which says that the total energy of a system is constant.[3]

Expansion – Poststenotic Dilatation

When the arterial cross-sectional area increases, as is the case in the poststenotic segment, the average velocity decelerates and the flow field is destabilized. This may lead to flow separation and turbulence development, both causing energy losses that cannot be recovered as pressure.

Pressure Drop and Energy Losses

The factors contributing to pressure drop and energy losses resulting from an arterial stenosis are complex. They consist of viscous, inertial, turbulence, or flow separation effects which vary individually with stenosis geometry, blood velocity, and arterial size. By applying the linear momentum equations to a stenotic segment, the instantaneous pressure drop across the stenosis, Δp, can be expressed as a function of these hydrodynamic effects (Eq. 18) [31]:

$$\Delta p = \underbrace{\frac{K_v \mu}{D} \langle V \rangle}_{\text{viscous effects}} + \underbrace{\frac{K_t}{2}\left(\frac{A_0}{A_1} - 1\right)^2 \varrho |\langle V \rangle| \langle V \rangle}_{\text{turbulence and separation effects}} + \underbrace{K_u \varrho L \frac{d\langle V \rangle}{dt}}_{\text{inertial effects}} \tag{18}$$

where K_v is a coefficient characterizing the degree of stenosis. For example, K_v is of the order of 800, 1700, and 6500 for 60%, 75%, and 90% symmetrical area reductions, respectively. K_t and K_u are parameters characterizing the stenosis geometry. They are almost independent of the degree of stenosis and are on the order of unity. $\langle V \rangle$ and $d\langle V \rangle/dt$ are the instantaneous velocity and time acceleration of the flow averaged over a cross section in the unobstructed tube. D is the diameter of the unobstructed tube, A_0 the area of the unobstructed tube, A_1 the minimum cross-sectional area of the stenosis, and L the length over which the pressure drop is measured.

It can be seen from Eq. 18 that viscous effects are linearly dependent on the flow velocity whereas turbulence and flow separation effects vary nonlinearly

---

[3] $E_{system} = E_{pressure} + E_{kinetic} + E_{potential} = \text{constant}$.

Table 1. Mean pressure drop (mm Hg) at different mean blood velocities with reference to Eq. 18[a]

Area reduction (%)	Mean blood velocity (cm/s)					
	10		20		40	
	Visc.	Turb.	Visc.	Turb.	Visc.	Turb.
20	0	0	0	0	1	0
50	0	0	0	0	1	2
70	0	1	1	2	2	9
80	1	2	1	6	3	25
85	1	3	2	13	5	51
90	2	8	4	32	7	129

[a] D = 1 cm; typical values for μ and ϱ have been chosen.

with the velocity. Also, the inertial effects which characterize blood acceleration are canceled out when the mean pressure drop across the stenosis is calculated over an entire cardiac cycle. The importance of viscous and turbulence/flow separation effects on mean pressure drops and the effects of the severity of the stenosis and of the blood velocity are illustrated in Table 1. The trans-stenotic pressure drops were calculated using the viscous and turbulence components of Eq. 18. In all examples, the pressure drop is dominated by turbulence effects. The calculation assumes that the blood velocity is constant even though for peripheral arteries in the human body this condition is not realized because the central perfusion pressure (the blood pressure) remains constant and a trans-stenotic pressure drop leads to a reduced blood flow velocity when the arteriolar resistance cannot be further reduced. Thus, the reduced blood flow velocity will result in diminished trans-stenotic pressure drop.

A critical stenosis is defined as a degree of stenosis at which a small change in the residual lumen area will produce a dramatic fall in poststenotic pressure energy. As seen from Table 1, for a given severity of stenosis, pressure losses are highly dependent on blood flow velocity.

Poststenotic Flow Field

Three fluid dynamic characteristics may be identified in a poststenotic flow field (Fig. 5) [32]:

1. Jets, which can be qualified as deterministic
2. Coherent (nonturbulent) disturbances
3. Turbulence (completely random fluid motion)

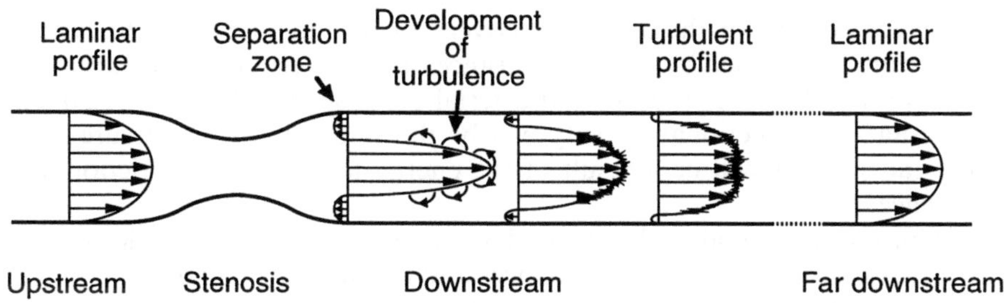

Fig. 5. Peak systolic velocity profiles at a stenosis

Jet and Separation Zone

The immediate poststenotic events in arteries are related to the behavior of confined jets [5]. As blood emerges from the constriction, it forms a jet which increases in size slower than the cross section of the artery does. Therefore, the jet eventually separates from the vessel wall, leaving a separation zone between the central forward flow and the vessel wall (Fig. 5).

Deterministic Disturbances

The flow field may become disturbed even at moderate degrees of arterial stenosis. For example, vortex shedding and development of coherent structures have been identified downstream of arterial cross-sectional area reductions of only 25%–50% [44]. The coherent structures can be found immediately downstream of the stenosis during the accelerating phase of the heart cycle and further downstream later in systole. The shear layer between jet and recirculation zone may give rise to vortex shedding. The spatial and timely location, the size and frequency of the coherent structure and vortices will in principle remain the same from beat to beat. Thus, these disturbances are deterministic.

Turbulence

When the stenosis is more severe or the blood flow velocity higher, viscous forces are not high enough any more to dampen the vortices developed in the shear layer and turbulence may develop. As noted earlier, turbulence is defined as a flow condition in which the various hemodynamic variables, such as flow velocity, become random in time and space.

Initially the turbulent velocity components are located at peak systole and in the deceleration phase of the systole. When *Re* increases, or when the severity of a stenosis increases, the turbulence may extend throughout the entire heart cycle.

Close to the stenosis, turbulence is found where the jet and the flow separation region meet. Farther downstream, turbulence has spread to span the entire cross section of the artery. Recovery of a laminar flow field farther downstream is preceded by the reappearance of the deterministic disturbances (vortices and eddies). If the original arterial cross-sectional area is regained, the average cross-sectional velocity will be the same, and the pressure drop across the stenosis will be a result of the energy loss in viscous friction and turbulence.

It is important to realize that these deviations from the prestenotic velocity flow field (jets, deterministic disturbances and turbulence) may be seen even in mild obstructions and subcritical stenosis, before any significant flow reduction can be detected. Indeed, in large arteries, poststenotic jets and deterministic disturbances only cause minimal pressure loss. It is the poststenotic turbulence associated with more severe stenoses which is the primary cause of pressure loss downstream of stenoses. However, when the peripheral vascular bed has the capacity to reduce its vascular resistance, a normal blood flow can be maintained and a severe lesion can remain undetected even if the energy losses are significant. By contrast, low-amplitude turbulence, even those involving only part of the cardiac cycle, can be detected in the velocity signals even though they do not result in significant energy losses [33]. Therefore, the development of velocity disturbances downstream of arterial stenoses can be used to detect and evaluate early subclinical occlusive arterial disease.

Predictors for Poststenotic Velocity Disturbances

Even though the size of a stenosis, simplified as the degree of cross-sectional arterial reduction, is an essential determinant of the disturbed post-stenotic flow, no simple relationship between poststenotic flow field and the severity of the stenosis exists. This is due to: first, a certain degree of stenosis must be reached before jet, flow separation or velocity disturbances develop; second, the relationship between the severity of the stenosis and the degree of post-stenotic disturbances is nonlinear; third, the poststenotic flow field changes with the different phases of the cardiac cycle. It is therefore important to consider factors that will most consistently predict the development of post-stenotic jets, recirculation zones, deterministic disturbances (vortices and eddies) and random turbulence. For the purpose of this discussion, three general groups of variables have to be considered:

1. Geometry of the stenosis
2. General hemodynamic variables
3. Biologic variables

Geometry of the Stenosis

The frequently complex geometry of a stenosis can be reduced to a simplified two-plane description. In cross section, the aperture may be symmetric circular or asymmetric irregular [34, 35]. In a longitudinal (axial) view, the length of the stenosis may vary [36], multiple stenoses may be present [37], and the stenosis inlet and outlet region may be gradual or sharp-edged [38]. The main direction of the stenosis with respect to the course of the artery may be oblique, leading to a skewed jet. In addition, the vessel wall in the stenosis may be smooth or rough [39]. These local geometric factors all affect the development, direction and size of poststenotic velocity field components and therefore the corresponding energy losses.

In general, the minimum energy losses are to be expected from a symmetric, centrally located, short, gradual stenosis with a smooth wall. Any deviation in lesion geometry will increase the trans-stenotic pressure losses and compromise even more distal perfusion. The effects of multiple serial stenoses, as observed, for example, in multilevel peripheral arterial disease, will be primarily determined by the most severe constriction [37]. At the same time, a severe critical stenosis will reduce the blood flow, thereby masking the effects of other stenoses which would be critical for normal blood flow. The salutory effect of eliminating the most severe constriction by an endovascular therapeutic procedure will therefore depend upon eventual additional stenoses in the artery, which now may become significant as the blood-flow-compromising effect of the most severe stenosis has been eliminated. However, determination of severity, localization, and significance of multilevel arterial stenoses is possible by comparing velocity measurements within and after these stenoses to velocity measurements in the proximal segment [40].

General Hydrodynamic Variables

An arterial stenosis destabilizes normal blood flow, and its effect depends on whether blood flow upstream is laminar or turbulent. The most important general hydrodynamic predictor of poststenotic flow disturbance appears to be the Reynolds number measured upstream of the stenosis, in the normal arterial segment. For example, let us consider the flow field downstream of a 50% stenosis in a steady flow in vitro experiment when *Re* is varied. When *Re* upstream of the stenosis is equal to 2000, the flow field displays vortices and turbulence. At *Re* equal to 1000 no turbulence can be seen, and at *Re* equal to 500 not even vortices can be seen [41].

In pulsatile flow, the added effects of Womersley's parameter, α, and the pulsatility of blood (U_{peak}/U_{mean}) have to be considered. Increasing α within the physiologic range in large arteries reduces the tendency for turbulence to develop (i.e., the critical Reynolds number is increased) and decreases the distance between the stenosis and the site where turbulence develops [38]. An isolated increase in the pulsatility of the flowing blood will elevate the peak Reynolds number and increase the tendency for poststenotic flow disturbances to appear in the decelerating phase of the systole.

Biologic Variables

Biologic responses to abnormal flow also need to be considered. For example, distal to a critical arterial stenosis the perfusion pressure declines, the peripheral vascular bed dilates, and the peripheral resistance decreases. The variability in the magnitude of ischemic responses among individual vascular beds may be explained in part by the considerable differences in the biologic effects of stenoses upon variables such as tissue perfusion [43]. In addition, the opening of preformed collateral channels and the development of new collateral flow pathways will also alter the effect of a stenosis significantly. Our understanding of biologic responses to altered blood flow at a cellular and biochemical level within the vessel wall is only at the beginning. Major progress in our knowledge of the biology of the vessel wall diseases is to be expected from studies on cellular flow effects.

It is important to recognise that the effect of the mentioned predicting and modifying variables has primarily been evaluated in terms of flow reduction and pressure drop, and their significance for regional velocity profiles and velocity disturbances remains to be investigated.

New hemodynamic indices for the evaluation of arterial stenoses are rapidly evolving [45]. The refinement of color Doppler ultrasound [46] and NMR velocimetry [47] will likely enhance our ability to detect an early arterial disease and to determine its clinical significance.

Venous Flow

Even though the venous system resembles the arterial tree, there are several fundamental differences. For example, the pressure in a vein is normally much lower than that in an artery at the same location, and some veins contain valves to prevent backflow. Also, veins have thin walls and may be collapsed in normal function.

In fact, the controlling factor for venous flow is the transmural pressure. When the pressure is higher inside the vein than it is outside, the flow is similar to that in an artery. However, the flow is greatly reduced when the vein is collapsed due to an outside pressure greater than that inside the vein. The intermediate case occurs when the entrance section of the vein is open (transmural pressure positive) and its exit section collapsed (transmural pressure negative). Then, the flow reaches a limiting steady state as it oscillates between a near-zero flow state caused by the choking of the vein and a near-"normal" flow state created by the entrance section.

Finally, while the flow in the arterial tree is mostly determined by the contraction of the heart and the impedance of the body, the flow in the venous system is more complex. Indeed, it is affected by factors such as the arterial pulse (very attenuated), right heart contraction, respiratory function, or even the action of muscles located close to the vein [48].

In conclusion, flow in the venous system is very complicated and although it is possible, in some cases, to make some analogies with arterial flow, much remains to be understood.

Conclusion

In this chapter, we have discussed some basic hemodynamic principles of blood flow in the large arteries of the body. The topics were discussed in general terms to provide vascular specialists with a basic understanding of arterial hemodynamics. For more detailed information on the subject, the reader is referred to the reference list.

References

1. Whitaker S (1986) Introduction to fluid mechanics. Krieger. Malabar, FL, USA
2. Bird RB, Stewart WE, Lightfoot EN (1960) Transport phenomena. Wiley, New York
3. Caro CG, Pedley TJ, Schroter RC, Seed WA (1978) The mechanics of the circulation. Oxford University Press, Oxford
4. Schlichting H (1978) Boundary-layer theory. McGraw-Hill, New York
5. Yoganathan AP, Cape EG, Sung HW, Williams FP, Jimoh A (1988) Review of hydrodynamic principles for the cardiologist: applications to the study of blood flow and jet by imaging techniques. J Am Coll Cardiol 12:1344–1353
6. White FM (1979) Fluid mechanics. McGraw-Hill, New York
7. Friedman MH, Deters OJ, Mark FF, Bargeron CB, Hutchins GM (1983) Arterial geometry affects hemodynamics – potential risk factor for atherosclerosis. Atherosclerosis 46:225–231
8. Bargeron BC, Hutchings GM, Moore GW, Deters OJ, Mark FF, Friedman MH (1986) Distribution of geometric parameters of human aortic bifurcations. Arteriosclerosis 6:109–113
9. Gosling RG, Newman DL, Bowden NLR, Twinn KW (1971) The area ratio of normal aortic junctions: aortic configuration and pulse-wave reflection. British Journal of Radiology 44:850–853
10. Feuerstein IA, El Masry OA, Round GF (1976) Arterial bifurcation flows – effects of flow rate and area ratio. Can J Physiol Pharmacol 54 (6):795–807
11. Siouffi M, Pelissier R, Farahifar D, Rieu R (1984) The effect of unsteadiness on the flow through stenoses and bifurcations. J Biomech 17(5):299–315
12. Ku DN, Giddens RP (1983) Pulsatile flow in a model carotid bifurcation. Arteriosclerosis 3(1):31–39
13. Lutz RJ, Hsu L, Menawat J, Zrubek J, Edwards K (1983) Comparison of steady and pulsatile flow in a double branching arterial model. J Biomech 16:753–766
14. El Masry OA, Feuerstein IA, Pound GF (1978) Experimental evaluation of streamline patterns and separated flows in a series of branching vessels with implications for atherosclerosis and thrombosis. Circ Res 43 (4):608–617
15. Cho YI, Back LH, Crawford DW (1985) Experimental investigation of branch flow ration, angle, and Reynolds number effects on the pressure and flow fields in arterial branch models. J Biomed Eng 107:257–267
16. Walburn FJ, Stein PD (1980) Flow in a symmetrically branched tube simulating the aortic bifurcation: the effects of unevenly distributed flow. Ann Biomed Eng 8:159–173
17. Moravec S, Liepsch D (1983) Flow investigations in a model of a three-dimensional human artery with Newtonian and non-Newtonian fluid. Part I. Biorheology 20:745–759
18. Liepsch D (1986) Flow in tubes and arteries – a comparison. Biorheology 23:395–433
19. Zarins CK, Giddens DP, Bharadvaj BK, Sottiurai VS, Mabon RF, Glagov S (1983) Carotid bifurcation atherosclerosis: quantitative correlation of plaque localization with flow velocity profiles and wall shear stress. Circ Res 53:502–514
20. Friedman MH, Bargeron CB, Hutchins GM, Mark FF, Deters OJ (1980) Hemodynamic measurements in human arterial casts, and their correlation with histology and luminal area. J Biomech Eng 102:247–251

21. Stein PD, Sabbah HN (1980) Hemorheology of turbulence. Biorheology 17:301–319
22. Brech R, Bellhouse BJ (1973) Flow in branching vessels. Cardiovasc Res 7:593–600
23. Fukushima T, Azuma T (1982) The horseshoe vortex: a secondary flow generated in arteries with stenosis, bifurcations, and branchings. Biorheology 19:143–154
24. Fukushima T, Komma T, Azuma T, Harakawa K (1987) Chracteristics of secondary flow in steady and pulsatile flows through a symmetrical bifurcation. Biorheology 24:3–12
25. Fox JA, Hugh AE (1966) Localization of atheroma: a theory based on boundary layer separation. Br Heart J 28:388–399
26. Walburn FJ, Blick EF, Stein PD (1979) Effect of the branch-to-trunk area ratio on the transition to turbulent flow: implications in the cardiovascular system. Biorheology 16:411–417
27. Karino T, Goldsmith HL (1985) Particle flow behavior in models of branching vessels II. Effects of branching angle and diameter ratio on flow patterns. Biorheology 22:87–104
28. McDonald DA (1974) Blood flow in arteries, 2nd edn. Arnold, London
29. Dean WR (1927/1928) The streamline motion of fluid in a curved pipe. Phil Mag 4 (7):208 and 5:673
30. Rodkiewicz CM (ed) (1983) Arteries and arterial blood flow, biological and physiological aspects. Springer, Vienna New York
31. Khalifa AMA, Giddens DP (1981) Characterization and evolution of poststenotic flow disturbances. J Biomech 14:279–296
32. Hinze J (1975) Turbulence, 2nd edn. McGraw-Hill, New York
33. Thiele BL, Hutchison KJ, Greene FM, Forster FK, Strandness DE (1983) Pulsed Doppler waveform patterns produced by smooth stenosis in the dog thoracic aorta. In: Taylor DEM, Stevens AL (eds) Blood flow – theory and practice. Academic, London, pp 85–104
34. Solzbach U, Wollschlager H, Zeiher A, Just H (1987) Effect of stenotic geometry on flow behavior across stenotic models. Med Biol Eng Comput 25:543–550
35. Young DP, Tsai FY (1973) Flow characteristics in models of arterial stenoses. I. Steady flow. J Biomech 6:395–410
36. Kindt GW, Youmans JR (1969) The effect of stricture length on critical arterial stenosis. Surg Gynecol Obstet 128:729–734
37. Vonrudden WJ, Blaisdell FW, Hall AD, Thomas AN (1964) Multiple arterial stenoses: effect on blood flow. Arch Surg 89:307–315
38. Yongchareon W, Young DF (1979) Initiation of turbulence in models of arterial stenoses. J Biomech 12:185–196
39. Neuwirth JG (1977) Pressure and velocity fluctuations associated with the flow through a stenosis with upstream roughness. IEEE Trans Biomed Eng BME 24:269–277
40. Neumyer MM, Thiele BL (1988) Evaluation of lower extremity occlusive disease with Doppler ultrasound. In: Taylor KJW, Burns PN, Wells PNT (eds) Clinical applications of Doppler ultrasound. Raven, New York, pp 317–337
41. Ahmed SA, Giddens DP (1983) Velocity measurements in steady flow through axisymmetric stenoses at moderate Reynolds numbers. J Biomech 16:505–516
42. Nerem RM, Seed WA (1972) An in vivo study of aortic flow disturbances. Cardiovasc Res 6:1–14
43. Evans DH, Barrie MJ, Bentley S, Bell PRF (1980) The relationship between ultrasonic pulsatility index and proximal arterial stenosis in a canine model. Circ Res 46:470–475
44. Talukder N, Fulenwider JT, Mabon RF, Giddens DP (1986) Post-stenotic flow disturbance in the dog aorta as measured with pulsed Doppler ultrasound. J Biomech Eng 108:259–265
45. Rittgers SE, Fei D-Y (1988) Flow dynamics in a stenosed carotid bifurcation model – part II: derived indices. Ultrasound Med Biol 14:33–42
46. Rittgers SE, Shu MCS (1990) Doppler color-flow images from a stenosed arterial model: interpretation of flow patterns. J Vasc Surg 12:511–522
47. Maier SE, Maier D, Boesiger P, Moser VT, Vieli A (1989) Human abdominal aorta: comparative measurements of blood flow with MR imaging and multigated Doppler US. Radiology 171:487–492
48. Lundbrook J (1962) Functional aspects of the veins of the leg. Am Heart J 64:706–713

Part II

Chapter 3

Basics of Color Doppler Imaging

E. G. CAPE, H-W. SUNG, and A. P. YOGANATHAN

Introduction

Color Doppler flow imaging (CDFI) is a relatively new ultrasound method to assess vascular function and morphology. Primarily introduced to echocardiography in the early 1980s [1–3], it has been more recently applied in vascular diagnostics [4]. The objective of this chapter is to describe the basic concepts applied in CDFI.

Retrospective

The medical applications of ultrasound span more than a quarter of a century; a concise historical review of its development has been provided by others [5] and only a brief retrospective will suffice for the purpose of this chapter.

Ultrasound was introduced into medical diagnostics in the early 1960s as one-dimensional (M-mode) imaging. In this mode the moving solid structures were intercepted perpendicularly by the ultrasonic beam and the motion was displayed as a function of time. In cardiology, M-mode imaging was used primarily to assess cardiac wall and valvular motion, as well as to determine the linear dimensions of the cardiac chambers. Later in the 1960s, new transducers were developed, allowing the M-mode line to sweep across a plane sector at a high speed. The rapid data acquisition and data processing allowed tomographic imaging in real time, i.e., with a frame rate > 16 images/s, and this was called B-mode imaging.

In the same decade Doppler frequency shifts of the echoes reflected by the blood corpuscles were utilized to measure flow velocities in the arteries. In this mode the blood flow velocities which are parallel to the ultrasound beam are displayed as a spectrum of reflected frequencies. The conventional Doppler ultrasound is performed either in the continuous wave (CW) mode, where all velocities along a line of ultrasound are measured, or in the later-developed pulsed wave (PW) mode, where only velocities within a discrete spatial location are determined. The spatial resolution in the PW mode is achieved by alternation of pulsing and receiving of ultrasound bursts.

In the 1970s, duplex imaging was introduced, combining the advantages of B-mode and echo-Doppler imaging. In duplex imaging, which was originally designed for carotid artery imaging, the Doppler frequency shifts are displayed simultaneously and in superposition on the gray-scale B-mode image.

CDFI, an extension of duplex imaging, was introduced in the 1980s. In this mode, a series of PW sample volumes were arranged along a line by multigating the Doppler bursts and by incremented monitoring of the returning backscattered signals. To facilitate the display, color encoding of data was implemented; for example, the color could represent the flow direction toward or away from the transducer. The color intensity, or brightness, defined the velocity magnitude. This color display of flow could then be superimposed on the standard image. Advancement of the data processing from a fast Fourier transform (FFT) to an autocorrelation method enabled sampling from a greater number of sample volumes along each line, thus allowing a real-time display of one-dimensional flow velocity maps. The next step was analogous to the evolution from M-mode to B-mode imaging. Multiple crystals mounted in the face of the transducer allowed sweeping of the Doppler scanning line, producing a pie-shaped array of sample volumes throughout the imaging sector. With the B-mode and Doppler scanning lines both sweeping at a high frequency compared with the frequency of cardiocirculatory events, color-coded blood flow maps superimposed on tomographic images in real time were obtained. The simultaneous display of color-coded flow maps and sector images became known as color Doppler flow imaging (CDFI).

Basic Principles of Color Doppler Flow Imaging

As discussed above, a color Doppler flow image is a composite of two images:
a) a color-coded image of the velocity flow field and
b) a two-dimensional B-mode image of internal structures.

B-mode ultrasonography is widely used clinically and its principles and applications have been described in a number of standard textbooks [6–8] and therefore will not be repeated in this chapter. In this context we shall assume that the tomographic ultrasound image is available and that we wish to superimpose flow information upon it. In the following sections the sequence of steps required to produce a color-coded Doppler flow image will be reviewed.

Conventional Doppler Ultrasound

The velocity information used for color encoding and display in CDFI is obtained according to the same principle utilized in PW Doppler ultrasound. Understanding of this principle is critical for a proper interpretation of color Doppler flow images. Therefore, the Doppler principle will be briefly summarized. Readers interested in a more in-depth treatment of Doppler physics and signal processing are referred to the standard textbooks on these topics [9, 10].

The Doppler Principle and Continuous Wave Mode

The Doppler effect was first described in the middle of the nineteenth century by an Austrian physicist, Christian Johann Doppler (1803–1853). Doppler found that when a light wave strikes a moving target the frequency of the reflected light changes; the change in frequency (shift) is proportional to the velocity of the target. The frequency increases with target motion toward the source and decreases with target motion away from the source. If the velocity vector is at an angle (θ) to the beam, this shift must be corrected by the cosine of that angle to produce an accurate measure of velocity.

It later became evident that the Doppler effect on apparent shift in frequency applies to any type of wave motion, including that of sound. Mathematically, the Doppler shift is calculated by Eq. 1:

$$F_D = \frac{2f_1}{C} V \cos\theta \tag{1}$$

where F_D is the Doppler frequency shift, f_1 is the frequency of the incident ultrasound, i.e., transducer frequency, C is the speed of sound in the medium, and V is the velocity of the target, i.e., blood flow velocity.

The speed of sound in tissue and blood is approximately 1540 m/s and remains constant. Therefore, Eq. 1 can be arranged to produce a measurement of the velocity of a red blood cell (Fig. 1):

$$V = \frac{F_D C}{2 f_1 \cos\theta} \tag{2}$$

In the CW Doppler mode, ultrasound is continuously emitted and received. Consequently, the returning signals arriving at the transducer cannot be assigned to any specific point along the ultrasonic beam. The frequency shifts are converted to velocities without spatial discrimination and displayed in spectral form. The number of signals within small intervals of frequency shift determines the intensity of the spectrum at that interval. The large numbers of returning data in conventional Doppler are usually processed and assigned into the discrete intervals by FFT methods.

Since the signals in CW-Doppler are received from all points along the beam, maximum flow velocity along the beam can be measured. The CW Doppler is valuable, for example, to assess the severity of valvular stenoses; the maximum velocity across the stenosis is measured and converted into a pressure gradient via the simplified Bernoulli equation.

The precise *location* of the maximum velocity along the beam, however, cannot be detected by CW Doppler. Not only is the location of the maximum velocity not known, but embedded beneath the peak envelope of the sepctrum is a rich array of frequency signals not accessible to useful interpretation.

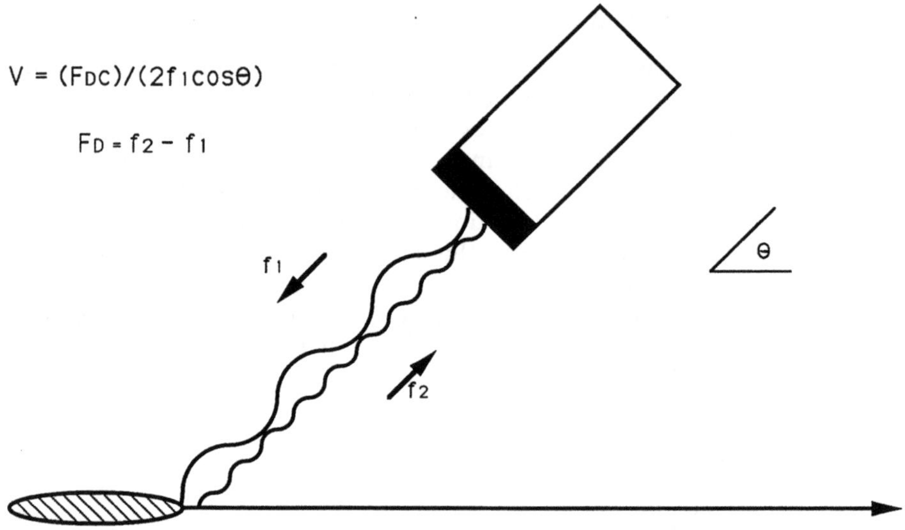

Fig. 1. The Doppler principle. An ultrasound beam of emitted frequency f_1 strikes a moving blood cell traveling at an angle θ to the beam. The beam returns to the transducer with a shifted frequency f_2. Blood cell velocity can then be calculated using the equation shown, with C = the speed of sound in the medium (approximately 1560 m/s)

Pulsed Wave Doppler

The limitations of CW Doppler are overcome by PW Doppler (Fig. 2). In this modality, ultrasound bursts are emitted at a set frequency called the pulse repetition frequency (PRF). By emitting a burst of ultrasound, waiting a precise time Δt, then monitoring the returning signal, it is possible to calculate the depth at which the returning signals must have been reflected by the moving target:

$$\text{depth} = \frac{C(\Delta t)}{2} \tag{3}$$

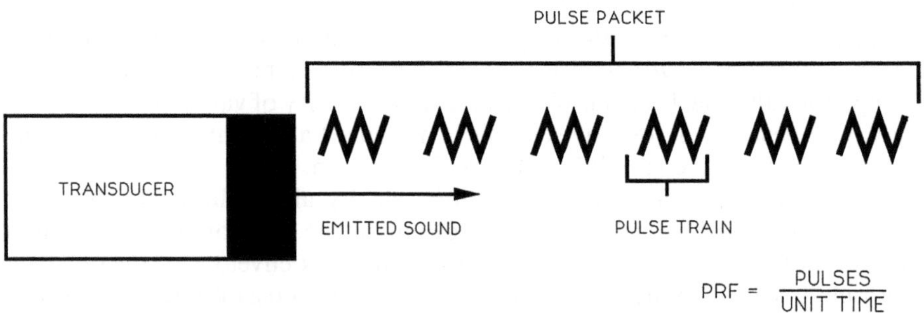

Fig. 2. Short bursts of ultrasound called pulse trains are emitted at a frequency called the pulse repetion frequency (PRF). A group of 3–16 pulse trains are averaged as a pulse packet to get a velocity sample

In conventional PW Doppler, the resulting frequency-shift signals from discrete sample volumes are processed and displayed in the same spectral format used for CW Doppler.

While PW Doppler does allow discrimination of returning signal depth, or range gating, the nature of the pulsing technique presents one very important limitation, namely, aliasing. Aliasing occurs when the Doppler shift frequency exceeds half the PRF, the *Nyquist limit,* resulting in "wrap around" of the spectrum. Thus, the PW Doppler has a limited capability to measure high velocities accurately. This is also true for CDFI.

Color Doppler Flow Imaging

Data Acquisition

In CDFI, the amplitude, frequency, and phase of the backscattered B-mode and Doppler echoes are processed to generate a composite B-mode and Doppler flow image. To produce such a composite image, the objective can be stated as follows: How can pulsed wave sample volumes be superimposed on the tomographic image in what appears to be real time?

We are initially faced with the problem of extending the single sample volume of PW Doppler to multiple PW volumes throughout the entire sector. We can obtain multiple sample volumes of Doppler data along a line by using the concept of multigating [11]. In conventional range-gated PW Doppler, Eq. 3 is used to determine the depth of returning signals for a single sample volume. To obtain multiple signals along a line, a burst of ultrasound is emitted and the returning signals are monitored at a number of incremental times equal to the number of sample volumes desired along the line. Equation (3) still holds, since to obtain these depths it must simply be multiplied or divided by factors reflecting the various depths.

Extension to Two-Dimensions

The nature of the transducer allows extension to two dimensions by "sweeping" the single line of sample volumes throughout the sector. After velocities along a line are obtained through multigating, the line is simply shifted to the next adjacent location (Fig. 3). The distance moved is, of course, determined by the sector width divided by the number of lines per sector. The number of lines is determined mainly by PRF, packet size, and frame rate. In the far field, where significant defocusing of the beam can occur, spaces are left *between* beams and must be filled in. To do this, instruments employ "line-averaging" techniques, which add to the mean velocity inaccuracies already present within individual gates and are discussed below. The *interplay* between these variables will be discussed in the section "Instrument Settings."

The lateral movement of the scanning line is accomplished with use of either a phased-array or a mechanically moving scan head transducer. The more

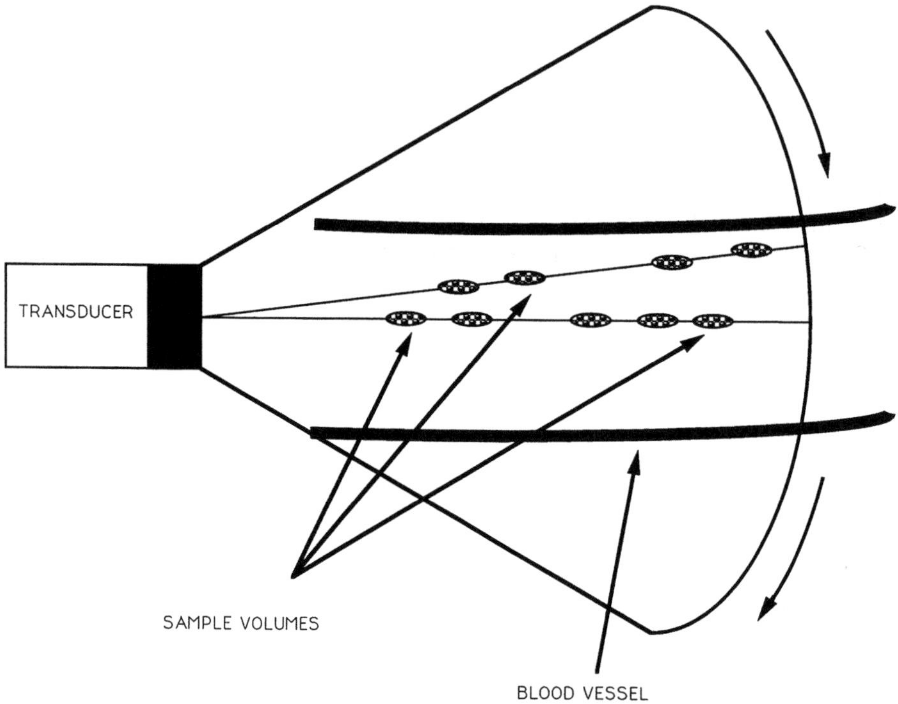

Fig. 3. Two-dimensional measurement of blood flow. Using the multigate concept, multiple sample volumes are placed along a line. The multiple-crystal configuration of the transducer then allows sweeping of the line across a pie-shaped sector at a high frequency for two-dimensional velocity measurement in quasi-real time. Individual sample volume velocities are then color coded for superimposition onto the two-dimensional echocardiographic image

popular *phased-array transducers* consist of multiple piezoelectric crystals mounted along the face of the device. Alterations in the output of individual crystals allow focusing of the scanning line at different angles, which eventually cover the sector angle of the imaging field.

Data Processing

Since typical color flow images have many gates along a single line (and subsequently this line will be swept across the sector), the data processing load is increased far beyond that of conventional Doppler and the techniques for data processing must be modified. The large number of data are handled in two-timensional color flow by using *mean frequency estimators* and changing from FFT techniques to *autocorrelation* methods.

Mean Velocity Estimation. Ultrasound is emitted from the transducer in a form known as "pulse trains." These pulse trains consist of short bursts of ultrasound waves with a wave frequency equal to the carrier frequency. These pulse trains are emitted at a frequency equal to the PRF setting of the instrument (Fig. 2). Several pulse trains form a pulse packet; the size of the pulse packet characterizes the accuracy of the estimation of mean velocity within each individual sample volume at a single point in space and time. For example, the conventional Doppler typically uses 128 pulse trains to characterize the velocity in a single sample volume. Because of the large number of sample volumes required to span the entire sector, the packet sizes in CDFI are much smaller than those in PW Doppler and typically consist of only 3–16 pulse trains. The velocities obtained from this limited number of pulse trains are then averaged to produce one mean value per sample volume, i.e., per picture element within the two-dimensional flow map. This mean velocity technique invokes an important limitation to CDFM. In laminar flow, instantaneous velocities within a sample volume will not deviate far from the mean velocity calculated from the ultrasound packet. In turbulent flow, however, fluctuations in velocity can be very significant and the true mean velocity can be as small as half of the peak fluctuating values. This problem is then compounded by the fact that the true mean is most likely not obtained in most cases, especially for those packets with a small number of pulse trains, e.g., three. This limitation should be kept in mind when interpreting clinical color flow data in a quantitative manner.

Autocorrelation Methods. The large number of data required for color flow mapping also requires a change from the spectral fast Fourier analysis to an autocorrelation technique. The large number of pulse trains within a packet and the time available for data processing in PW Doppler allows precise calculation of velocities via Eq. 2. In CDFI, individual returning frequencies are not compared directly with the carrier frequency; rather, the phase shift between successive signals is used to calculate velocity. The simplest way to think of this autocorrelation technique is as follows. Imagine a blood cell which moves a distance ΔZ over the time, Δt, during which two successive bursts of ultrasound strike it. The velocity of the cell is constant within this extremely small interval, and that is the value desired. The two successive signals will return to the transducer with a frequency which differs from the carrier frequency in proportion to the velocity of the target. A velocity value could therefore be calculated for each signal using Eq. 2. However, a faster way to determine the velocity is to simply measure the phase shift of two successive signals, referred to as the autocorrelation function. The waves should be shifted out of phase by a distance ΔZ. The known period between signals, Δt, then allows more rapid calculation of target velocity V:

$$V = \frac{\Delta z}{\Delta t} \tag{4}$$

Data Display (Color Encoding)

By using the multigating concept, streamlining the data processing technique, and using either phased-array or mechanically moving scan head transducers, we can now understand how velocity data are obtained throughout the imaging sector in quasi-real time.

Digital Processing. The data obtained through the autocorrelation technique are first processed by what is called the quadrature detector. This circuit detects the shift in phase of the returning signals and assigns a *digital* value to the shift. This digital value is used for calculation of individual velocities in the color flow processor which are then averaged to obtain the mean value discussed above. Since only one color can be assigned per gate, only this final mean value receives an assigned color and it carries with it the limitations (especially in the case of turbulence) discussed above.

The individual colors within gates are assigned according to a color bar displayed alongside the image. Velocities below a certain high-pass threshold U_c will not be assigned a color. Velocities between U_c and the Nyquist limit U_N (determined by PRF and depth as for conventional Doppler) will be assigned colors of varying intensity as shown on the bar. Dull to bright *red* corresponds to velocities *toward* the transducer, increasing between U_c and U_N. Dull to bright *blue* corresponds to velocities *away* from the transducer, ranging between $-U_c$ and $-U_N$.

The aliasing is reflected in CDFI by a transfer from the brightest of one color to the brightest of the opposite color, in analogy with the wrap-around effect observed in conventional Doppler. The intersection of the two bright colors allows easy detection of the location of aliasing, which is important since these high-velocity regions often correspond to locations of cardiovascular pathology.

Variance. As stated above, the mean velocity estimation technique used to assign one color to a given gate can result in significant underestimation of maximum velocity, especially in the presence of turbulent flow where fluctuations about the true mean are high. Unfortunately, the large number of data required to construct a two-dimensional image does not allow us to overcome this problem with presently available technology. An option available on many currently marketed instruments may at least allow us to identify areas where such inaccuracies are most likely to occur. From the same data used to calculate the mean velocity within a gate, a variance value can also be obtained by standard statistical definition with little additional computation time. Variance values are then color encoded using the third main color, green. A variable-intensity (dull to bright) process such as that used for velocities is also used for variance. Note that variance values calculated from either positive or negative velocities are positive due to the squared nature of the variance statistic, and therefore only one additional color (green) is required for mixing on the image.

By its definition, triggering of the variance algorithm should alert the user to regions where deviations from the mean are high and therefore the validity of individual colors should be questioned. If quantitative maximal velocity information is desired in such areas, a conventional pulsed wave sample volume should be steered in for measurement.

During the time in which variance-encoding algorithm first became available, it was suggested that green areas could be used to signal flow turbulence. Such associations have more recently been deemed inaccurate. Turbulence is generally assessed in traditional fluid mechanics by a quantity called turbulent shear stress:

$$\mathrm{TI} = -\varrho\,\overline{U'V'}$$

where U' and V' are fluctuations about the mean axial and radial velocities, respectively, ϱ is the density of the fluid, and the overbar denotes an average. The analogy to the statistical definition of sample variance (used in the color instruments) is clear.

$$\mathrm{var} = \frac{\sum_{i=1}^{n} (U_{\mathrm{i}} - U)^2}{n-1}$$

where n is an assigned number of samples to be averaged. Unfortunately, the physical size of color sample volumes (especially in the far field where defocusing can occur) compared with the geometry of flow patterns within the vessel (for example, velocity profiles at the exit of a stenosis have steep gradients) can result in calculation of significant spatial variance in the absence of any true flow turbulence. Furthermore, depending on the frame rate or PRF settings of the instrument, the pulsatile nature of physiologic flow fields can cause a significant temporal variance in the absence of any true flow turbulence. Hopefully, more sophisticated algorithms will soon allow true allocation of flow turbulence. Such capabilities would be very important since flow turbulence has been shown to damage and/or alter the function of the blood cells and endothelium. In the meantime, however, the variance algorithms are still of values in alerting the user to regions of maximum flow velocity underestimation.

Interleaving of Color and B-Mode Images

Color Doppler and B-mode data are generally acquired in separate sweeps. The data are stored in the digital scan converter, then laid down on the video screen together. When superimposing color flow information, the resolution of the B-mode is decreased to allow more detailed processing of Doppler data. The time required for Doppler data acquisition depends on the various machine settings: depth, wall filters, scan rate, line density, and scan angle [12].

Angle Correction in CDFI

One difficulty in applying conventional CW or PW Doppler for quantitative measurement of blood velocities has been the angle of interrogation. The angle

θ in Eq. 2 is generally taken as zero in the data processing algorithms, and it is left to the user to maintain a beam direction parallel to the flow. If a parallel beam path is not maintained, underestimation of true velocity will occur. By visualizing flow, CDFI enables the operator to better adjust the orientation of the transducer or to properly correct the incident angle for calculation of velocities.

The estimation of true blood flow velocity is considerably more complex in CDFI than with conventional Doppler methods. Since Eq. 2 governs the measurement of velocities in every single sample volume throughout the imaging sector, placement of the transducer at an angle to flow results in underestimation of velocity at all points within the sector. Also, the different spatial locations of each sample volume with respect to the transducer result in underestimation to a different extent (different angle) for each gate.

Furthermore, an additional complication exists which is unique to color flow mapping and is related to the scanning nature of the transducer. Even if the transducer (and centerline of the sector) are placed parallel to the flow, all off-centerline measurements will underestimate velocities as the angle of deviation increases. Consequently, CDFI does not allow precise quantitative measurements of blood flow velocities and routine combination with PW or CW Doppler is usually implemented to perform these measurements. Figure 4 summarizes these angle correction considerations.

These limitations are not insurmountable and will probably be corrected in the near future. The off-centerline measurements could be corrected by adding a few lines to the processing algorithm. The location of each sample volume, and therefore the angle, is known, so the velocities could conceivably be corrected. More importantly, although it may complicate operation of the

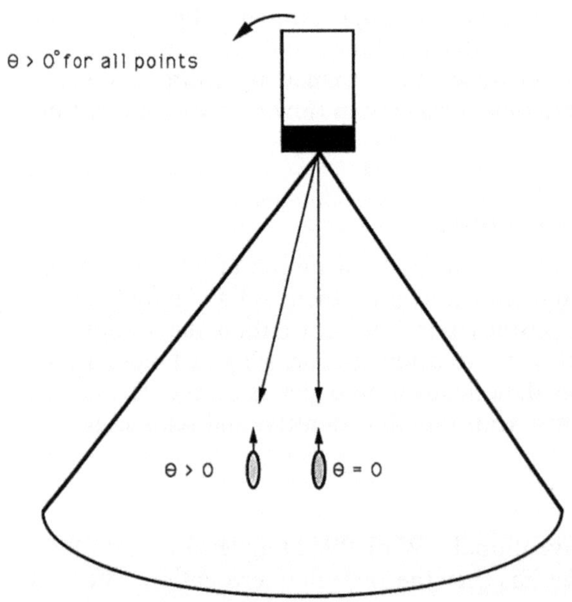

Fig. 4. Angle effects in CDFM. With the center scan line parallel to flow, no underestimation occurs along that line. As the scanning line sweeps away from the center, however, the angle θ in the Doppler equation becomes increasingly nonzero and velocity underestimation occurs. Rotation of the transducer as shown at the *top* of the figure superimposes additional contributions to this important limitation of CDFM

machine, the ability to enter an angle of transducer placement would result in a global correction for angle throughout the sector and would resolve the former limitation. In the meantime, it is important to keep in mind current limitations which account for the semiquantitative nature of blood flow velocity mapping using color Doppler imaging.

Instrument Settings

Having explained how velocity data are obtained and color coded for display in a Doppler image, we can now briefly discuss the various controls available on the instrument panel which are used for enhancement of the final clinical image. The present variability in instrument settings among manufacturers makes a comprehensive discussion difficult. We will at least discuss the qualitative functional importance of the basic settings to explain the general system functions.

Gain. Gain is generally defined as any amplification of a signal transmitted from one point to another. Varying gain in color flow imaging is generally reflected by increased or decreased "sensitivity" to a given flow field. Gain can affect the spatial extent of a flow as imaged by CDFI. For example, Fig. 5 shows a turbulent jet modeling mitral regurgitation recreated in an in vitro flow chamber. It is evident that for a constant orifice velocity, the jet spatial extent varies with gain settings. This is unfortunate, since jet spatial extent has been correlated with the severity of regurgitation in some studies. Furthermore, the precise gain setting at which the optimal image is obtained, varies from patient to patient. Therefore, care must be taken in making quantitative interpretations of flow jets, and the dependence of their spatial extent on gain setting and other settings discussed below must be taken into consideration.

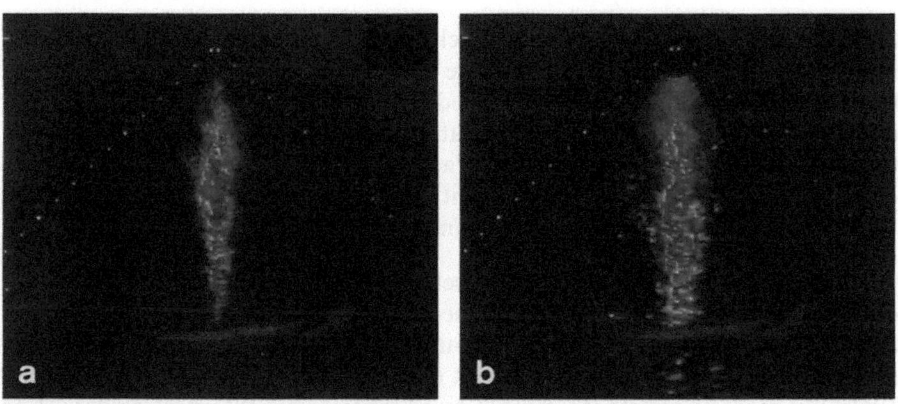

Fig. 5. Gain effects in CDFM. A steady-flow turbulent jet is imaged in vitro with two different gain settings but constant physical flow conditions. The difference in the quality and spatial extent of the image should caution against correlations of flow extent with disease

Pulse Repetition Frequency (PRF). PRF has its most obvious effect in setting the Nyquist limit, as discussed previously. It also affects depth settings; by increasing PRF, one increases the Nyquist limit. However, in PW Doppler, increasing the PRF beyond certain limits results in multiple sample volumes as the transducer is unable to distinguish signals returning from multiples of the desired depth. Since the spatial location of signals is precisely the idea of color flow imaging, such range ambiguity cannot be tolerated. Therefore, an increase in PRF beyond certain limits usually results in an automatic decrease in viewing depth on the screen.

Carrier Frequency. The frequency of the transducer plays an important role in the final image quality obtained. Because of higher absorption of high-frequency ultrasound in interrogation of deeper vessels (e.g., the abdomen or pelvis), the use of lower-frequency transducers is dictated. However, higher-frequency transducers produce better quality images with higher axial resolution. The ability to image *flow* fields also depends on the frequency of the transducer. Lower-frequency transducers allow accurate measurement of higher velocities by producing smaller frequency shifts (Eq. 1) and increasing the Nyquist limit.

Wall Filters. High-pass wall filters remove low-frequency signals and are particularly important in color flow imaging, since both blood cell and solid vascular structure movements are being interrogated simultaneously. Low-frequency, high-amplitude signals from the vascular walls can result in saturation of the data processing circuits, resulting in noise which makes the true flow indistinguishable. This instrument setting eliminates such noise through high-pass frequency filters, and their levels can be adjusted depending on the magnitude of flow velocities. Optimum adjustment is achieved when the entire dynamic range of the signal is spanned by the flow signal. Figure 6 shows turbulent jets obtained in the in vitro model of Fig. 5. The effect of wall filter settings on the spatial extent and general appearance of the jet is shown.

Frame Rate. The frequency at which the scanning line moves across the sector can also be varied. Slower frame rates allow more time for data acquisition and processing and generally produce better images. However, high-frequency cardiovascular events necessitate the use of high frame rates in order to avoid stroboscopic effects, frame fractures, or simultaneous frame display of nonsimultaneous events in different parts of the imaging sector. Typical frame rates utilized clinically are around 15 frames/s.

Interrelationships. The instrument settings are interrelated. The principles discussed in this chapter should lead the reader to certain intuitive conclusions as to the relationship between the various settings. These adjustments are often automatically made by the instrument (e.g., reduction in depth for increased PRF) but must be constantly kept in mind when one is interpreting data. For example, for a given frame rate and depth setting, increased packet size necessarily results in fewer lines to construct the image.

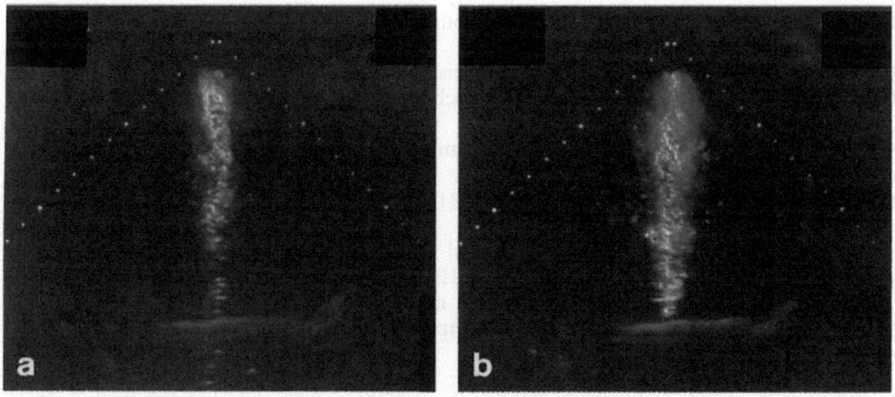

Fig. 6. Wall filter effects in CDFM. The same turbulent jet as in Fig. 5 with varying wall filter settings. Elevating the level of the high-pass filter can trim color from the lateral boundaries of a flow, but it also eliminates low-frequency, high-amplitude signals from cardiac structures which saturate the data processing circuitry, thereby producing cleaner images. Note that the wall filter frequency does not necessarily correspond to the high-pass threshold used as the lower limit for color encoding

Summary

Color Doppler flow imaging represents the highest level of clinical ultrasound imaging technology currently available. A detailed understanding of the principles involved requires an advanced level knowledge of the state-of-the-art signal processing imaging technology. Due to the exceedingly high technicality and complexity of this topic a more intuitive approach to the essential concepts of the CDFI engineering and instrumentation principles has been selected for the purpose of this medical textbook. Understanding of these basic concepts as described in this chapter sets the stage for a better utilization of CDFI in clinical practice and facilitates access to the advanced texts on CDFI technology.

References

1. Bommer WJ, Miller L (1982) Real-time two-dimensional color-flow Doppler: enhanced Doppler flow imaging in the diagnosis of cardiovascular disease. Am J Cardiol 49:944 (abstract)
2. Miyatake K, Okamoto M, Kinoshita N, et al. (1984) Clinical applications of a new type of real-time two dimensional Doppler flow imaging system. Am J Cardiol 54:857–868
3. Switzer DF, Nanda NC (1985) Doppler color flow mapping. Ultrasound Med Biol 11:403–416
4. Merritt CRB (1987) Doppler color flow imaging. J Clin Ultrasound 15:591–597
5. Goldberg B, Kimmelman B (1988) Medical diagnostic ultrasound: a retrospective on its 40th anniversary. American Institute of Ultrasound in Medicine and Eastman Kodak, Washington, JC, and Rochester

6. Kremkau FW (1989) Diagnostic ultrasound: principles, instruments, and exercises. Saunders, Philadelphia
7. Goldberg BB (ed) (1984) Abdominal ultrasonography. Wiley, New York
8. Feigenbaum H (1986) Two-dimensional echocardiography. Lea and Febiger, Philadelphia
9. Kremkau FW (1990) Doppler ultrasound: principles and instruments. Saunders, Philadelphia
10. Taylor KJW, Burns PN, Wells PNT (eds) (1988) Clinical applications of Doppler ultrasound. Raven, New York
11. Vieli A, Jenni R, Anliker M (1986) Spatial velocity distributions in the ascending aorta of healthy humans and cardiac patients. IEEE Trans Biomed Eng BME 33:28–34
12. Lee R (1989) Physical principles of flow mapping in cardiology. In: Nanda NC (ed) Textbook of color Doppler echocardiography. Lea and Febiger, Philadelphia

Chapter 4

Color Doppler Instrumentation

S.H. MASLAK and J.G. FREUND

Introduction

In the last several years, color Doppler imaging has grown rapidly in clinical application. It was introduced for echocardiography instrumentation in 1984, for peripheral vascular imaging in 1986, and for general radiology use in 1987, and is now included on the vast majority of high-performance ultrasound systems placed in the United States and Europe today.

What are the criteria by which high-performance color instrumentation should be judged? Ultrasonographers who have spent years building expertise in B-mode imaging may feel uncomfortable evaluating color images that they cannot assess by the standard criteria used in B-mode: detail resolution, contrast resolution, and image uniformity. [1] They may feel equally uncomfortable judging color Doppler instrumentation whose performance they are not trained to evaluate using objective, easily learned criteria.

This reluctance is understandable, but unnecessary. Physicians familiar with B-mode ultrasound – even those unfamiliar with spectral Doppler – can quickly develop the expertise needed to evaluate high-quality color Doppler images and instruments. The reason is that while color does rely on Doppler information, it is essentially an *imaging* modality, and can be evaluated by imaging criteria.

Intuitively, for example, the clinician realizes that the color Doppler image of a normal kidney in Fig. 1 a or the transesophageal cardiac images in Fig. 18 are of high quality. In Fig. 1 a, for example, the fine structure of vasculature branching from renal hilum to renal cortex is directly reminiscent of the vascu-

[1] B-mode ultrasound images are often referred to by various other names, principally "2-D" and "gray scale." All are appropriate, and may be used interchangeably. In many ways, we believe B-mode is the least confusing term given the characteristics of the different modes available in ultrasound today. For example, the term "2-D", often preferred by cardiologists, might be confusing because color Doppler images are also two-dimensional images. Further, the term "gray scale" might be confusing as well, because "gray-scale" information can now be mapped in colors (e.g., "B-color") as well as in shades of gray (appropriately called gray scale).

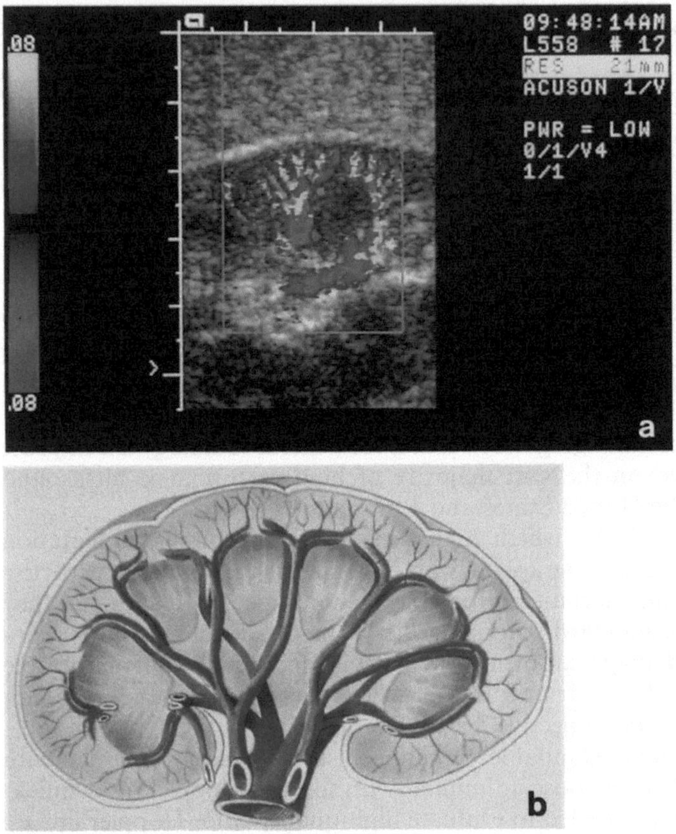

Fig. 1 a, b. Color Doppler image (**a**) compared to anatomic drawing of renal vasculature (**b**)

lar anatomy presented in anatomical drawings (Fig. 1 b). The B-mode image underlying the color has excellent detail resolution, contrast resolution, and image uniformity as well.

However, while most clinicians are familiar with the criteria used to judge images of anatomy, criteria used to judge blood flow imaging are not readily available. No textbook, for example, immediately tells the clinician what the fine details of an eddy current *within* a particular diseased vessel should look like. Therefore, the clinician needs a framework that describes the components that make Fig. 1 a excellent, and allows him to apply this framework to evaluate color images and color Doppler instrumentation. This chapter establishes such a framework. Below, we examine the parameters of color Doppler image quality, and set forth objective and easily evaluable criteria by which color images and equipment can be evaluated in any clinical laboratory.

Contributions of Color Doppler

Color Doppler is valuable to the clinician not because it utilizes color *per se,* but because it adds new physiologic information – images of blood flow – to B-mode ultrasound images of anatomy. These blood flow images represent a new dimension for ultrasound, and provide novel, real-time, and simultaneous diagnostic information that is not easily accessible with other imaging modalities such as X-rays, nuclear medicine, computed tomography (CT), magnetic resonance imaging (MRI), or position emission tomography (PET).

Obviously, the quality of this new color Doppler information is determined by the system's sensitivity to blood flow. Therefore, in evaluating color Doppler instrumentation it is most important to understand its *sensitivity* to flow and its ability to display it accurately in a proper relationship to the underlying anatomy.

But what exactly *is* color Doppler sensitivity? How should the clinician judge it? We begin by examining several common misconceptions about what color flow sensitivity is, proceed to define it properly, and then discuss how it can be evaluated in a clinical setting.

Color Doppler Sensitivity: Correct and Incorrect Definitions

What *is* sensitivity in color Doppler, and what is it not? What is it comprised of?

Sensitivity in color Doppler does not refer to the *amount* of color in an image, the *vibrancy* of the color, or a particular *shade* of robin's egg blue in the physician's favorite color map. These descriptions refer to aesthetic *image presentation* factors established by personal preference and habit: they reveal little about the underlying quality of the information that an instrument can extract from B-mode and Doppler echoes. Experience has shown that clinicians vary considerably in their initial preference for image presentation and readily adapt to alternative methods that offer equivalent or better information.

By contrast, sensitivity in color Doppler does refer both to an instrument's ability to extract the maximum flow information from echoes and its ability to allow the clinician to visualize it as physiology. The instrument must accomplish both these tasks reliably and accurately, in faithful relationship to the underlying B-mode image and anatomy, without introducing artifacts.

Sensitivity should not be confused with depth of penetration. Ultrasound beams, whether for B-mode or color Doppler, are attenuated by tissue; this attentuation ultimately defines the maximum usable depth of ultrasound in any given patient. However, because color Doppler echoes are weaker than B-mode echoes, penetration depth for color Doppler at a given frequency is generally somewhat less than B-mode penetration (Fig. 2). This is based on physical principles, and is therefore true for all instruments. *Sensitivity,* on the other hand, refers to the quality of B-mode or color Doppler information

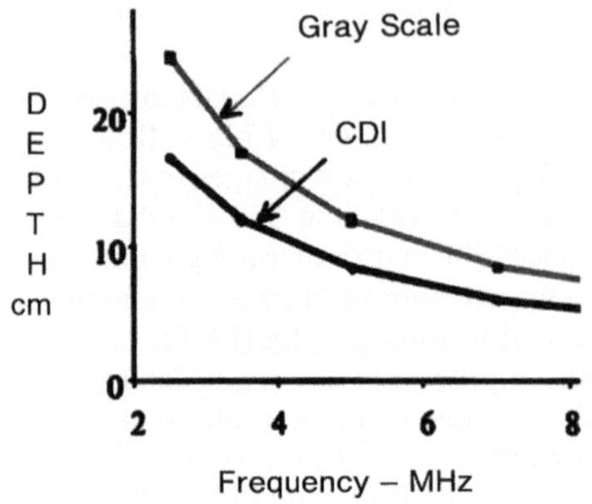

Fig. 2. Graphic representation of the relative penetration of grayscale and color Doppler ultrasound beams (actual penetration may vary for different patients and different types of tissue)

within the usable viewing depth allowed by ultrasound penetration at a given frequency. It is a powerful indicator of the instrument's overall performance.

The Five Key Elements of Color Doppler Sensitivity

Now that we have defined what sensitivity refers to, the next steps are to establish the key building blocks of sensitivity in color Doppler images, and to discuss how each can be assessed clinically.

Many factors are responsible for an instrument's color Doppler sensitivity. However, in this context we shall discuss the five most important ones:

1. motion discrimination,
2. temporal resolution,
3. spatial resolution,
4. image uniformity, and
5. high-level computer control.

The first four of these elements determine the maximum information content possible in a given color image or instrument, while the fifth is required to make a sensitive color Doppler instrument practical for use in an everyday clinical setting. Each of these elements can be measured and evaluated with tools commonly available in a clinical laboratory.

Of these factors, the first two – motion discrimination and temporal resolution – are based on the Doppler effect, but can be understood far more easily by analogy to two-dimensional B-mode imaging than by analogy to spectral Doppler. The third and fourth factors – spatial resolution and image uniformity – will be directly familiar to clinicians by analogy to their experience in evaluating B-mode images.

The fifth factor – high-level computer control – describes an instrument's ability to optimize multiple parameters simultaneously in response to simple commands given by the clinician. This is critical because color Doppler imaging requires not only the adjustment of all the parameters used in B-mode imaging and spectral Doppler, but also requires new adjustments specific to color Doppler. Without an ability to translate a few intuitive commands entered by the clinician into the hierarchy of multiple parameters required to control the color Doppler instrument, flow sensitivity would be compromised: the system would be too complicated to be optimized routinely.

Motion Discrimination

Motion discrimination refers to the ability of a color Doppler system to discriminate between different types of motion through simultaneous analysis of Doppler-shifted and B-mode echoes. The need for motion discrimination is unique to color Doppler imaging, and has no direct analog in B-mode. Nevertheless, its presence or absence in a color Doppler instrument can be easily assessed by the clinician, as will be discussed.

Motion discrimination has two aspects, each critical to an instrument's clinical sensitivity to blood flow information: (a) ability to discriminate moving blood from tissue (whether moving or stationary) *within* a given pixel of the image, and (b) ability to discriminate and resolve flow patterns comprised of subtly different velocities within a vessel or chamber.

Discrimination of Blood from Tissue

As is well known, Doppler shifts occur when a reflecting object within the body moves with respect to the sound generator in an ultrasound transducer. Thus, the Doppler shift caused by *tissue* moving within a given velocity range is indistinguishable from the Doppler shift caused by *blood* moving within that same range.

Since many tissues in the body move constantly – due to transmitted cardiac motion, vessel pulsation, respiratory motion, and involuntary tissue movement – this motion can cause Doppler shifts falsely attributed to blood flow. This was a significant problem in all color systems until recently, and is usually referred to as "flash artifact" or "motion artifact."

An example appears in the kidney image in Fig. 3a. Here, the flash artifact appears as a large flashing pool of red on the right side of the image; it obviously obscures any renal anatomy and vasculature that might be present, compromising sensitivity.

The conventional approach to reducing this type of artifact is to use a velocity cut-off filter, which filters out from the color image any velocity below an adjustable critical value. This improves the artifact, because the tissue motion causing the flash artifact is typically slow. Unfortunately, renal flow includes velocities lower than this threshold, which are also removed by the filter. Consequently, this type of filter can reduce flash artifact, but at the expense of eliminating the visualization of low flow velocities in the kidney parenchyma, as shown in Fig. 3b. This is especially noticeable near the renal cortex, since the arcuate arteries are not visualized in color.

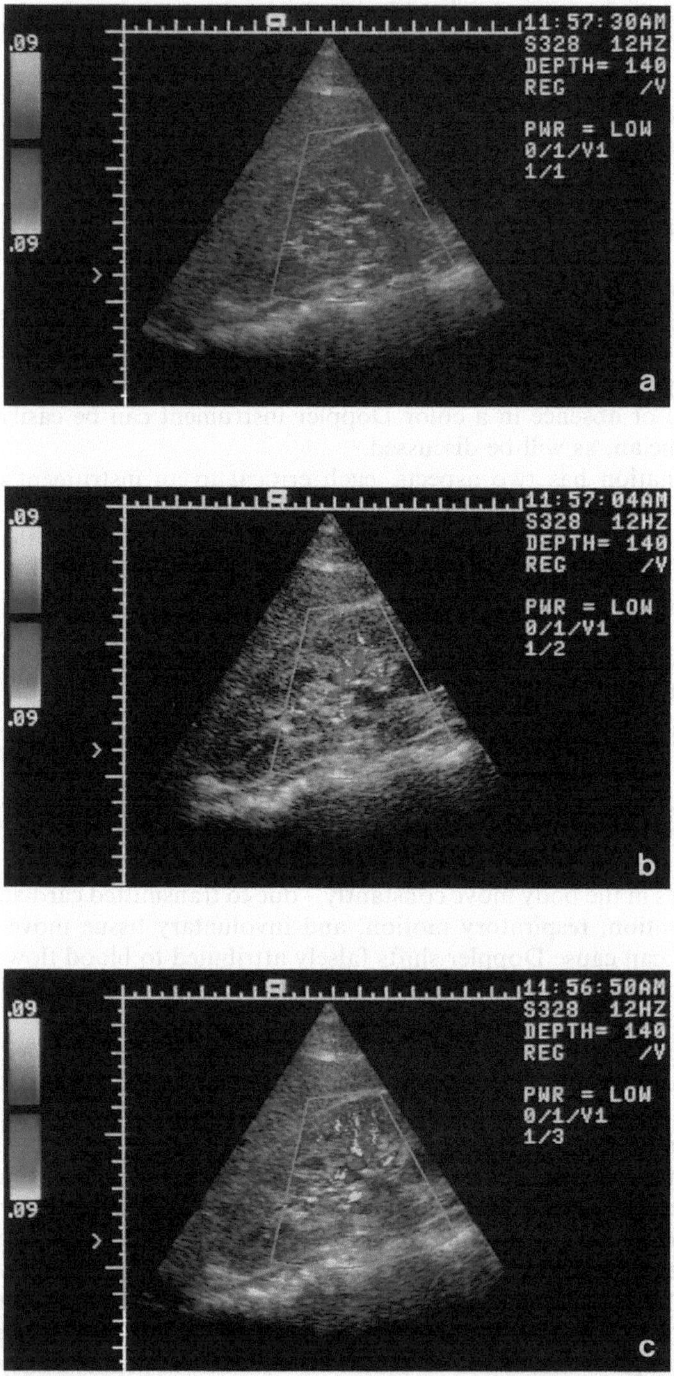

Fig. 3a–c. Effect of motion discrimination on renal image. **a** Appearance of flash artifact. **b** Use of conventional velocity cut-off filter. **c** Use of multivariate motion discriminator

This tradeoff is now unnecessary, due to the development of the multivariate motion discriminator.[2] In this new approach, sophisticated motion discriminators were designed to be capable of analyzing not only Doppler shifts due to velocity, but *multiple* aspects of returning echoes and their Doppler spectra, simultaneously and in real time. The imaging results are impressive particularly in real time, but can also be appreciated in frozen images. In Fig. 3c, the multivariate motion discriminator removes the flash artifact while allowing flow to be visualized all the way to the renal cortex. The result is increased effective color Doppler sensitivity.

It is important to note that good motion discrimination is inherently more practical in electronic scanning than it is in motor-driven scanning, such as is found in mechanical or annular array transducers. The reason is that the Doppler effect caused by a moving motor-driven transducer element is indistinguishable from similar Doppler effects caused by moving blood or tissue. As a result, all-electronic transducers (phased arrays), without moving elements, are used by the vast majority of manufacturers to perform color Doppler.

However, all-electronic transducers do not automatically employ multivariate motion discrimination. The presence or absence of a multivariate motion discriminator in a color Doppler system or transducer can be easily evaluated on a normal native kidney, especially if the patient is given a glass of fruit juice or other fluid shortly before the examination to hydrate the renal vasculature. The system should be set up to provide maximum color spatial resolution while maintaining adequate frame rate. If adequate motion discrimination is present, the system should be capable of imaging renal flow all the way to the cortex with minimal or no flash artifact, as in Fig. 3c.

The normal kidney is a good motion discrimination test object for three primary reasons. First, the normal kidney is readily available and constantly moves within the patient from cardiac and respiratory motion. Second, fine kidney vasculature is woven within the renal parenchyma, so that returning echoes in virtually every pixel contain a mixture of Doppler-shifted echoes from anatomical structures and Doppler-shifted echoes from flow. Finally, examination of the kidney is relevant not only to radiology ultrasound specialists but to an increasing number of echocardiographers or vascular specialists who are mindful of the renal origins of some cardiovascular disease.

The need for motion discrimination is acute when imaging flow under demanding circumstances irrespective of anatomical location. These demanding circumstances arise, for example, with extremely ill patients who cannot easily hold their breath or lie still. Demanding circumstances also arise unpredictably whenever blood and tissue echoes are intermixed, such as in areas adjacent to blood clot, adjacent to vessel walls or cardiac chambers, or that pick up secondary motion echoes from the heart or diaphragm. In these circumstances, the multivariate motion discriminator can act effectively as an enhancer of flow contrast – eliminating or reducing artifact without removing actual flow.

---

[2] Developed by Dr. J. W. Allison and colleagues, 1265 Charleston Road, Mountain View, CA (private communication).

Conversely, motion discrimination is least needed where the field of view contains relatively large areas that contain only blood or only tissue, away from tissue-blood interfaces.

Discrimination of Flow Patterns

The second aspect of motion discrimination essential to clinical sensitivity is the ability to resolve flow patterns comprised of subtly different velocities within a vessel or chamber. This is analogous to contrast resolution in gray scale, where a system must resolve subtle tissue characteristics that lie adjacent to each other, allowing the clinician to visualize abnormal pattens in echo texture.

A good example of the reason it is important to image subtle flow velocity differences is shown in Fig. 4, which deliberately displays no motion discrimination at all. In this image, a bright red uniform color is shown throughout the lumen. This suggests that flow is a single velocity. But blood actually moves at different velocities throughout a vessel; it is slower along the walls than in midstream, with subtle variation throughout. Thus, the uniform red color in Fig. 4 may initially appear striking, or may appear to display "a lot of color," while the experienced user recognizes that the illustration exhibits no subtle gradations in flow and therefore has virtually no color information content.

By contrast, good motion discrimination allows the system to display all the subtle velocities that are present. This is demonstrated in Fig. 5, which displays subtle gradations in color in contrast to the uniform color in Fig. 4. The slowest-moving blood in Fig. 5a is found near the vessel wall, as expected (as tagged in white), while the fastest streamline – counterintuitively – is skewed somewhat away from the exact center (as shown by the white tag in Fig. 5b).

Under physiologically unperturbed conditions, the nonturbulent laminar flow in vessels is organized into streamlines and patterns; the flow contrast

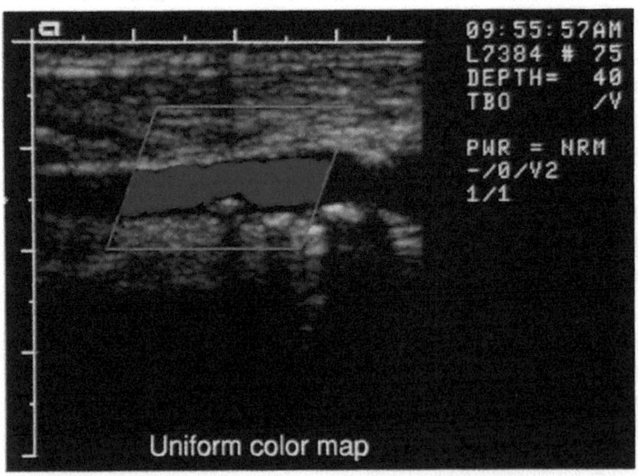

Fig. 4. False impression of color sensitivity created by use of an inappropriately uniform color map

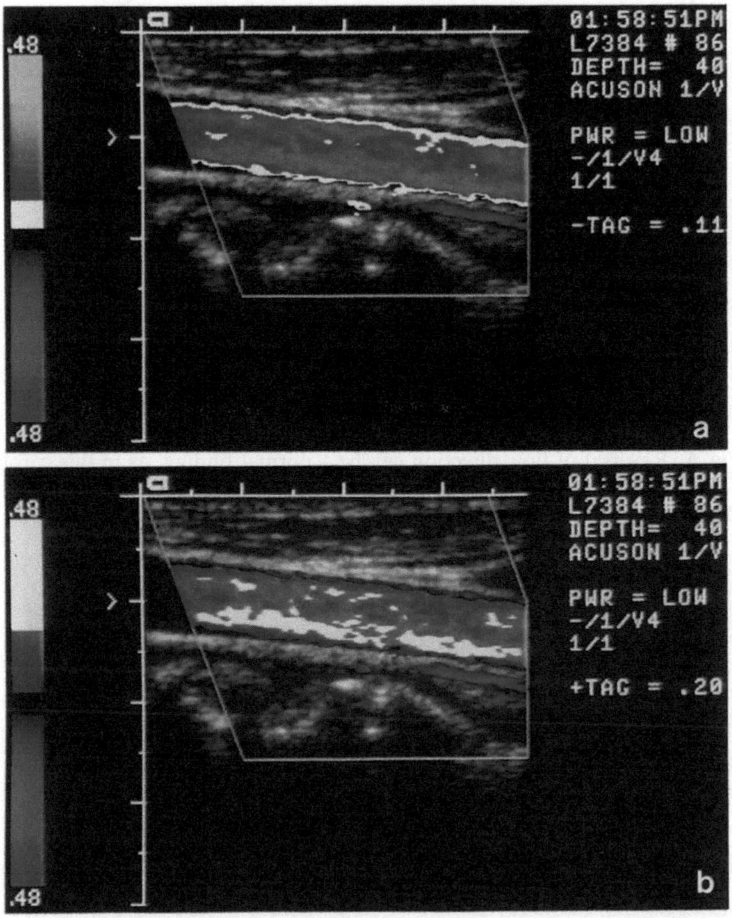

Fig. 5. Motion discrimination demonstrated within a nonturbulent vessel. **a** The *lowest* flow velocities have been tagged with *white*. **b** The *highest* flow velocities have been tagged with *white*

made possible with good motion discrimination brings these streamlines and patterns out and more accurately represents vessel hemodynamics. As the clinician gains experience with such characteristic patterns, he can often recognize their absence as pathological or abnormal. An example of one of these typical patterns is the reversed flow normally found in the carotid bulb on a sensitive color Doppler instrument (Fig. 6). In other situations, reversed flow can represent pathologic changes.

Similarly, with good motion discrimination a system can image the changes in blood flow velocity profiles that correspond to different phases of the cardiac cycle. In many vessels these profiles are typically blunt during systole, and become more parabolic during diastole. In other words, not only is the highest velocity found during peak systole, but the velocity difference between

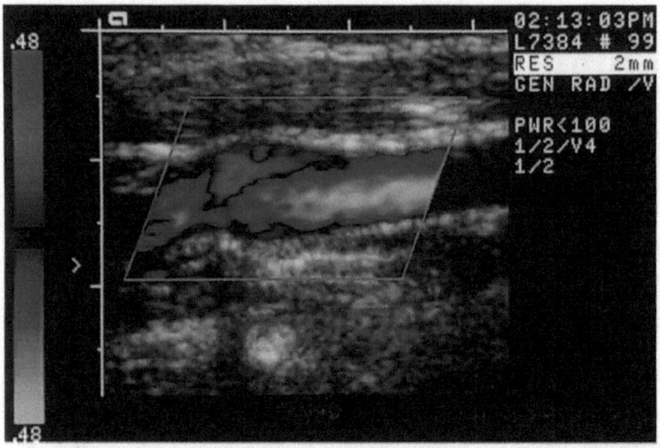

Fig. 6. Reversal of flow direction can be a normal finding, as in this carotid bulb

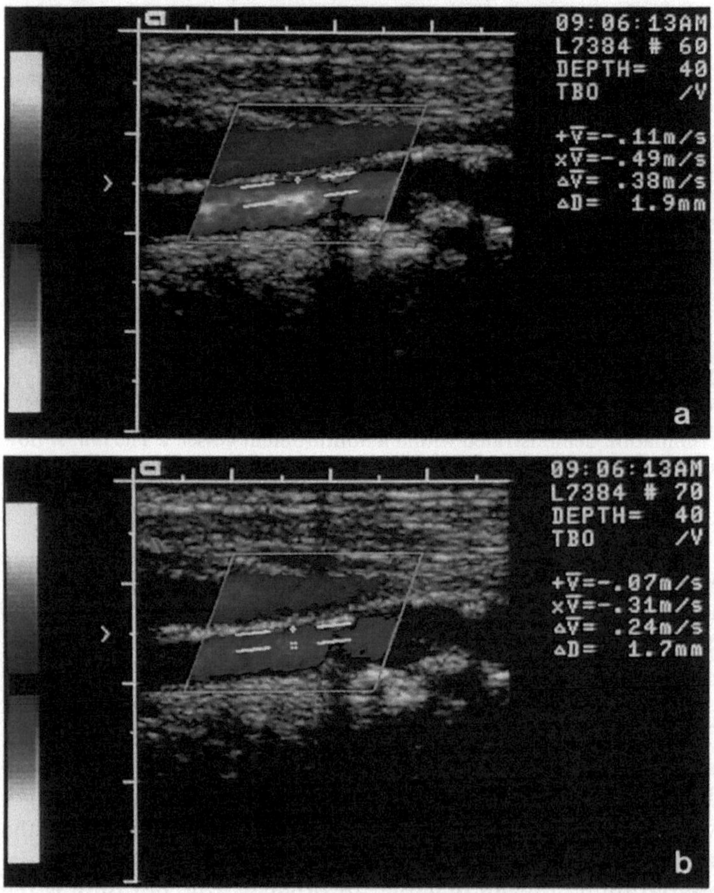

Fig. 7 a, b. Flow profiles vary during cardiac cycle. **a** Systole. **b** Diastole

the center and periphery of the vessel generally varies during the cycle. In Fig. 7a, for example, in systole, the color calipers show that velocity in the center of the vessel is 0.49 m/s, with a difference of 0.38 m/s between the center and the periphery. In Fig. 7b, on the other hand, obtained during diastole, the velocity in the center has dropped to 0.31 m/s, while the difference between center and periphery has dropped to 0.24 m/s.

Again, the clinician using a sensitive color Doppler instrument comes to expect certain profile changes that are characteristic of different vessels, and can suspect pathology when these changes are not present. These expectations become intuitive with clinical color Doppler experience, as the clinician's eyes become trained. By contrast, users of equipment without good motion discrimination may be unable to see such patterns consistently or reliably, which may impede accurate diagnosis, as well as slow the learning curve of the clinician.

In summary, motion discrimination has no direct analog in B-mode imaging, but, as we have seen, it critically impacts the clinical sensitivity of color Doppler. The closest analogy familiar from B-mode imaging is that of contrast resolution, which after being poorly understood for many years was significantly improved in B-mode systems with the introduction of dynamic apodization [3, p. 12], a technology which removed stray reflected echoes from the B-mode image in real time. Similarly, the work of Dr. J. W. Allison et al. has only recently allowed sophisticated motion discrimination to be introduced. Its impact on clinical color Doppler imaging may parallel the impact of improved contrast resolution on B-mode imaging.

Temporal Resolution

Temporal resolution refers to the ability to visualize changing physiological parameters in real time. It is determined primarily (but by no means exclusively) by frame rate. The effect of frame rate on the ability to assess motion is familiar from B-mode imaging, but high frame rates are much more difficult to achieve in color Doppler.

The reason for this greater difficulty comes from basic physical principles. B-mode systems require only one pulse-echo cycle for each scan line. However, each color scan line formed analyzes the movement of blood *between* pulse-echo cycles. Therefore, the system must "dwell" on each color Doppler scan line for several pulse-echo cycles before moving on to form the next line. The "uncertainty theorem" in Fourier analysis [see 1 p. 160, 5 p. 485], which explains the tradeoff between accuracy and frequency of sampling, shows that the longer the dwell time on a given scan line, the greater the accuracy of motion analysis for that line. The result of this longer dwell time is inherently lower frame rates – often only half as great for a similar image set-up – for color Doppler when compared to B-mode imaging.

However, sensitivity in clinical use requires a color Doppler instrument to achieve frame rates adequate to capture the physiology under examination, which can vary from organ to organ. Further, in normal circulation, the

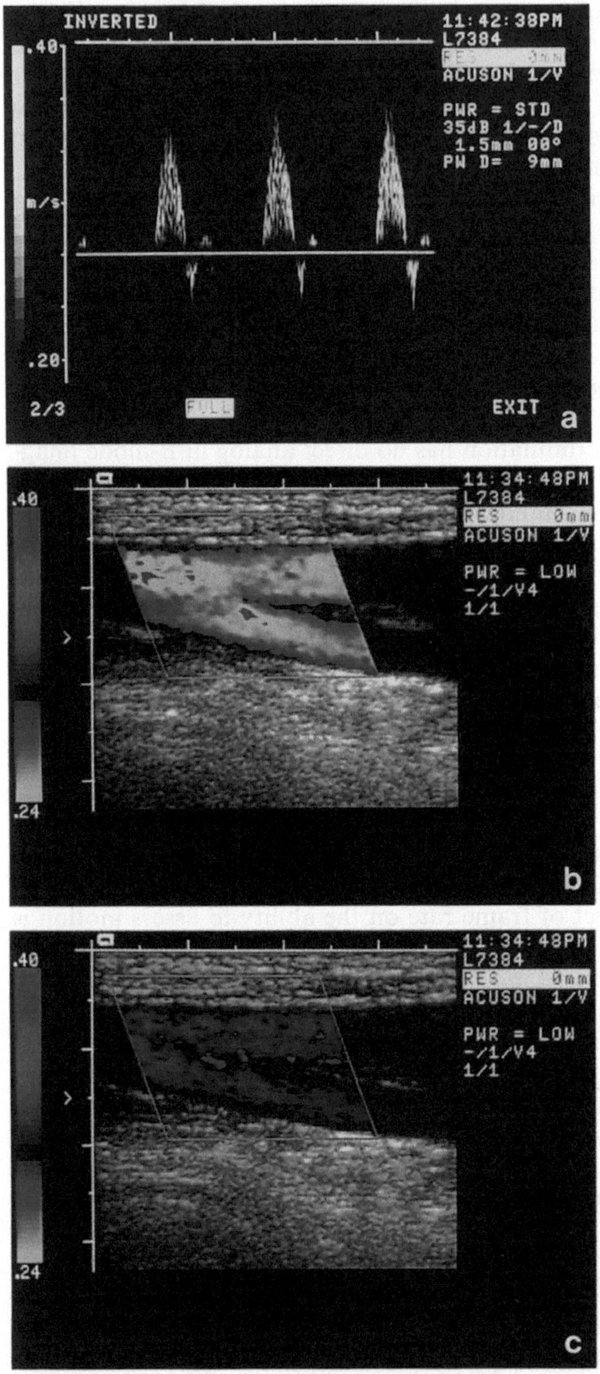

Fig. 8 a–c. Triphasic flow demonstrated in a normal femoral artery. **a** Spectral Doppler. **b** Systole. **c** Diastole

pattern of flow can change dramatically throughout the cardiac cycle, and deviations from expected patterns may represent pathology. Examples that require moderately high frame rates between 10 and 20 Hz include:

- The femoral artery, which normally exhibits triphasic flow. This is demonstrated in the spectral Doppler image shown in Fig. 8a; it can also be visualized in the corresponding systolic color image (Fig. 8b) and the diastolic flow reveral demonstrated by the blue color in Fig. 8c.
- The kidney parenchyma. Highly pulsatile flow in the parenchyma of the kidney (or lack of diastolic flow) indicates high vascular resistance to flow and can be diagnostic of an inflammatory process, such as transplant rejection.

Much higher frame rates are required for echocardiographic color Doppler imaging, where jets may be very transient and present only during a brief part of the cardiac cycle. This is shown in Fig. 9, an image of a pediatric atrioventricular septal defect where the patient had a heart rate of 145 beats/min. In this sequence, the dual-origin jet is visible for only a fraction of a second in Fig. 9b, as demonstrated by Fig. 9a, formed before the jet appears, and Fig. 9c, formed after the jet has disappeared. Clearly, temporal resolution adequate for each organ under examination qualifies as an essential component of clinical color Doppler sensitivity.

In general, a clinician can assess the adequacy of a system's color Doppler temporal resolution by a simple test. First, select an appropriate organ of interest and obtain a high-quality spectral Doppler signal. Note the phases of flow throughout the cardiac cycle, then image the same location with color Doppler. The system's temporal resolution for color Doppler in that organ is adequate if the clinician can visualize each phase that was visualized in spectral Doppler. For example, in Fig. 10a, a spectral Doppler image of a renal artery, the spectrum clearly demonstrates that flow occurs in both systole and diastole. A color Doppler image of the same artery (Fig. 10b) shows the peak systolic flow corresponding to the spectral Doppler, while Fig. 10c demonstrates that the instrument is capable of imaging the end-diastolic flow in color. Performing this test on several organs in the normal patient makes for an excellent overall assessment of an instrument's temporal resolution.

Knowledge that a system can image with adequate temporal resolution should increase the clinician's diagnostic confidence, and vice versa. For example, if inadequate temporal resolution in a color Doppler system makes it unable to visualize normal renal diastolic flow on a patient, the clinician might misinterpret the absence of this flow as pathologic when in fact the real reason is inadequate equipment performance.

We have shown that the clinical need for a high frame rate is at odds with the need (based on Fourier theory) for long dwell time and accuracy of motion analysis. Considerable engineering effort has been directed at this dilemma, and some important solutions are now available. For example, most color Doppler instruments now offer the operator some method to select a "window" or "region of interest" outside of which no color Doppler information

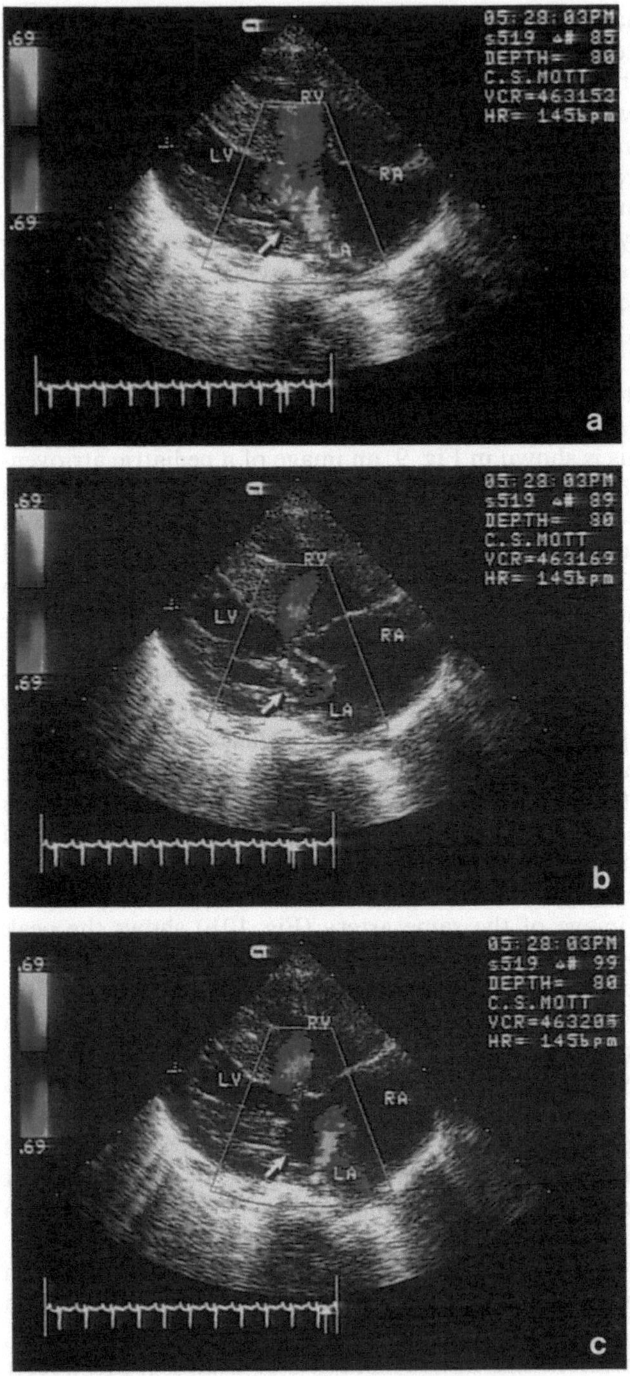

Fig. 9 a–c. Temporal resolution in a pediatric atrioventricular septal defect. **a** Just prior to jet. **b** During jet. **c** Just after jet

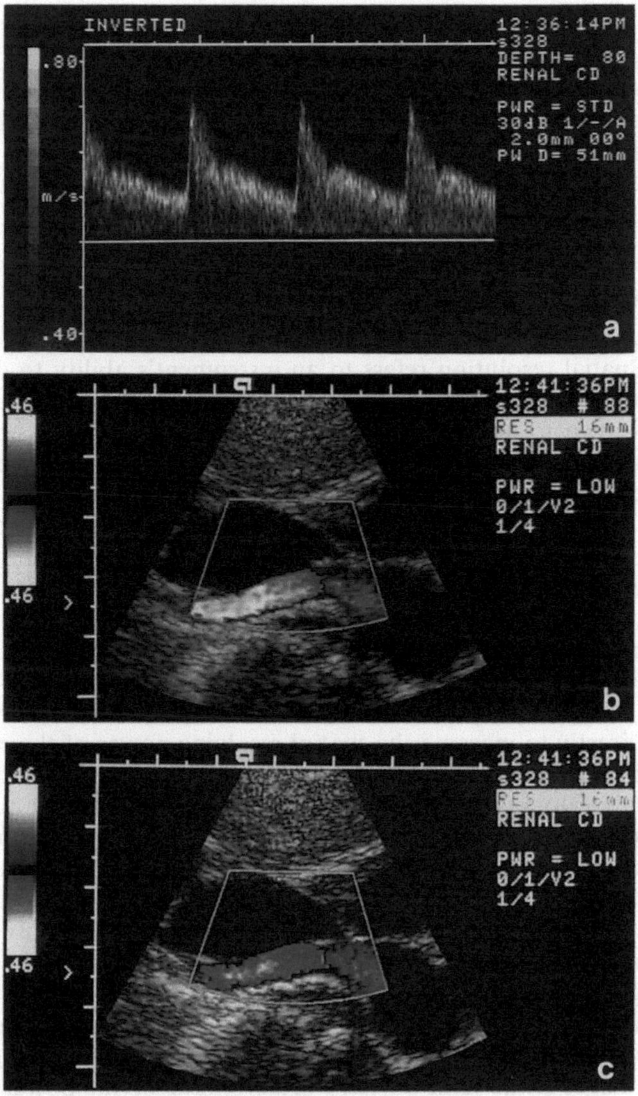

Fig. 10 a–c. Temporal resolution demonstrated in a normal renal artery. **a** Spectral Doppler. **b** Color Doppler, peak systole. **c** Color Doppler, end diastole

is presented: by reducing the color Doppler field or view, the frame rate is increased. As a second example, in some demanding circumstances sophisticated ultrasound systems display two color Doppler frames for every grayscale frame, increasing the amount of temporal resolution that can be visualized. The underlying engineering sophistication and resulting clinical performance of these and other methods vary greatly from instrument to instrument.

The complex tradeoffs involved in optimizing frame rate for a given clinical situation involve controlling many components of the color Doppler image simultaneously, in real time. Different organs and types of vasculature require different tradeoffs and optimization. The level of control required to maximize effective sensitivity throughout the body would be impossible to achieve in clinical settings without high-level computer control technology, as discussed below.

Spatial Resolution

Spatial resolution refers to an instrument's ability to detect small structures and display them in their correct anatomical locations. It should be well known to any sonographer familiar with B-mode, since it is the aspect of B-mode imaging that allows an instrument to distinguish small structures with clarity.

B-mode spatial resolution can be assessed visually and measured analytically by using an ultrasound tissue-equivalent phantom [3 p. 5]. These tissue-equivalent phantoms are readily available for use in clinical laboratories, and if used properly are a convenient way to analytically assess the spatial resolving power of ultrasound systems, most typically in the lateral and axial dimensions.

Since tissue phantoms do not contain actual blood flow, they cannot be used directly to measure the spatial resolving power of the color Doppler imaging beam. Such a direct measurement must be made in a sophisticated engineering laboratory. However, the B-mode spatial resolution achieved on a phantom while the color Doppler mode is active is an excellent indirect measure of the theoretical maximum resolving power of the system used to form the color Doppler image.

The validity of using a B-mode phantom to measure color Doppler spatial resolution can be easily understood by analogy to photography. The photographic test object in Fig. 11 is a standardized tool for measuring the spatial resolving power of a camera lens. A high-resolution lens gives high-resolution images of the test object, while a low-resolution lens gives lower-resolution images of the same object. Further, the resolving power of the lens is unaffected by whether the camera is loaded with black and white film (Fig. 11 a) or color film (Fig. 11 b). For similar reasons, the spatial resolution achievable with a given color Doppler capable transducer can be no greater than that achievable with that same transducer in B-mode: as in B-mode, maximum lateral resolution achievable for color Doppler is a function of the number of independent electronic channels an ultrasound system and transducer use simultaneously (on transmit and receive) to form an image line. [3]

[3] As has been well described, a pin target in a phantom at a distance R from an array ultrasound transducer with elements spaced 1/2 wavelength apart will be imaged with a lateral dot size inversely proportional to the number of independent electronic channels used simultaneously [see 3 p. 11].

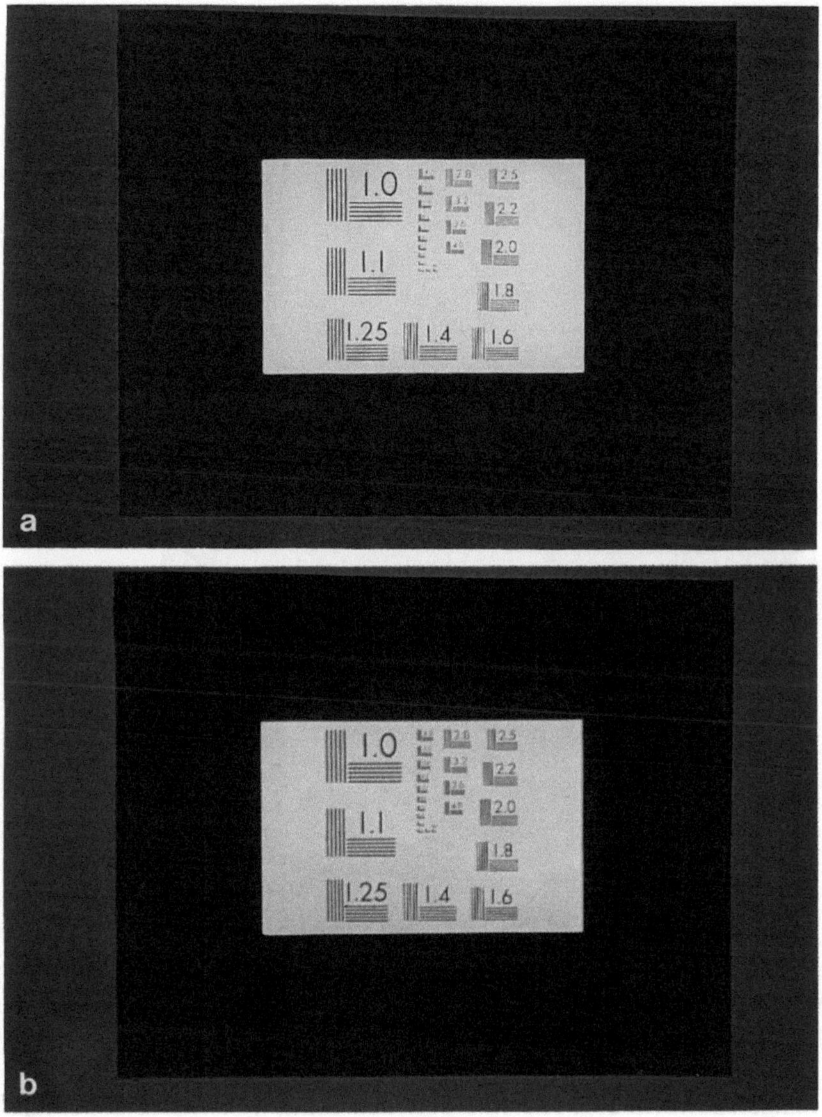

Fig. 11 a, b. The same test object demonstrates similar spatial resolving power in photography using black and white film (**a**) and color film (**b**)

This relationship between channels and resolving power can be illustrated in gray scale in several ways. Figure 12 a compares a liver and kidney imaged with 128 channels to an image of the same liver and kidney using the same system and sector transducer with only 64 channels turned on (Fig. 12 b). A tissue-equivalent phantom allows a more objective measurement. Figure 13 represents the type of assessment that can be easily performed by a sonographer comparing systems or transducers in any clinical laboratory. The phantom

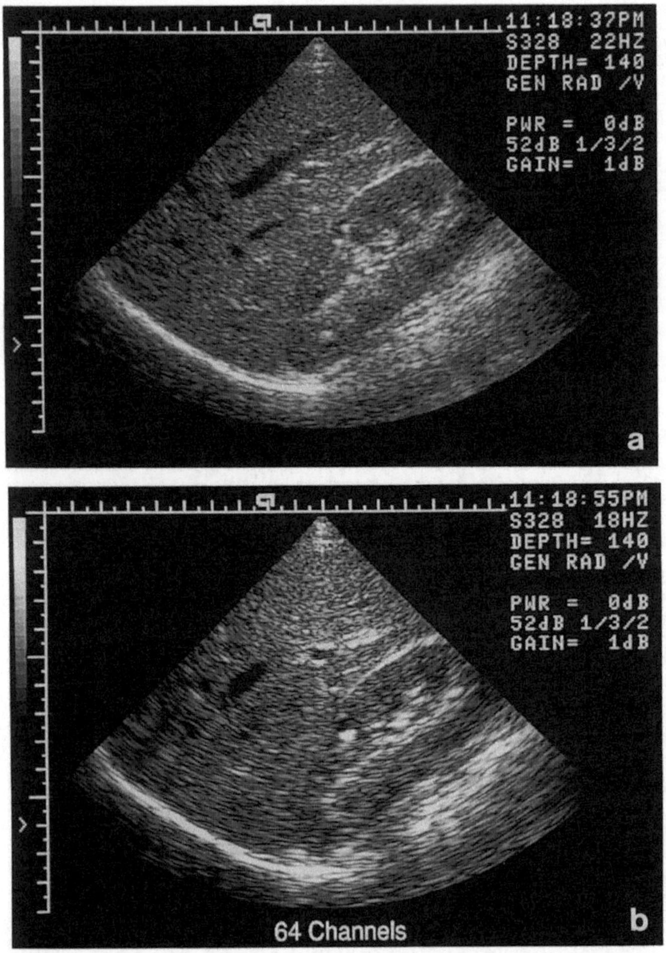

Fig. 12 a, b. Images of the same normal liver and right kidney obtained with the same ultrasound system and sector transducer **a** with 128 channels active and **b** with only 64 channels active

image shown in Fig. 13a was obtained using a 3.5-MHz abdominal sector transducer with 128 channels, while in Fig. 13b the number of simultaneously active channels on the same acoustic lens has been reduced to 64 in order to demonstrate the decreased lateral resolution that results from fewer channels, especially in the far field. Note that in both phantom images, the color bar is present, demonstrating that the color Doppler mode was enabled. Clearly, the 128-channel acoustic lens displays the pins with finer lateral resolution than the lens with fewer channels.

It is important to realize that the number of channels being used by a system at any time can be no greater than the number of channels used simultaneously by the active transducer. A 128-channel system using a transducer with 64

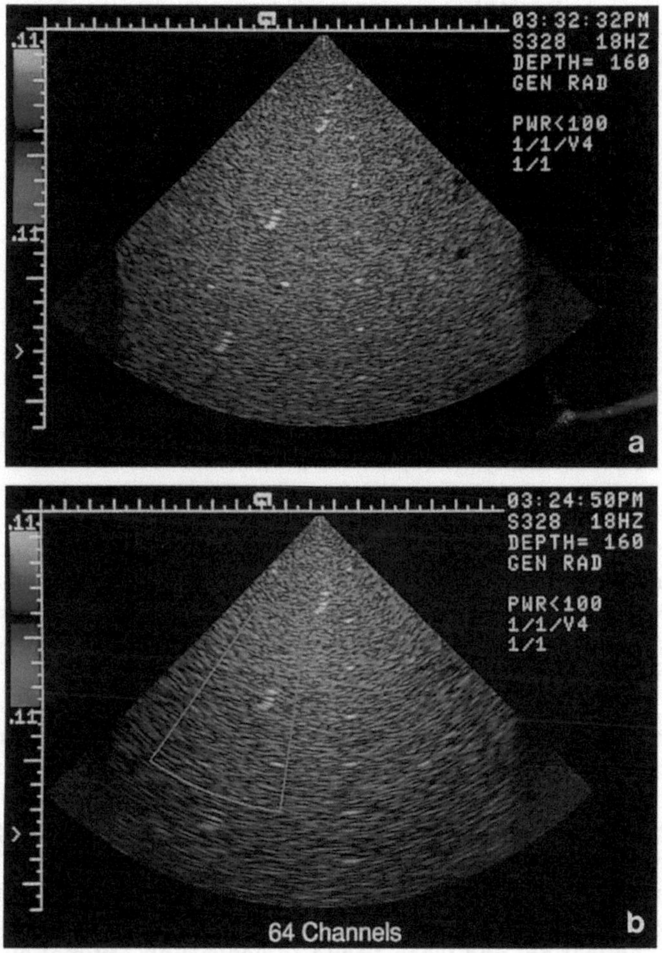

Fig. 13 a, b. Images of the same tissue-equivalent phantom obtained with the same ultrasound system and sector transducer **a** with 128 channels active and **b** with only 64 channels active. Note that the color Doppler mode is active in both images

channels performs as if it had 64 channels, not 128, regardless of the total number of channels that the system may offer with other transducers.[4]

[4] Two terms – "elements" and "channels" – are often confused by manufacturers. "Channels" refers to the number of independent electronic circuits used to transmit and receive ultrasound information from an electronically scanned transducer. "Elements" refers to the number of separate piezoelectric elements in the transducer that generate and receive sound. Each channel may be connected to more than one element; however, additional elements per channel do not fundamentally allow an increase in the information content of the image, while increased numbers of channels do. It is always worthwhile to ascertain the number of simultaneously active independent *channels* a transducer or ultrasound system offers before evaluating it, and not be confused by discussions regarding the numbers of *elements*.

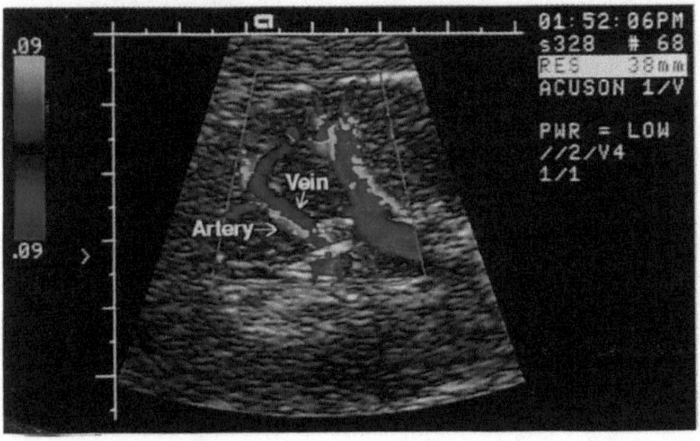

Fig. 14. Spatial resolution allows visualization of the interlobar artery and vein in this study of a normal kidney

As in B-mode, greater spatial resolution equates with more diagnostic information on clinical examinations. With excellent color spatial resolution, for example, the small interlobar artery and vein lying next to each other in Fig. 14 can be more clearly resolved, demonstrating the contribution lateral resolution makes to color Doppler sensitivity.

When performing the type of phantom image comparison discussed here, it is also important to compare transducers that require similar anatomical access, in order to ensure that the comparison is based on transducers that would be used in similar types of clinical situations. The physical size of the transducer face (or transducer "footprint," as it is often called) generally determines how much access is required. The physical access available for a given examination on a patient is often limited, as in intercostal or endocavity scanning, when attempting to image around bandages or bowel gas in an abdominal examination, and when trying to achieve an optimal angle to flow. All else being equal, the highest B-mode and color Doppler spatial resolution obtainable with a given ultrasound system will result from choosing the color-Doppler-capable transducer with the largest footprint (assuming it also has the largest active aperture) that is suitable for a particular application.

In other words, even if the number of channels and frequency is unchanged, if all else is equal a larger footprint size should be capable of proportionately better lateral resolution.[5] This can be appreciated by comparing phantom images using the techniques described earlier.

[5] Conversely, two transducers with similar footprint sizes and different numbers of channels may offer the same theoretical lateral resolution, if all else is equal, since lateral resolution is proportional to the width of the acoustic aperture. In this type of comparison, however, a greater number of channels should offer substantial benefits other than lateral resolution. For example, a 2.5-MHz sector transducer with 128 channels spaced across a 19-mm footprint will offer approximately the same lateral resolution as a 2.5-MHz sector transducer

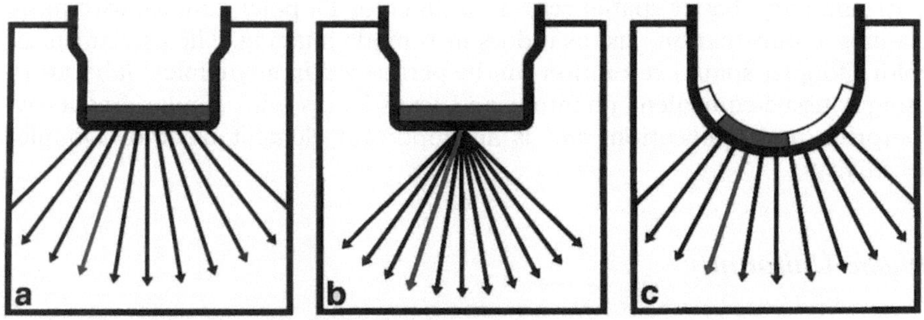

Fig. 15a–c. Graphic representation of the simultaneously active apertures used by different transducer designs with the same size footprints and similar fields of view. **a** Vector Array. **b** Sector. **c** Tightly curved array

Finally, some transducer geometries inherently limit the number of channels or elements that can be used simultaneously, limiting lateral resolution. Electronic sector, Vector Array®, linear array, and gently curved array transducers all have geometries that, if the designer elects, allow them to use all their channels and elements simultaneously, permitting lateral resolution to be maximized. By contrast, the extremely convex geometry in a tightly curved array transducer prevents elements on one side of the transducer from being used simultaneously with elements on the opposite side, compromising lateral resolution, especially in the far field. This is illustrated in Fig. 15, which graphically represents the simultaneously active apertures used by Vector Array, sector, and tightly curved array transducers with the same size footprints and similar fields of view. The illustration explains why Vector Array transducers can produce color images with uncompromised resolution while allowing much larger fields of view than sector transducers at all depths: both sector and Vector Array transducers use their entire footprint as a simultaneously active aperture. By contrast, tighty curved array images may offer large fields of view but may compromise lateral resolution if an extreme curvature does not permit them to use their entire footprint simultaneously as an active aperture.

---

with 64 channels spaced across the same 19-mm footprint. In this case, geometry dictates that the spacing between the channels on the 128-channel transducer is twice as dense as on the 64-channel transducer, and the more densely packed channels offer very significant increases in contrast resolution. In other comparisons between transducer designs with similar footprint sizes but different numbers of channels, the increased channels may result in greatly diminished imaging artifacts such as grating lobes, which can appear in the image and masquerade as anatomical or flow information where none actually exists. Regardless of which benefits may be offered by increased channels in a given transducer – increased lateral resolution, increased contrast resolution, or decreased artifacts – an increased number of channels offers significant clinical benefits if properly utilized. With respect to evaluating the spatial resolution of color Doppler imaging, the impact of channels on lateral resolution for a given transducer design can be easily assessed using the phantom imaging technique described.

In summary, better spatial resolution in color Doppler equates with more diagnostic information, just as it does in B-mode imaging. The assessment of color Doppler spatial resolution can be performed in any clinical laboratory using a tissue-equivalent phantom and transducers with similar frequency, footprint, and application, and is an important element of color Doppler sensitivity.

Image Uniformity

As experienced clinicians have learned from B-mode imaging, an ultrasound system must strive for uniform performance throughout its field of view. The same is true for color Doppler. Within the field of view defined by the depth of color Doppler penetration, an ideal color Doppler instrument should be uniformly sensitive to blood flow. The reason is simple: when performing an ultrasound examination in any mode, the ultrasonographer cannot know in advance where in the field of view the pathology he is looking for might be found.

A powerful tracking lens system [3 p. 13] – an electronic ultrasound lens capable of real-time dynamic focusing, aperture, and apodization – is perhaps the most essential contributor to image uniformity in B-mode imaging. This is equally true for a color Doppler system. As discussed above, any clinician can use a tissue-equivalent phantom with the color Doppler mode active to assess image uniformity. In Fig. 16a, the phantom image shows a high degree of uniformity because it was obtained on a color Doppler system using dynamic focusing. In Fig. 16b, the same color Doppler system and transducer are used with the dynamic focusing portion of the tracking lens system disabled to produce an image with significantly less uniformity, as evidenced by poorer resolution away from the center of focus. Note that the color Doppler mode is enabled in both phantom images.

True tracking lens systems are sometimes confused with a less effective alternative correctly called "zone focusing," although both types are often called "dynamic focusing" by manufacturers. In a true tracking lens, the focal length is continuously reshaped during the receive portion of the pulse-echo cycle to maintain precise focus for returning echoes along each scan line. This allows for a sharply focused image from skin line to maximum depth, and only requires a single transmit pulse to form each ultrasound line. By contrast, in zone focusing the lens cannot change its focal length in real time during the receive portion of the pulse-echo cycle. Instead, each ultrasound line is actually a composite of several lines obtained sequentially with separate fixed electronic lenses. Each portion of the line in the composite requires its own pulse-echo cycle; thus, obtaining an entire image with five zones requires five times as many pulse echo cycles as the same image with a single zone. These additional pulse-echo cycles require additional time, decreasing frame rate.

The result is that zone focusing achieves some improvement in B-mode imaging, but at a sacrifice to frame rate. Since frame rate is so critical in color

Doppler (for reasons discussed above), instruments generally disable zone focusing while in the color Doppler mode. Thus, systems that rely on zone focusing instead of a true tracking lens generally achieve less uniformity for color than for B-mode imaging.[6]

Reliance on zone focusing by a system can easily be determined with a phantom, by comparing image uniformity in B-mode with image uniformity with the color mode active. If the phantom image shows acceptable image uniformity in B-mode while using several zones (see Fig. 16c, which uses zone focusing but with dynamic focusing disabled), but defaults to a single zone with sharply poorer lateral resolution away from the center of focus when color Doppler is turned on (as shown previously in Fig. 16b), it may be indicative that the system utilizes zone focusing as a substitute for a true tracking lens. Note that Fig. 16a uses dynamic focusing, achieving uniformity in color Doppler mode even while using only a single zone.

High-Level Computer Control

A fully capable and optimized color Doppler imaging system requires the simultaneous adjustment of all the parameters used for B-mode imaging, many of the parameters used for spectral Doppler, and a number of new parameters unique to color Doppler. Increasing the difficulty, many of the raw color Doppler parameters are not at all intuitive. Thus, manual optimization of all of these parameters in real time, while imaging an actual patient, would be extremely burdensome, if not impossible. Clearly, then, the ability to optimize these parameters more easily and intuitively is important to clinical color Doppler sensitivity.

However, this optimization cannot be fully automatic: there is no substitute for a clinician's expertise in evaluating the quality of an image and adjusting the system setup to improve it. Thus, for practicality, sensitive color Doppler instruments must be able to optimize many parameters simultaneously in response to one or two intuitive commands given by the clinician. We call this ability "high-level computer control," and it is the fifth key element of color Doppler sensitivity. The major objective of high-level computer control is simplicity and user friendliness. Thus, the objective is to use more computer sophistication to produce *less* operator complexity: with high-level computer control, the system can make it easy for newcomers to develop color Doppler expertise, and can also make the system easier to optimize for color Doppler experts.

---

[6] Systems with a true tracking lens on receive may also use zone focusing on transmit in B-mode. These systems use zone focusing not to increase lateral resolution, but to improve contrast resolution in B-mode. Use of zone focusing on transmit in B-mode, while using a true tracking lens for receive, should not be confused with the inferior solution of relying on zone focusing on receive (when a true tracking lens is unavailable) in order to achieve acceptable lateral resolution away from the focal center of the image.

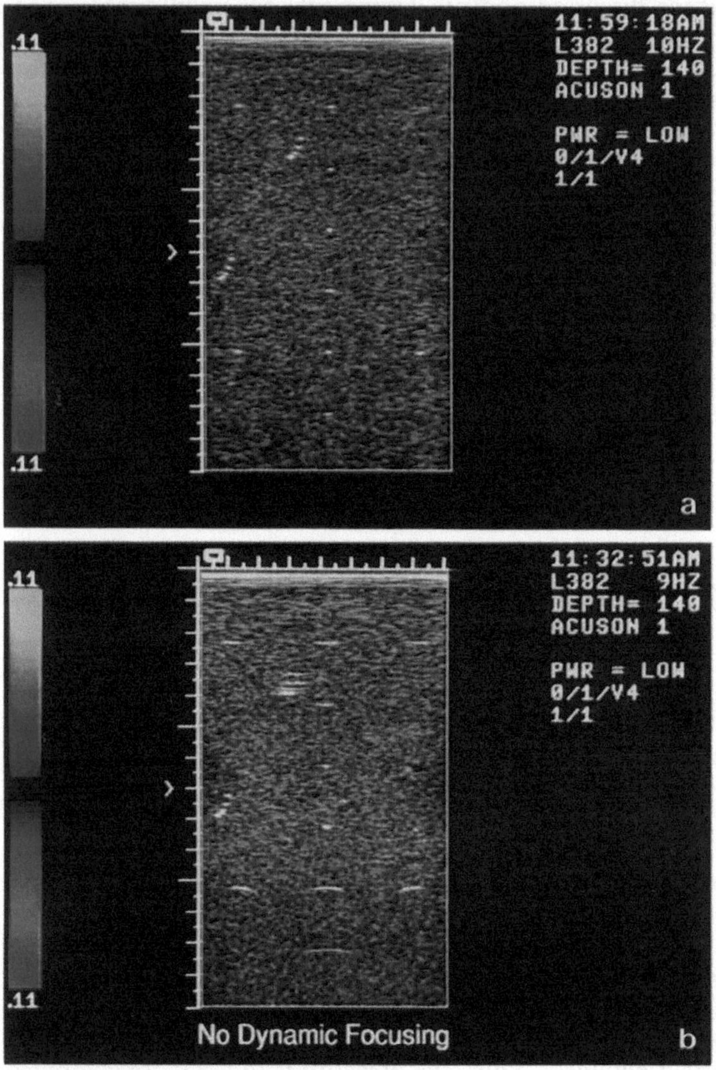

Fig. 16a–b. The effect of dynamic and zone focusing. **a** Dynamic focusing enabled, single zone. **b** Dynamic focusing disabled, single zone

Two types of high level software programs are necessary to control a color Doppler instrument with high sensitivity: a) *user-selected* programs, and b) *embedded* programs.

User-Selected Programs

The reason for user-selected controls is virtually self-explanatory. The simplest way to illustrate how they can be implemented using principles of high-level

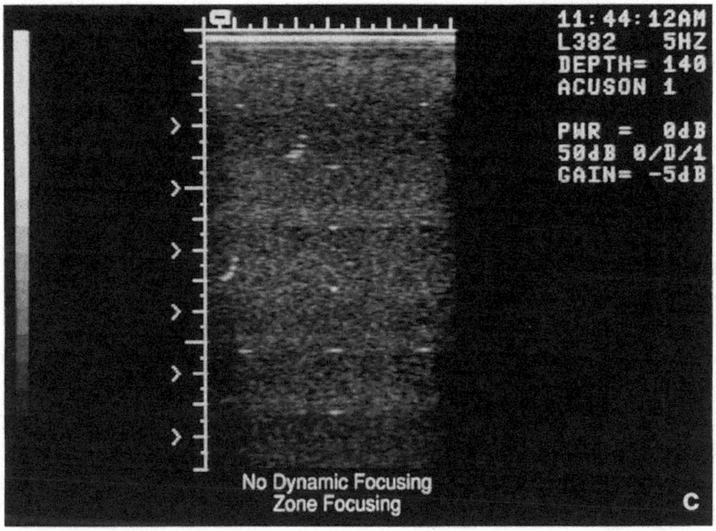

Fig. 16c. Dynamic focusing disabled; multiple zones can be used if color is not required and offers some improvement in uniformity if dynamic focusing is unavailable

computer control is to consider a few specific examples. In the first example we will use the function called *RES*®.[7]

RES. In the section on temporal resolution, we discussed the fact that the clinical need for a high frame rate is at odds with the need for a long "dwell" time in color Doppler. The result is that it is virtually impossible to maximize spatial resolution, accuracy of motion analysis, and frame rate simultaneously while performing color Doppler throughout the entire field of view. As a result, most instruments now offer the operator some method to limit the field over which color Doppler is performed, by selecting a "window" outside of which no color information is acquired.

This concept is implemented with varying degrees of sophistication in different systems. For example, suppose that the clinician wants to increase frame rate and color spatial resolution, at the expense of a smaller color field of view. In a less sophisticated system, the clinician might first restrict color Doppler scanning to a selected window, saving the time required for the system to scan outside the window. This increases the frame rate, but typically has no impact on color spatial resolution. Thus, in a second step, the clinician must separately increase color spatial resolution by adjusting an additional control that trades off frame rate for color resolution. Perhaps, in additional steps, the clinician

[7] The examples of the RES, Application, and MultiHertz functions, and the specific high-level computer controls described, are taken from the Acuson® 128XP color Doppler imaging system because of the authors' familiarity with it. The user interface, sophistication, and degree of high-level computer control can differ considerably for other clinical systems.

might separately have to adjust dwell time or sample volume size in order to optimize temporal and spatial resolution at this new frame rate.

By contrast, the RES mode accomplishes the clinician's goal at a higher level of computer control, requiring fewer and more intuitive adjustments. The result is greater effective clinical sensitivity and greater ease of use. When the clinician changes the size of a color RES box, the system automatically balances frame rate, line density, and dwell time, as well as many other nonintuitive parameters such as transmitter focus, sample volume size, and apodization. Further, because different clinical applications require different optimal parameter combinations for similarly sized RES boxes, the operator uses an intuitive "preprocessing" control to give weighted preference to temporal resolution, as in cardiac imaging, or to give preference to spatial resolution, as in much radiology imaging.

The weightings accomplished by these programs and the operation of the RES program itself are "preprocessing" activities that must be accomplished during real-time image acquisition: they control the way the ultrasound beam is formed and the way the subsequent image is acquired, and therefore determine the information content which results. Proper optimization cannot be performed post-detection, during freeze-frame or cine review, because the information available in the frozen or cine review mode is predetermined during its real-time acquisition.

To maximize color Doppler sensitivity, a system should allow the user to effectively optimize the examination using the smallest possible set of intuitive high-level controls. In addition to RES, these controls should include some version of:

- *Filter*. Selects appropriate motion discriminator (or simple velocity filter for systems without sophisticated motion discriminators).
- *Scale*. Selects appropriate velocity range or "velocity scale" of Doppler information to be acquired and displayed.
- *Preprocessing*. Selects desired preprocessing strategy (e.g., emphasizing spatial resolution for the kidney, vs emphasizing temporal resolution for the heart).

Operator controls at this high level can be quite logical, especially when their relationship to the normal flow characteristics of specific organs is understood. A table such as the abbreviated one in Table 1 can then be developed as an intuitive guide to matching control settings to the clinical situation being addressed.

Tables such as this are useful in developing the intuitive understanding clinicians need to get the most out of color Doppler, and are helpful as teaching tools. However, a more complete table might include 15 or 20 columns describing different clinical situations, not simply four or five, making it impractical for use in optimizing the instrument for every different organ during every examination. Instead, a system with high-level computer control can obviate the need for such a table by effectively including it automatically as a small part of what in one manufacturer's system is called the "Application™" function.

Table 1. An example of one abbreviated table that matches high-level control settings to the clinical situation addressed

	Cardiac	Carotid	Renal	Venous
Normal flow state	High velocity High volume	Medium velocity High volume	Medium velocity Low volume	Low velocity Medium volume
Motion discrimination setting (or velocity filter)	High	Moderate	High	Very low
Velocity scale	Very high	Moderate	Low	Very low
Preprocessing strategy	Maximize frame rate	Balance frame rate and resolution	Maximize resolution	Maximize resolution

Application. The Application function is controlled by a dedicated key. When using the Application function with color Doppler enabled, the key allows the user to select from a number of application-specific optimization setups, such as "Adult Heart", "Transesophageal Echo," or "Adult Kidney," etc. These setups can be set by the manufacturer, or easily modified by the user based on clinical experience with his own patients. Selecting an "Adult Heart" program, for example, would in a single keystroke appropriately reset not only the Filter, Scale, and Preprocessing settings according to an overall strategy appropriate for that organ, but would use that strategy to reset other color Doppler-related parameters as well. The result is a greatly shortened learning curve for novice color Doppler users, as well as far greater ease of use and clinical sensitivity for experienced users.

Two final examples illustrate the impact of high-level computer control. The first is the need for fully duplex color Doppler imaging.

Fully Duplex Imaging. We define this as the capability of directing the B-mode beam in one direction while simultaneously steering the color Doppler beam in a different direction. This steering requires much internal reoptimization of color Doppler parameters, but can be accomplished very simply on a system with high-level computer control.

The need for fully duplex imaging arises from the differing physical laws and constraints applicable to Doppler and B-mode imaging. These constraints have long been recognized, as shown in the following quotations regarding cardiac and peripheral vascular Doppler imaging, respectively:

"Doppler and imaging systems actually make competing demands on an ultrasound device.... The optimal echocardiographic image is obtained by orienting the transducer perpendicular to the surfaces of interest. As we have previously noted, optimal Doppler information is obtained in a position parallel to flow." [2 pp. 25–26]

"Inherent to the combinations of Doppler imaging and B-mode imaging is a conflict of basic imaging approaches. B-mode imaging traditionally seeks out the specular reflections of a structure. That often places the ultrasound beam perpendicular to the vessel walls. It also places the beam perpendicular to the flow pattern within the vessel. Doppler needs an angle away from 90°, preferably less than 80° for greatest accuracy. Scanning for flow will systematically take us away from conventional imaging approaches to vessels and organs." [4]

Note that for vascular studies, a sophisticated fully duplex color Doppler system should actually be capable of achieving much more favorable angles of 70° or even 60°.

Fully duplex imaging is particularly useful for evaluating peripheral arteries and veins because of the need for high-resolution B-mode imaging combined with the problematic (for color Doppler) orientation of the vessels parallel to

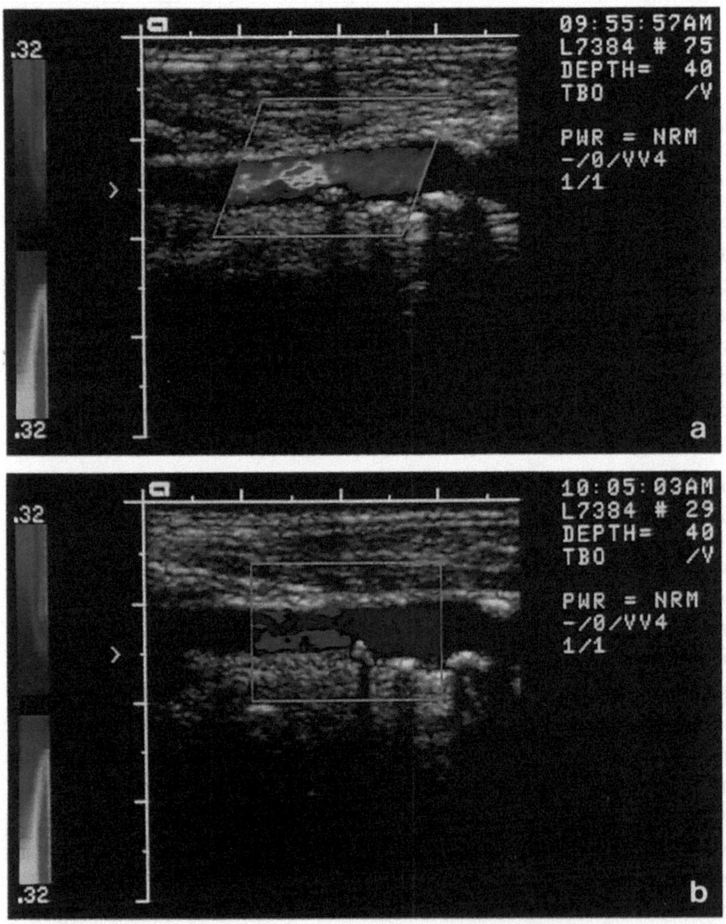

Fig. 17 a, b. Images of a mildly diseased common carotid obtained while steering color **a** with an appropriate angle to flow and **b** with an inappropriate angle

the skin surface. With fully duplex imaging the clinician can easily steer the color Doppler beam in several different directions without disturbing the B-mode image or any anatomical reference. For example, in Fig. 17, the color Doppler image obtained from an area of mild carotid disease is significantly changed when steering the color with a favorable angle to flow (Fig. 17a) vs an inappropriate angle (Fig. 17b). While common sense usually gives the correct angular orientation for color Doppler imaging, exceptions can be caused by exactly the kind of pathological flow the color Doppler examination is intended to discover. A check of at least two different color Doppler angular orientations is quick and prudent.

Echocardiographers generally examine jets from different angles by moving small-footprint, highly maneuverable transducers between long-axis, short-axis, parasternal, and other views. The way cardiac imaging can employ different views to achieve different Doppler angles is shown in Fig. 18, which images paravalvular and valvular cardiac jets from two different angles to flow, using a biplane transesophageal transducer with Vector Array. These vastly different views do not really exist for carotids or many other peripheral vascular structures. As a result, fully duplex imaging is highly useful in peripheral vascular work, but is not required or useful in echocardiography.

Finally, the last example addresses the need of a sensitive color Doppler system to easily switch between different frequencies without changing transducers.

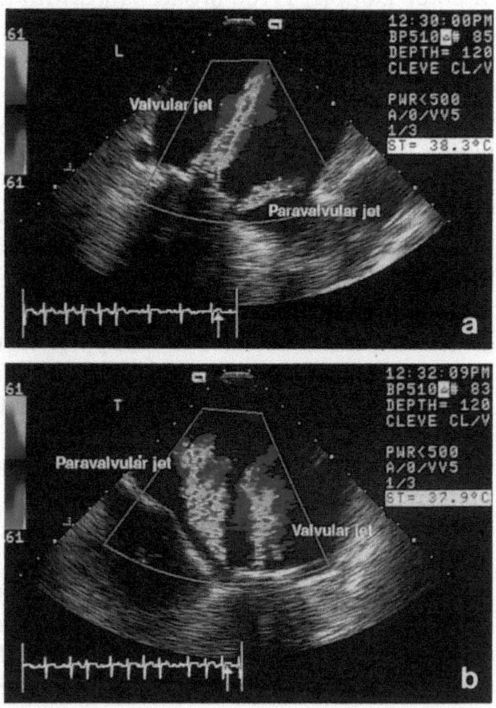

Fig. 18a, b. Different angles to flow in a biplane transesophagal study. **a** Sagittal (or "longitudinal") view. **b** Transverse view

MultiHertzⓉⓜ. Ease of changing imaging frequencies is critical, for several reasons. First, the best frequency for color Doppler is often somewhat lower than the best frequency for B-mode. This conclusion is driven mainly by two observations:

1. Doppler signals from blood are weaker than B-mode echoes from tissue, and
2. The clinician often wishes to use a lower color Doppler frequency to reduce aliasing, while he wishes to use a higher B-mode frequency to maximize resolution.

However, there are many circumstances where a lower color Doppler frequency is *not* optimal. For example, many subtle flows or jets in the near or midfield can be visualized better at a higher color Doppler frequency, as long as penetration is adequate. In the far field portions of these same examinations, the clinician may require a lower frequency for greater B-mode and/or color Doppler penetration. Further, some lesions (for both B-mode and color) have optimal contrast resolution or motion discrimination at a particular transducer frequency, but not at other frequencies.

The result is that use of any single transducer frequency is not always optimal and the best possible frequency to use cannot accurately be predicted. Further, changing transducers and reacquiring the image is so time consuming that most clinicians avoid it in busy practices.

One possible solution that does not properly solve this problem for Doppler or color Doppler is the broad bandwidth technique, a feature used by a number of systems under several different trade names in an attempt to gain better penetration for a given transducer frequency. On transmit, a broad-bandwidth system attempts to emit ultrasound with a broad spectrum of frequencies (broad bandwidth). Then, on receive, the system's receive electronics listen for higher frequencies in the near field, and listen for lower frequencies in the far field. However, this approach is inappropriate for Doppler and color Doppler, and in many circumstances may not be used in these modes even in systems that employ broad bandwidth for B-mode. The reason is simple: Doppler techniques measure the shift of the receive frequency relative to the original transmit frequency, so the original transmit frequency must be known with great accuracy. As a result, Doppler is most accurate when the range of transmit frequencies is kept deliberately narrow. Thus, for Doppler or color Doppler, broad bandwidth is inappropriate, as it actually decreases Doppler accuracy instead of increasing it. Finally, broad bandwidth techniques do not allow the clinician the ability to change transmit frequencies, which is the preferred solution to this entire dilemma.

The problem is solved optimally, by contrast, with a technology called MultiHertz. With MultiHertz, the clinician can easily switch B-mode and color frequencies (both on transmit and receive), at the push of a button, without changing transducers. This allows the operator to choose a higher frequency for part of an examination, and a lower frequency for other parts, without changing transducers. As a result, the user gains the ability to optimize the frequency for the clinical application at hand, and to easily switch this optimization several times during a single examination. Further, MultiHertz optimizes the image at each frequency selected while employing the narrow bandwidth for Doppler and color Doppler that is required for accuracy in these modes.

This switching of a transducer's frequency requires special transducer technology and system hardware, and requires extremely wide bandwidth transducers that can transmit at either of the frequencies the operator selects. However, while operating in any transmit frequency for Doppler and color Doppler, MultiHertz uses a narrow spectrum in order to maximize Doppler accuracy and sensitivity while avoiding the problems with broad bandwidth noted above.

MultiHertz would be completely impractical without high-level computer control. No user would change frequencies so often unless the software automatically reoptimized key system parameters when the operator pushed the MultiHertz button.

Embedded Programs

As we have already described, functions like RES, Application, and MultiHertz are selected and controlled by the clinician. By contrast, embedded programs are critical for excellent clinical performance, but are invisible to the user. Embedded programs tailor the system for each specific transducer and its intended spectrum of clinical applications.

For example, the range of motion discriminators accessible while using an abdominal transducer on a system should ideally be different from the range available when using a cardiac transducer on the same system. Many systems without sophisticated high-level control do not make this distinction, and apply a "one size fits all" approach which can produce disappointing results. When a transducer on a system with high-level computer control is activated, an appropriate embedded program can automatically modify the detailed functioning of high-level controls and imaging algorithms. Complexity is reduced because the operator need not memorize a battery of adjustments unique to each transducer, and clinical sensitivity is improved because the operation of such critcal components as motion discriminators are optimized for the clinical application.

Fig. 19 illustrates the dangers of the "one size fits all" approach. Fig. 19a shows excellent perfusion in a normal kidney, obtained with an appropriate renal program, while Fig. 19b shows poor perfusion in the same kidney, in an image obtained using inappropriate cardiac software. Similarly, Fig. 19c shows an excellent cardiac image demonstrating a mitral regurgitation jet, obtained with an appropriate cardiac program, while Fig. 19d shows a virtually useless image of the same heart, in an image obtained using inappropriate renal software. These images illustrate the need both for system flexibility (so that the embedded programs have sufficient parameter options to choose from) and for high-level computer control (for ease of use).

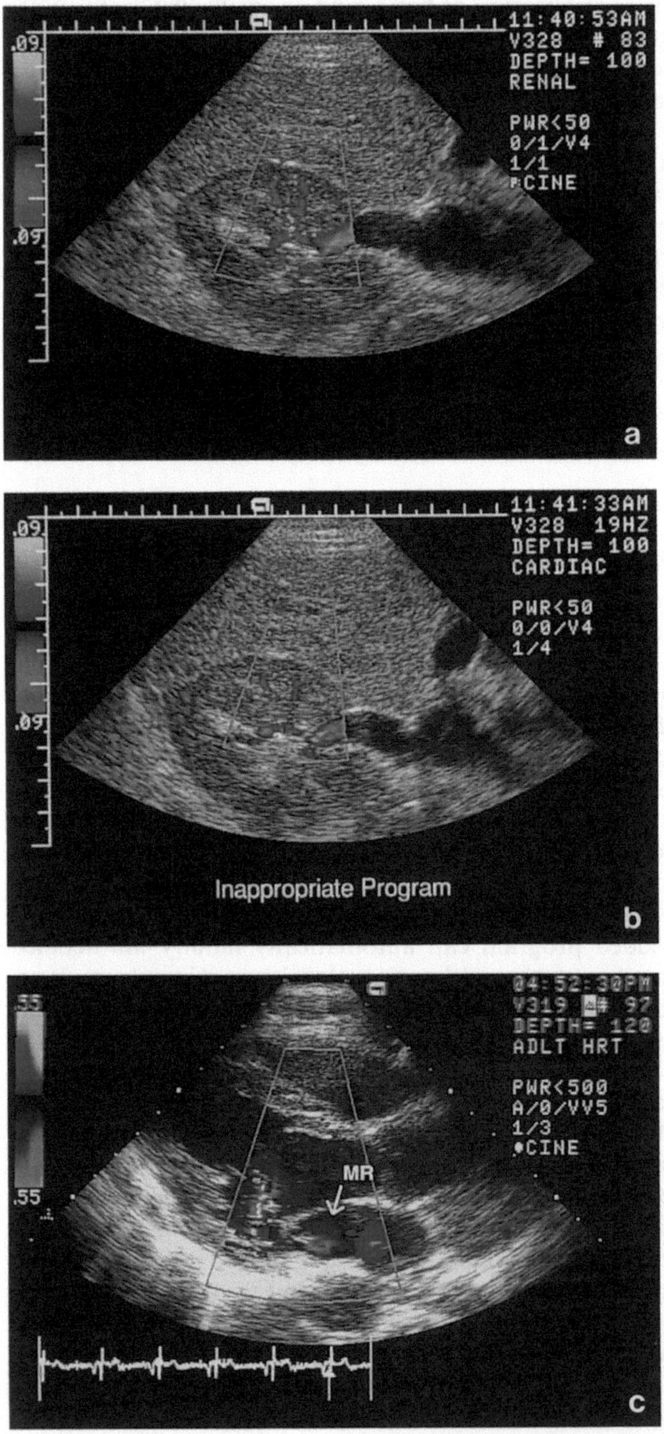
11:40:53AM
V328 # 83
DEPTH= 100
RENAL
PWR<50
0/1/V4
1/1
CINE
.09
.09
a
11:41:33AM
V328 19HZ
DEPTH= 100
CARDIAC
PWR<50
0/0/V4
1/4
.09
.09
Inappropriate Program
b
04:52:30PM
DEPTH= 120
ADLT HRT
PWR<500
A/0/VV5
1/3
CINE
.55
.55
MR
c

Fig. 19 a–c.

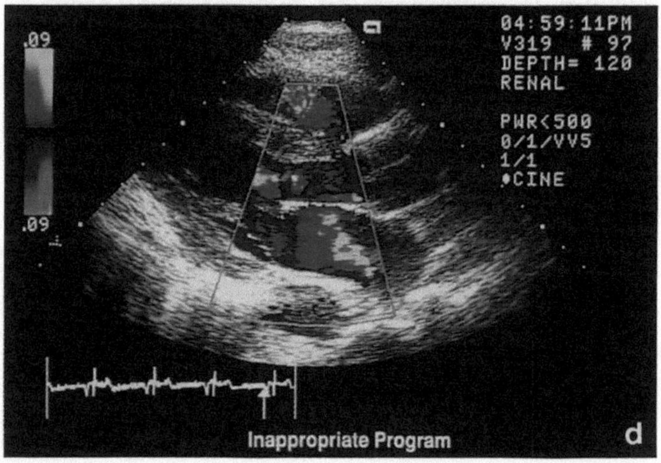

Fig. 19 a–d. Appropriate vs. inappropriate software. **a** High-quality image of a normal kidney, obtained with a renal program. **b** Same kidney imaged poorly, using inappropriate cardiac software. **c** High-quality cardiac image, obtained with a cardiac program. **d** Same heart imaged poorly, using inappropriate renal software

In summary, well-designed high-level computer controls make color Doppler far more sensitive, and far more practical. Available clinical instruments respond to this need with varying success, and with various degrees of convenience: for example, major color controls are far more useful if directly accessible on the keyboard, instead of being buried several layers deep in a scheme of menus as they are on some systems. Perhaps as important as the design a system offers today is whether its parameters are optimized with software, and eventually field upgradeable to new capabilities at reasonable cost, or whether it is hardwired and upgradeable only at great expense or by means of a system swap. A hardwired implementation of embedded or user-selectable programs can be successful for a limited range of clinical situations, but these same hardwired programs may be difficult to upgrade as clinical knowledge evolves or new clinical applications need to be supported.

Image Aesthetics

The five key components of color Doppler performance – motion discrimination, temporal resolution, spatial resolution, image uniformity and high-level computer control – determine the underlying information content in a color Doppler image and the operational complexity of the clinical examination. Other instrumentation-based presentation factors that affect the perceived or "aesthetic" quality of the final image are, to a large extent, subjective. They relate to how the image is presented, rather than how much information it contains.

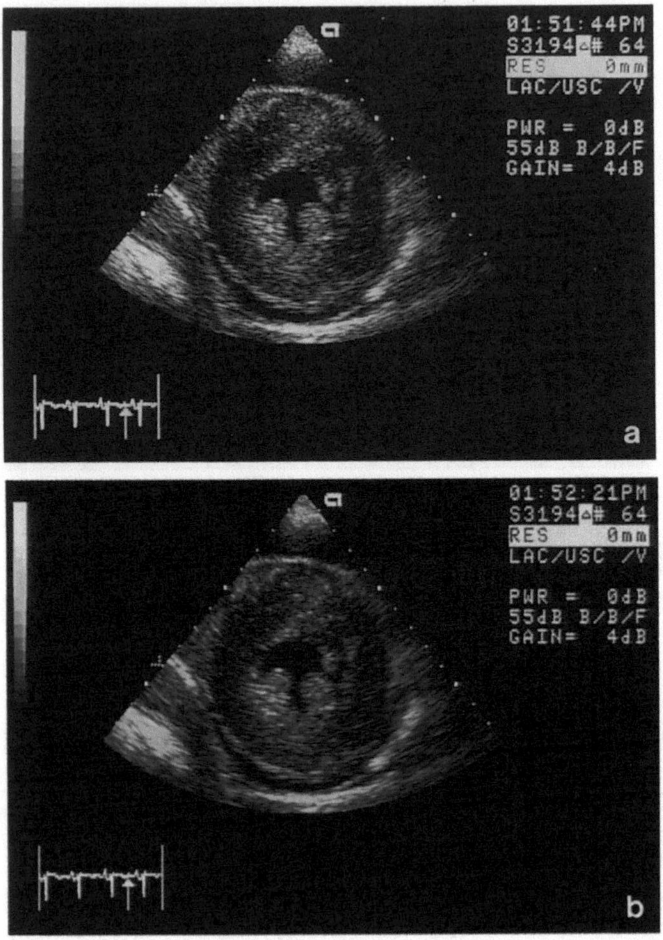

Fig. 20 a, b. B-color image presentation – cardiac. **a** Ventricular hypertrophy mapped in gray scale. **b** The same information mapped in B-color

Image presentation is clearly important, and its importance increases with the degree of information content of an image. Presentation cannot increase the underlying information content, but it can make the information easier to *perceive* and help bring out the subtleties in a good image. Further, this is true not only for color Doppler. For example, mapping B-mode information in color using a *B-color* function can increase the amount of anatomical information the eye can perceive compared with the same information mapped in shades of gray.

This is illustrated in Fig. 20, an image of left ventricular hypertrophy where the endocardial margin and myocardial fill-in are easier to perceive when mapped with B-color (Fig. 20 b) instead of with gray scale (Fig. 20 a). A similar comparison is shown in Fig. 21, where the full extent of cholecystic empyema

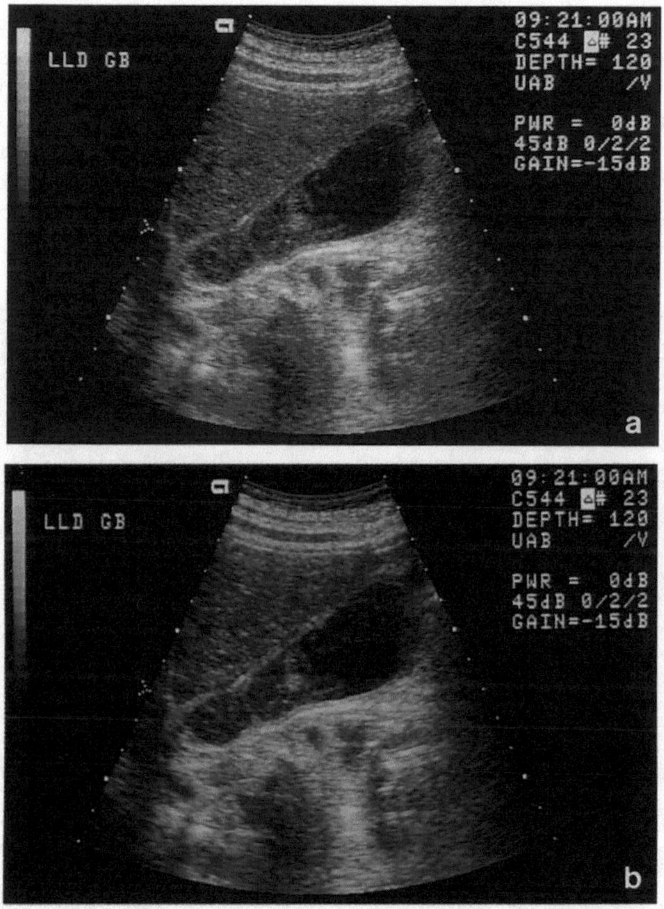

Fig. 21 a, b. B-color image presentation – abdominal. **a** Cholecystic empyema mapped in gray scale. **b** The same information mapped in B-color

is easier to perceive when mapped with B-color (Fig. 21 b) than with grayscale (Fig. 21 a).

It is in the area of color Doppler image presentation that individual preferences for vibrant colors, or for robin's-egg blue or fluorescent red, are appropriate and legitimate considerations. These qualities of color are *not* determined by the information content in the image. Instead, they are determined by the color map (analogous to the "B-mode postprocessing curve") that translates velocity (or velocity and variance) into the hue, saturation, and brightness of the image (or equivalently, into some combination of red, green, and blue components). Color maps cannot increase the information content in a color image: they only map the information after it has been derived. Choice of an appropriate color map can increase the amount of information the user can *perceive*, however, especially if the underlying information content is high.

Most color Doppler instruments incorporate a variety of color maps to accommodate a wide range of user preferences. These present the same information with different colors, which may affect perception. (Compare the carotid images in Fig. 22, where the identical color Doppler information looks quite different when presented with two different color maps.) Many instruments also provide the option of spatial persistence or "smoothing," and some users take strong positions expressing their preference for or against this capability. A small amount of smoothing can eliminate some artifacts and make

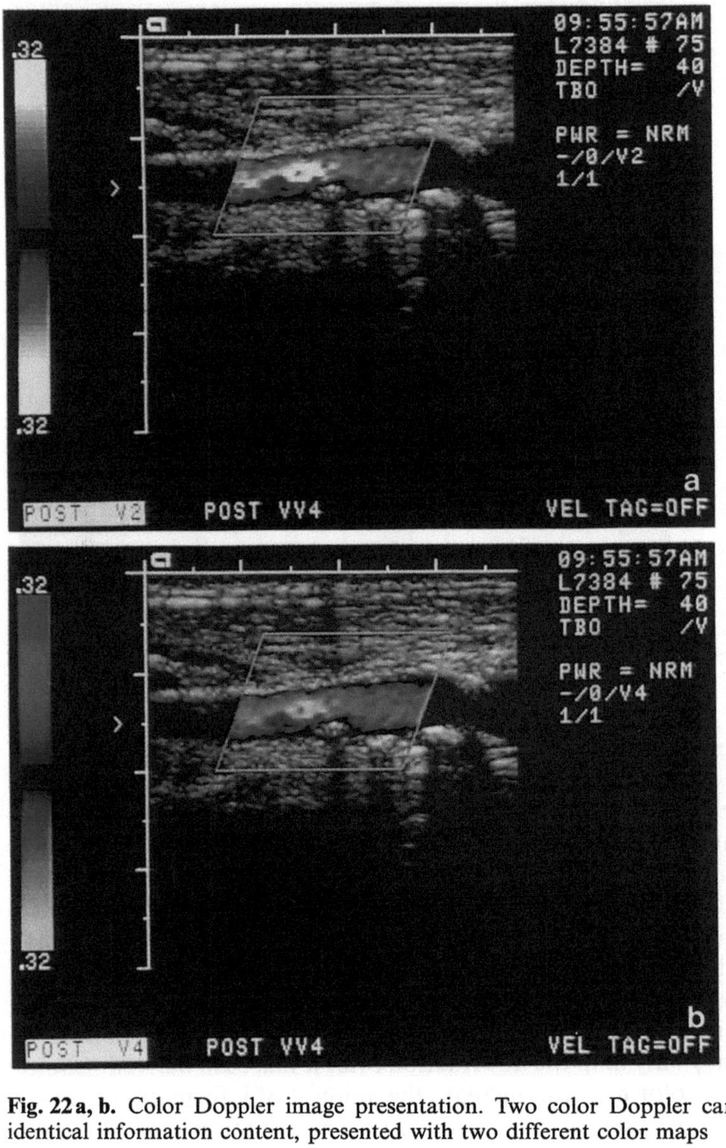

Fig. 22 a, b. Color Doppler image presentation. Two color Doppler carotid images with identical information content, presented with two different color maps

information easier to see with the human eye, but excessive smoothing can create a smooth, flow-like appearance where none really exists, perhaps leading to an incorrect diagnosis.

A user who is confident of having the maximum underlying information content can select the most pleasing color map and can choose or reject spatial smoothing in order to display color Doppler information according to his or her preference. However, the converse is not true: a more "vibrant" image may, upon knowledgeable examination, contain less diagnostic information than a similar image with more muted, but more finely graduated color. Bicolored or poorly continuous maps combined with heavy spatial smoothing can often use bright, overabundant color to hide low information content. When coupled with poor motion discriminators, such maps and spatial smoothing may initially look striking, but they defeat the objective of acquiring and presenting the maximum amount of physiological information.

Conclusion

When comparing ultrasound instrumentation in a clinical environment, ultrasound clinicians can teach themselves to see through aesthetic factors and learn to evaluate the underlying diagnostic information, simply by concentrating on the five elements we have outlined. Further, using the techniques presented, the clinician can learn to evaluate color Doppler instruments with his or her own expertise. Tissue-equivalent phantoms can be used to help assess spatial resolution and image uniformity objectively. The evaluation of motion discrimination and temporal resolution requires evaluation of such organs as the kidney, peripheral arteries and veins, and the heart, which can be done on normal patients in any clinical laboratory.

In the final analysis, diagnostic information content is usually the best measure of clinical imaging instrumentation. As important as image aesthetics are, a clinician can adjust his or her eyes to differences in image presentation after a few weeks, but information cannot be seen where none exists.

Acuson and RES are registered trademarks and MultiHertz, Vector and Application are trademarks of Acuson Corporation. Vector Array is Acuson's trademark for its omni-steerable omni-originating image formation technology (patent pending).

References

1. Bracewell R (1965) The Fourier transform and its applications. McGraw-Hill, New York
2. Kislo J, Mark DB, Adams D (1986) Basic Doppler echocardiography. Churchill Livingstone, New York
3. Maslak SH (1985) Computed sonography. In: Ultrasound annual 1985. Raven Press, New York
4. Powis RL (1988) Color flow imaging: understanding its science and technology. JDMS 4:236–245
5. Skolik MI (1962) Introduction to radar systems. McGraw-Hill, New York

Part III

Chapter 5
Principles of Magnetic Resonance Imaging

N. M. HYLTON and L. E. CROOKS

Introduction

Both ultrasound and magnetic resonance imaging (MRI) produce images of shoft tissue structures and perform their measurements noninvasively. Neither technique uses ionizing radiation and both are relatively hazard free. The major application of the two modalities has been for imaging soft tissue; however, both techniques have recently found an application to the detection and measurement of blood flow. The information in the ultrasound and the MR image is derived from different physical properties, and the two modalities can thereby provide unique and often complementary information. A general understanding of these processes is necessary for the effective interpretation and comparison of these images.

This chapter provides a review of basic MRI principles. An effort is made to help familiarize the reader with concepts and terminology unique to MRI. A glossary of conventional nuclear magnetic resonance (NMR) terms prepared by the American College of Radiology is recommended to the reader for standard definitions and conventions [1]. We approach this subject by first discussing general characteristics of digital images and then the phenomenon of NMR. The topics that follow include relaxation parameters, Fourier transformation, spatial localization, the effects of imaging parameters on contrast, instrumentation, quantitative image characteristics, and pulse sequences. The reader interested in a more thorough discussion on any of these subjects can refer to the texts referenced in the respective sections and listed at the end of the chapter.

The Digital Image

The information in an MR image is presented to us in the form of a *digital image* [2–4]. A digital image is a matrix of brightness values, each representing the numerical value of a measured property of the object being imaged. In a digital ultrasound image the response of a tissue to a vibrational wave is mapped whereas in MRI the response of the tissue to a magnetic stimulus is measured. The image is formed by sampling the property at regular spacings and displaying these values as a rectangular grid or matrix on paper, computer screen, or other media. The grid is so fine that the eye perceives a picture and not the individual elements. Under magnification, it becomes apparent that the image consists of a grid of square boxes, each with a different shade of gray (Fig. 1). Each picture element or *pixel* of the digital image represents a specific value of the sampled tissue property and can be described by its location (row and column number) and by an integer value called its *gray level*. The image is stored on the computer disk as a single stream of these values listed row by row.

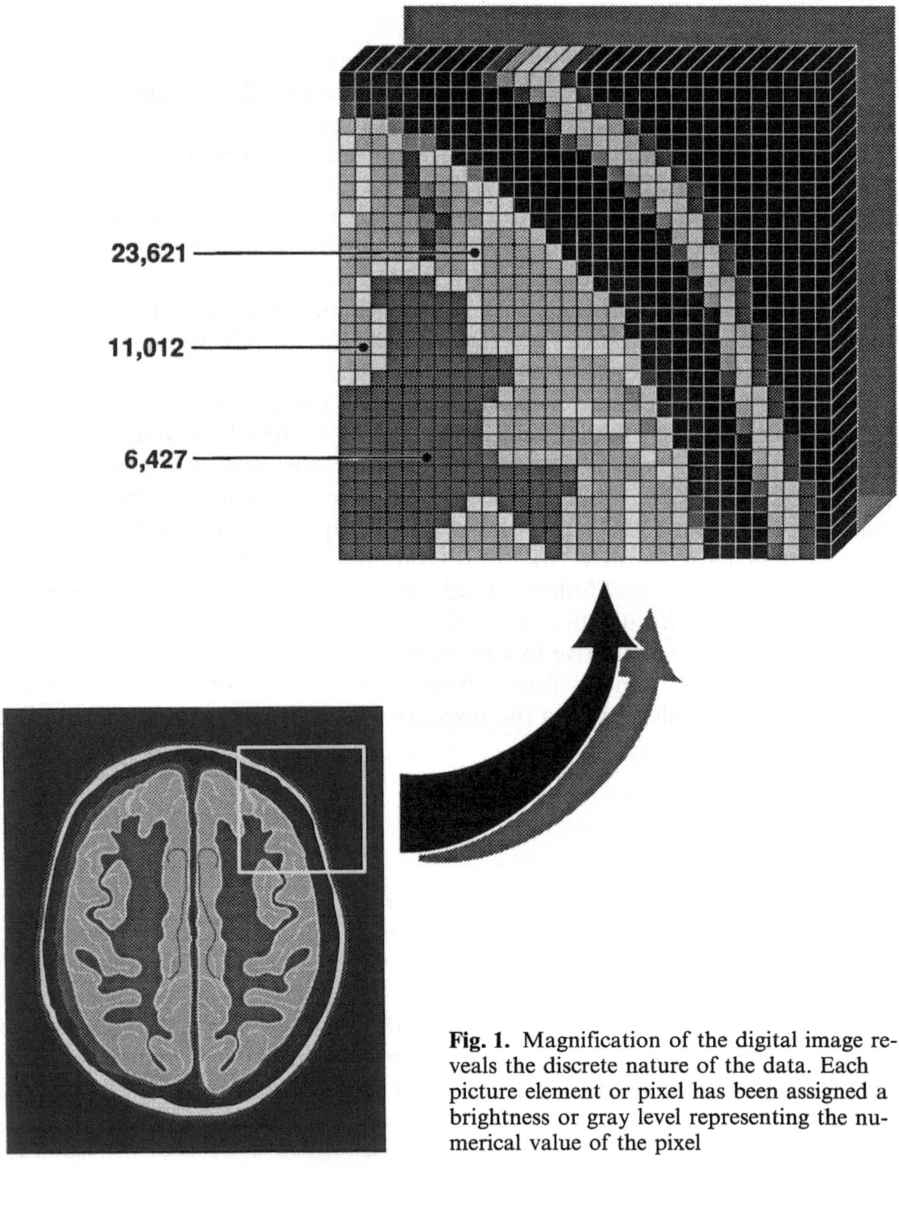

Fig. 1. Magnification of the digital image reveals the discrete nature of the data. Each picture element or pixel has been assigned a brightness or gray level representing the numerical value of the pixel

Digitization, Processing, and Display

The measurable quantity in MRI is the current induced in a receiving antenna by oscillating magnetic fields generated by the body tissues after stimulation. In this form, it cannot be visually interpreted and must first undergo three stages of processing: digitization, digital processing, and display. The input signal is first converted into discrete values by sampling and quantizing the signal at regular intervals, generating a string of integer values. Quantization is performed by an analog-to-digital converter. Computer processing then modifies the pixel values subject to operator control. In the case of MR images, the data are unrecognizable in their initial form and must undergo at least one stage of digital image processing to unscramble the input signal. Further processing is often used to correct for systematic errors or to improve the appearance of the image by smoothing or enhancement.

In the final display stage, a brightness is assigned to the corresponding location on the display screen based on the integer gray level value of the pixel. The appearance of the image can be further manipulated by adjusting brightness assignment, magnification, or other display time variables. These adjustments do not modify the stored data values.

Gray Scale and Windowing

Each pixel is represented by a shade of gray ranging from black for zero to white for maximum value. In an MR image, different tissues will appear with a characteristic range of gray level values. The gray shades are usually assigned to encompass the full dynamic range of the signal; i.e., black is assigned to the lowest signal found in the image and white is assigned to the highest value. These assignments can be changed, usually at display time, with controls at the operator's console. Increasing the minimum integer value and decreasing the maximum integer value corresponding to the full gray scale allows more gray shades to be used to represent a smaller range of values, giving better differentiation. However, all values below that range will now be black and all values above will be white. By adjusting the gray scale assignments, the operator is able to window the intensity range of interest. Figure 2 demonstrates the effect of changing the gray scale. The sagittal head image in Fig. 2a is displayed with the full gray scale corresponding to 100% of the image pixel values. In Fig. 2b the range has been narrowed to correspond to values between 25% and 60% of the maximum image value. All values below 25% are displayed as black and those above 60% are displayed as white. Since the total number of gray shades are now being used to represent a smaller range of values, greater differentiation is possible for tissues with pixel values within the range.

Color displays are also available on many display systems. A color or shade is assigned to each integer value in the digital image. One common color assignment utilizes the optical spectrum to represent the range of values, from red for the lowest value to violet for the highest value.

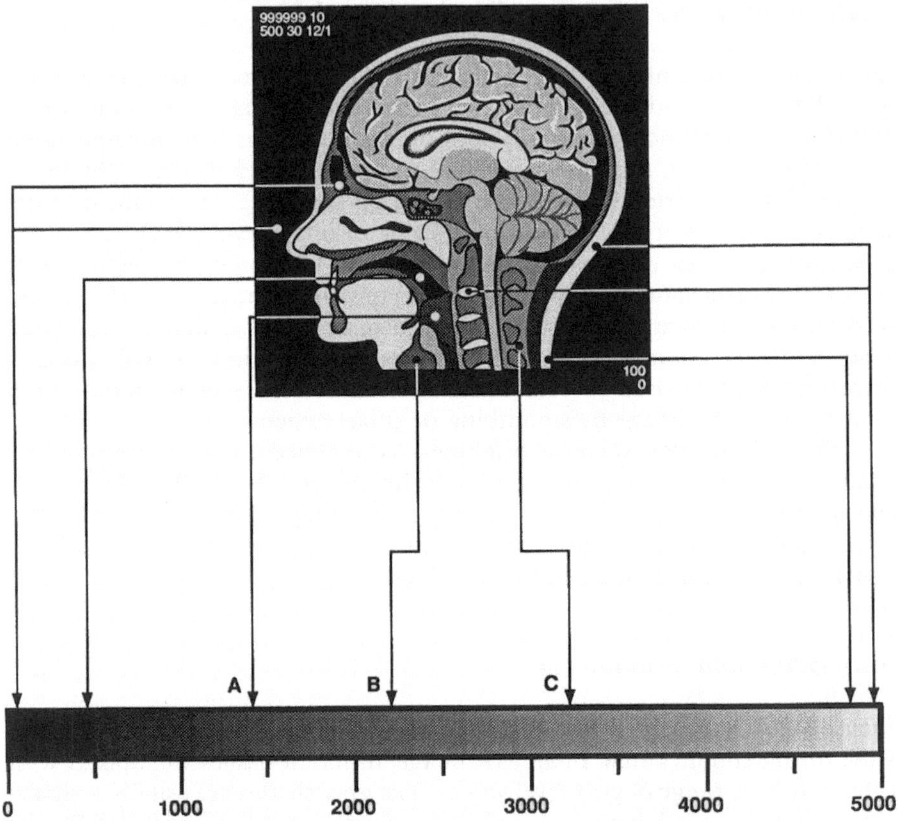

Fig. 2a, b. **a** An image with integer pixel values between zero and 5000 s displayed with full dynamic range of the gray scale. **b** The gray scale range has been reduced to span 25%–60% of the intensity values. Tissues above and below this range are displayed as white and black respectively. Relative contrast between tissues A, B, and C has increased

We tend to interpret a digital image as if we were looking at the object itself, but it is important to remember that it is only a representation. The fidelity of trueness of the image compared with the actual object will depend on the accuracy of the sampling, i.e., image data acquisition, and on the quality of image data processing. Clearly, the finer the spacing of the samples, the better the representation of the original object. When the sample region, or *voxel* (for volume element) is large and contains a mixture of tissues, the measured voxel value will represent a mean rather than a unique value. This is known as *partial volume averaging*. It is impossible to separate the components within the voxel once the image has been acquired. To reduce the effects of partial volume averaging it is desirable to improve the resolution by decreasing the dimension of the voxel. As we will discuss, there are some practical as well as fundamental limitations to reducing the voxel size. Another type of misrepresentation can

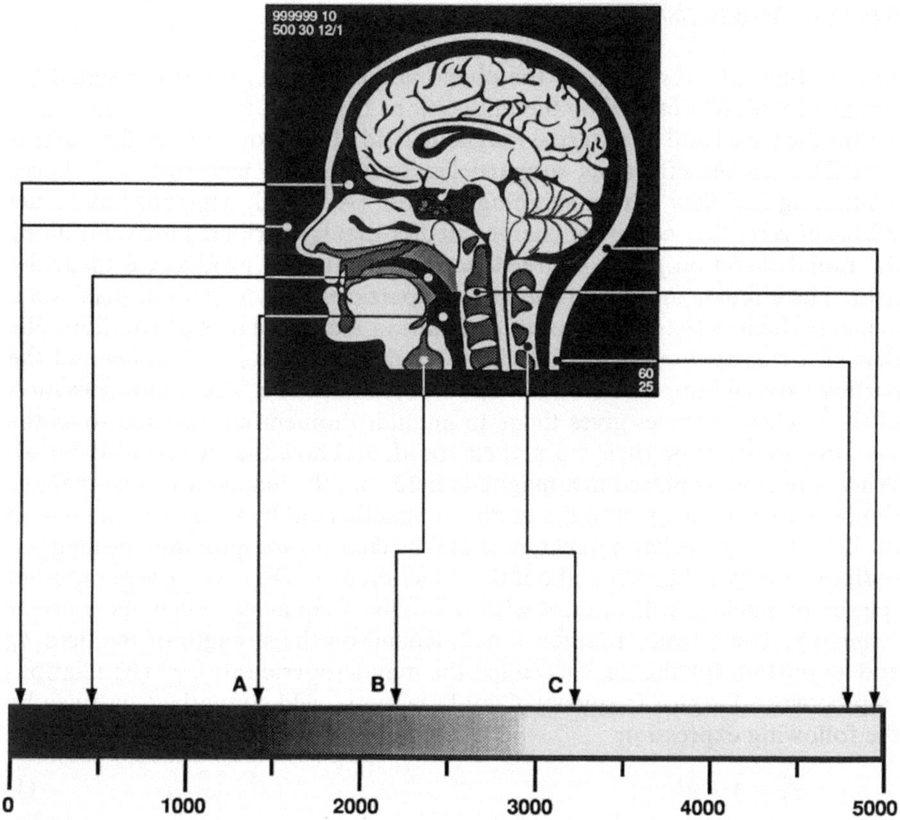

Fig. 2b

occur if the position of the samples is inaccurately mapped due to instrumental imperfections, such as field inhomogeneities or nonlinear gradients. This can result in image distortion or, in the worst case, a complete degradation.

Nuclear Magnetic Resonance – How the Signal is Generated

The number value assigned to each pixel in the image is a measurement of the amount of signal generated by a small voxel located at the corresponding position in the body. The magnitude of the signal is determined by properties of the tissue contained within the voxel, namely the number of nuclei of the element of interest (usually hydrogen), the T_1 and T_2 relaxation parameters that characterize how the nuclei respond to magnetic fields, chemical shift, diffusion, and motion.

Nuclear Magnetism

Nuclear magnetic resonance is the phenomenon that generates the signal for imaging [5, 6]. It is caused by the interactions between the nuclei of the atoms within the tissue and an external magnetic field. Two properties of the nucleus contribute to the effect: its subparticles, protons and neutrons, are charge containing and they have a spinning motion. (Although neutrons have a net charge of zero, they consist of charged matter that is distributed unevenly along the radial direction.) Any charged particle in motion produces a magnetic field. The circular motion of the nuclear particles causes them to generate a magnetic field with a north-south orientation along the axis of rotation, like that of a current loop. This field is referred to as a *magnetic dipole* and the particles are said to possess a magnetic dipole moment μ. The spinning motion of the nuclear particles gives them an angular momentum, referred to as the *spin*, that depends on their mass, their speed, and how the mass is distributed. When a nucleus is placed in a magnetic field, B_0, the magnetic dipoles behave similarly to a bar magnet in the earth's magnetic field by tending to align with the field. Unlike the bar magnet, because the charges are spinning, the magnetic dipole precesses like a top about the direction of B_0 (Fig. 3a). Every different species of nucleus will precess with a unique frequency, called the *Larmor frequency*. The Larmor frequency will depend on the strength of the field B_0 and a constant for the nucleus called the magnetogyric ratio, γ. The relationship between Larmor frequency f_L and magnetic field strength B_0 is given by the following expression

$$f_L = \gamma \times B_0 \tag{1}$$

The angular momentum of a particle can be positive or negative, depending on the two directions of rotation called spin up (aligned with B_0) and spin down (aligned opposite B_0) (Fig. 3b). Each proton and neutron contributes its angular momentum to the total angular momentum of the nucleus. In nature, systems of particles will orient themselves so as to lower the overall energy state of the system. In the nucleus this is done by pairing, in which every pair of like particles will occupy opposing spin directions, thus cancelling each other's angular momentum and magnetic dipole contributions (Fig. 3c). It follows that if the nucleus has an even number of protons and an even number of neutrons, the net angular momentum and net magnetic dipole moment of the nucleus will be zero. Only nuclei with a nonzero angular momentum and dipole moment can undergo NMR. The magnetic dipole of the nucleus is not free to assume any orientation with respect to the direction of the field b_0. It is restricted to occupying orientations that have certain exact energy levels. This implies that in order to change from one orientation to another, energy must be absorbed or released in an amount exactly equal to the difference between the two energy states (Fig. 3d). The magnetic component of an oscillating electromagnetic field can be used to transfer energy to the magnetic dipole. In order to be exactly equal to the difference between the two energy

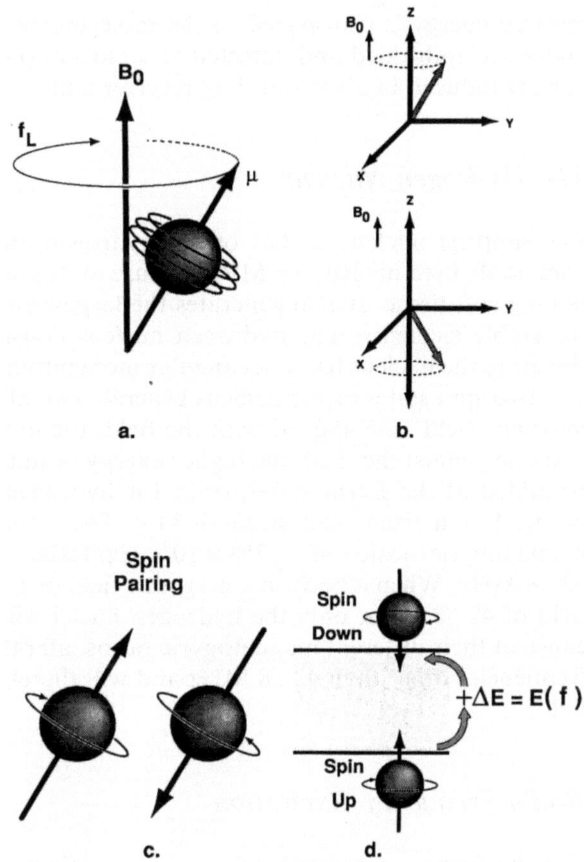

Fig. 3a–d. a A magnetic dipole μ in magnetic field B_0 will precess about the direction of B_0 at the Larmor frequency f_L. **b** The two allowed orientations for a spin in a magnetic field. When the spin is aligned along the direction of B_0 it is in the lower energy, spin-up state. When aligned opposite the direction of B_0, it is in the higher energy or spin-down state. **c** The spins in each pair will orient themselves in opposite directions, an energy-lowering process known as pairing. **d** A transition from the spin-up to the spin-down state requires the addition of an amount of energy exactly equal to the difference between the two energy states. This energy difference depends on the precession frequency of the spins, f_L. Electromagnetic energy must be at frequency f_L of the system to provide photons of this amount

states, the electromagnetic field must be oscillating at the Larmor frequency of the nucleus. This is the phenomenon of *nuclear magnetic resonance*. If energy is added at a frequency other than the Larmor frequency, transitions will not be made and resonance will not occur.

A group of nuclei in a field B_0 at a given temperature will form a natural distribution among their allowed orientations defined by the Boltzmann equation:

$$N_+/N_- = \exp(-\Delta E/kT) \qquad (2)$$

where N_+ and N_- are the relative numbers of spins in two allowed orientations, ΔE is the energy difference between the two states, T is the absolute temperature, and k is Boltzmann's constant. If energy is added at the Larmor frequency using an oscillating field B_1, transitions to the higher energy states will be made and the distribution of energy states of the spins will be shifted away from equilibrium. When B_1 is discontinued, nuclei will return to the lower state by releasing energy and equilibrium will be reestablished. The released energy is transferred to the molecular lattice structure in which the nuclei are embedded and detected as a current oscillating at the Larmor frequency induced in a surrounding receiver coil.

The Hydrogen Nucleus

The simplest nucleus is that of the hydrogen atom. Hydrogen is the most commonly used nucleus for MRI because of its great abundance in water-containing soft tissue. It also generates the largest amount of NMR signal of all the stable elements. The hydrogen nucleus consists of a single proton and therefore the nucleus has a net angular momentum equal to that of the proton. The two spin states of the nucleus generate two allowable energy states in the magnetic field, one aligned with the field, the lower energy or parallel state, and one against the field, the higher energy or antiparallel state. Energy must be added at the Larmor frequency for hydrogen in order for resonance to occur. For a fixed field strength of 1 Tesla, for example, and hydrogen's magnetogyric ratio of 4.258×10^7 Hz/Tesla, the Larmor frequency is 42.58 MHz. When a body in a magnetic field of 1 Tesla is irradiated with a B_1 field of 42.58 MHz, only the hydrogen nuclei will experience resonance. Because of their different magnetogyric ratios, all other nuclei will have Larmor frequencies other than 42.58 MHz and will therefore be unaffected by the B_1 field.

Radio Frequency Excitation

The B_1 field is generated using an oscillating electromagnetic field at the frequency f_L. The B_1 field must be directed perpendicular to the B_0 field and is turned on to create resonance and turned off to allow the system to return to equilibrium. Because we usually pulse the B_1 field, and since f_L is in the frequency range of radio waves for typical whole-body imaging, the use of the B_1 field is generally referred to as the radio frequency, or *RF pulse* [7, 8]. A common terminology with MRI is the phrase *90° pulse* (or any angle) to describe *RF* excitation sufficient in amplitude and duration to cause the net magnetic moment to rotate 90° away from the direction of B_0. A transmitter coil is usually placed in close proximity to the region being imaged and is designed to generate a B_1 field in the direction orthogonal to the main magnetic field.

Magnetization Vector in a Rotating Frame of Reference

To facilitate our understanding of what happens in a magnetic resonance experiment, it is convenient to use the net magnetic moment vector M to represent the vector sum of all of the spins in a single voxel. Consider vector

M in the frame of reference shown in Fig. 4a. The magnetic field B_0 is directed along the z axis, also called the longitudinal direction. The plane formed by the x and y axes is called the transverse plane. As we mentioned earlier, the magnetic moments of individual nuclei precess about the direction of the magnetic field B_0. If we summed up all of the magnetic moment vectors in the voxel, because of their distribution around the z axis their components in the transverse direction would cancel, leaving only a large longitudinal component M (Fig. 4a). If energy is added at the Larmor frequency using an oscillating

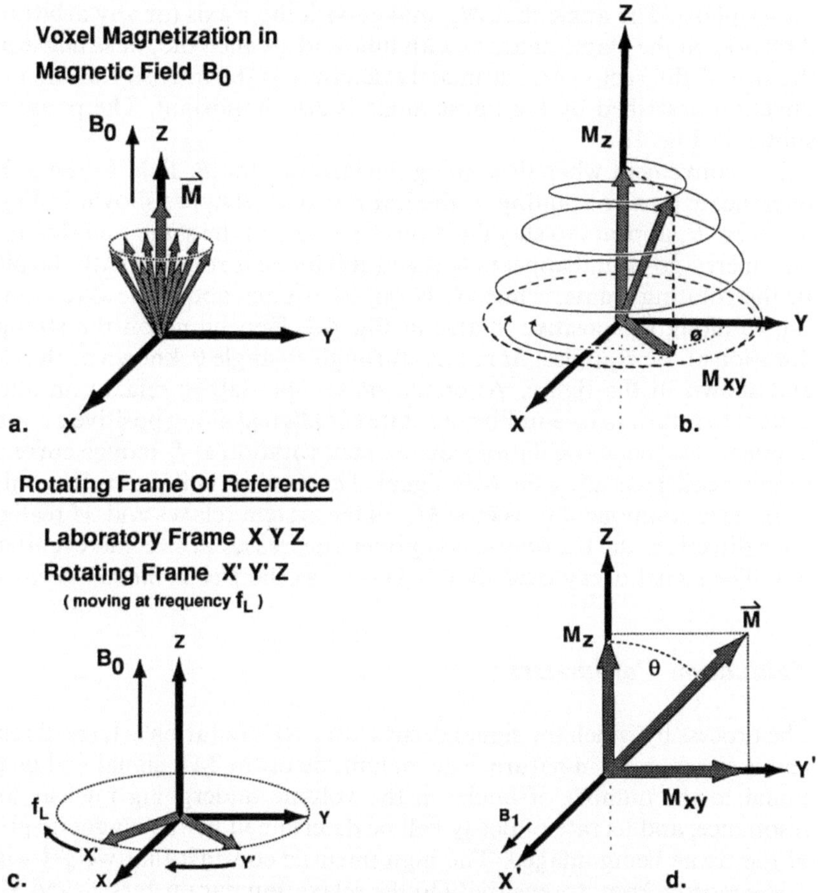

Fig. 4a–d. a Vector M is used to represent the net magnetization of all the spins contained in one voxel in the field B_0. While all precessing at the same frequency, they will be randomly distributed about B_0, causing cancellation of their transverse magnetization components and superposition of their longitudinal components. **b** When electromagnetic energy at the frequency f_L is added to the system in the form of an oscillating magnetic field B_1, vector M moves away from the direction of B_0 in a spiraling motion at frequency f_L. **c** Using a frame of reference $x'y'z$ moving at the frequency f_L simplifies the motion of M. **d** In the rotating frame of reference, when B_1 is applied along the x' axis, M rotates in the y'–z plane. Depending on the strength and duration T or B_1, M will rotate through an angle of θ degrees determined by $\theta = 2\pi\gamma \int B_1(t)\, dt$, with the integral taken over the duration T

magnetic field B_1 directed in the transverse plane, individual spins will absorb energy and make transitions to the upper energy state. The effect on the vector sum M is for it to begin to move away from the z axis toward the transverse plane while precessing about z. This spiraling motion is shown in Fig. 4b and will continue as long as B_1 remains on. As M begins to move away from the z axis, a component in the transverse plane appears. M can be represented by its two components M_z and M_{xy} as shown. We refer to the M_z component as the longitudinal magnetization and M_{xy} as the transverse magnetization. Because of the precessing motion of M, the transverse component M_{xy} rotates on the xy plane. The angle that M_{xy} makes with the y axis (or any arbitrary fixed direction on the plane) changes with time and is called the *phase angle* ϕ. While the size of the vector M_{xy} ultimately determines the intensity of the pixel, the direction described by the phase angle is also important. The phase angle is shown in Fig. 4b.

It is convenient when describing the effect of the B_1 field to use a frame of reference $x'y'z$ also rotating at the Larmor frequency, as shown in Fig. 4c. In this way, B_1 appears to stay fixed on the x' axis in the same way that an object on a merry-go-round appears to stand still if one is rotating with the platform. In the rotating frame, while B_1 is on, M rotates about the direction of B_1, shown along the positive x' axis in Fig. 4d. Depending on the strength and duration of the B_1 pulse, M rotates through an angle θ, known as the *flip angle* and shown in the figure. After the pulse, spin-lattice relaxation allows the system to return to its equilibrium state (M aligned along positive z). While this happens, magnetic oscillations due to spin rotation at f_L induce current in the receiver coil generating the MR signal. The receiver coil is sensitive only to the transverse component of vector M. As the system relaxes and M realigns with the z direction, the transverse component decreases in size and eventually dies out. The signal decay over time is known as the *free induction decay (FID)*.

Relaxation Parameters

The process by which the signal decays after *RF* excitation is termed relaxation and is exponential in nature. The magnitude of the MR signal will be proportional to the number of nuclei in the volume undergoing nuclear magnetic resonance, and its rate of decay will be determined by the magnetic properties of the tissue being imaged. The high intrinsic contrast that we get with MRI arises mostly from its sensitivity to the relaxation parameters T_1 and T_2 [9, 10]. Differences in T_1 and T_2 between neighboring tissues give rise to signal differences and may provide discrimination of disease states [11, 12].

T_1 Relaxation

Spin-lattice or T_1 relaxation is the process by which equilibrium is reestablished after the absorption of *RF* energy. Following absorption, energy can be released to the surrounding molecular structure or lattice. In the same way that *RF* energy is absorbed by a spin if the frequency matches its Larmor frequency, energy can be transferred from the spin to a neighboring process that is also oscillating at the Larmor frequency. As this happens, the excited spin system returns to its equilibrium distribution at a rate characterized by the relaxation constant T_1. The molecular makeup of different tissues gives them characteristic values of T_1. T_1 depends also on the frequency of the applied external field and is known to increase with frequency [13].

T_2 *Relaxation*

A second process, called *spin-spin* or T_2 *relaxation,* reduces the measureable transverse magnetization. To discuss T_2 and the idea of dephasing in general, we need to remember that the magnetization M is really the vector total of the magnetic moments of many individual nuclei. In a perfectly homogeneous magnetic field, all of the nuclei will precess at exactly the same frequency. They remain coherent, or in phase, meaning that as they rotate they always have the same phase angle. However, if slight variations exist in the strength of the field from location to location, slight variations will also exist in the frequencies at which individual nuclei are precessing. If this happens, faster precessing nuclei will begin to get ahead of slower nuclei, causing phase differences to appear (Fig. 5). As the phase dispersion increases, cancellation of individual spin magnetic moment vectors will cause a decrease in the net value M_{xy}, decreasing the signal detected by the receiver coil. When the nuclei are uniformly dispersed over 360°, no signal is measurable. There are several sources of this loss of phase coherence. The first is the intrinsic, irreversible process of T_2 relaxation. Individual spins being magnetic entities, they themselves make small contributions to the total magnetic field. Each spin experiences the main static magnetic field plus small contributions from each of the spins around it. Since the configuration is slightly different from the vantage of each spin, the total magnetic field felt by a spin is slightly different from that of its neighbor. This creates a variation in field values over the voxel, causing the value of M to decay over time at a rate described by the exponential constant T_2. For hydrogen in tissues, the rate of T_2 decay is much faster than the rate at which longitudinal magnetization recovers as characterized by T_1. In other words, after *RF* excitation and the rotation of the net magnetization M into the transverse plane, signal in the receiver coil dies out as transverse magnetization dephases. However, though no longer measurable, transverse magnetization still exists. The T_1 relaxation process, by which energy is transferred away from the spin system to the lattice and longitudinal magnetization is restored, is still occurring. Typical T_2 values for soft tissue in the human body range from 20 to 100 ms while typical T_1 values range from 200 to 1000 ms [14]. Differences in the T_1 and T_2 values among body tissues are largely responsible for the contrast in MR images. Blood has relatively long T_1 and T_2 values.

Many of the more successful angiographic techniques do not depend on the relaxation parameters to provide the contrast.

Phase Coherence and Dephasing

The other phenomena that can cause dephasing are field inhomogeneity and magnetic field gradients. They are reversible to some extent and can be corrected. Added to spin-spin relaxation, these effects cause the FID to decay at a rate faster than that given by T_2. The constant T_2^* is often used to describe the rate of decay due to the collective dephasing effects.

Inhomogeneities in the main magnetic field contribute frequency variations with a voxel in the same manner as described above. A spin located in a slightly

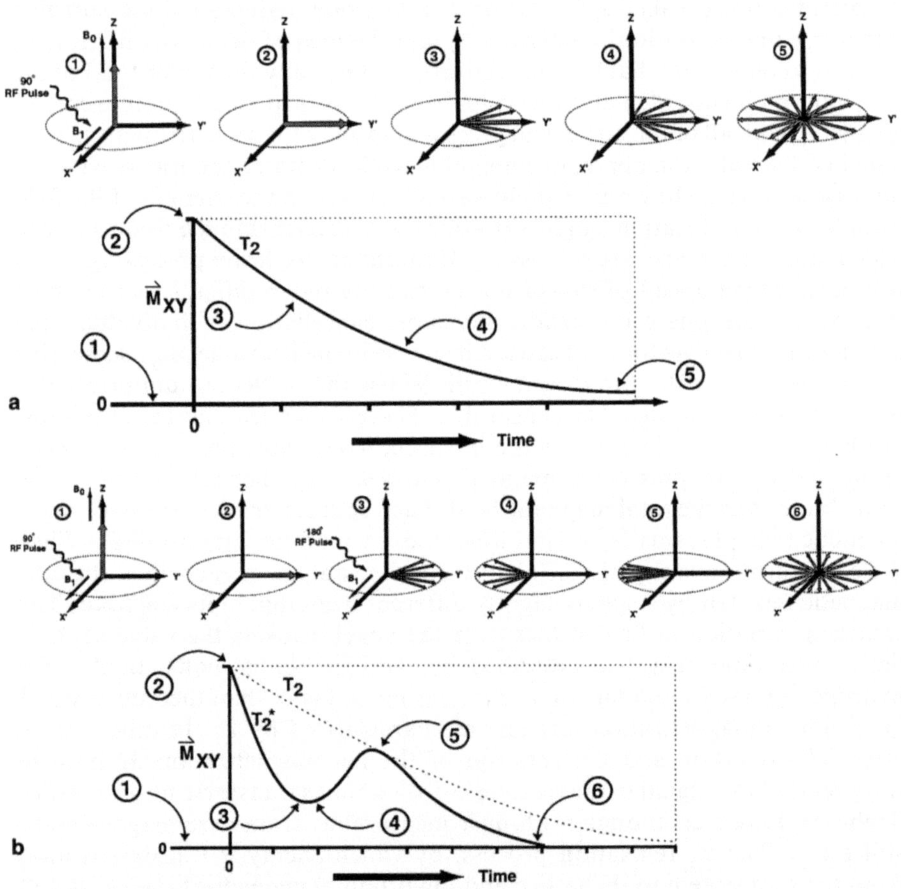

Fig. 5a, b. a After B_1 excitation, the free induction decay signal is measured in the receiver coil. The signal decays according to the relaxation constant T_2 as spin dephasing occurs. **b** Field inhomogeneity causes additional dephasing and results in signal decay at a faster rate by T_2^*. The effect of inhomogeneity can be overcome using at 180° refocusing pulse to reverse the spin system and refocus the spins. This is known as a spin-echo technique

higher field will develop a phase leader over a slower-moving spin as described in Fig. 5b. By using at 180° *RF* pulse about the x axis, the system of spins can be reflected about that axis as shown. The precession direction is still the same, but the faster-moving spin is now situated behind the slower-moving spin. In a time equal to the time to develop the phase lead, the faster spin will catch up with the slower spin. A regrowth of the signal will occur as the spins rephase, creating a signal echo. This is known as a *spin-echo* technique. Because of the dynamic nature of the T_2 dephasing process, it is not reversible like the dephasing caused by field inhomogeneity. The peak of the echo created by the refocusing pulse will still reflect T_2 decay.

Fourier Transform and Data Collection

The basic magnetic resonance experiment consists of a sample in a magnetic field. A transmitting coil excites the spin system with an *RF* pulse, causing the net magnetization vector to rotate into the transverse plane. When the pulse ceases, the spin system begins to return to equilibrium. The receiver coil is turned on and measures the FID. One way to create an image would be to repeat this process at every location within the object. This is impractical for many reasons, particularly because of the excessive amount of time it would take to complete an image with, for example, 256 by 256 elements. The process is expedited by taking advantage of a mathematical tool known as the *Fourier transform* [15]. The principle behind the Fourier transform is that any signal can be represented as a sum of sine and cosine waves of many frequencies and can be decomposed into its individual frequency component. This is expressed in the following Fourier transform equation

$$s(t) = C \int m(v) e^{ivt} dv \qquad (3)$$

where $s(t)$ is the signal at time t, $m(v)$ is the magnetization at frequency v, and C is a constant. In the case of MRI, if several voxel were placed in different field strengths and excited simultaneously they would each generate an FID at a different Larmor frequency (Fig. 6). The signal measured by the receiver coil is the sum of these signals. This signal is processed by Fourier transformation and yields the distribution and content of the frequencies from each of the contributing voxels. Furthermore, once we have the frequencies we can deduce the value of the field eat each voxel, due to the one-to-one relationship between frequency and field strength given in Eq. 1. If the spatial pattern of the field is known, the location of the voxel is known. In this way the signal values can be assigned to the appropriate location in the digital image. In practice, this is accomplished by superimposing linear magnetic field gradients on the main magnetic field and exciting many frequencies simultaneously.

Many methods for data collection have been used successfully including the sensitive point, line scan and projection reconstruction techniques [16]. The method found on most commercial imagers today encodes position along rows

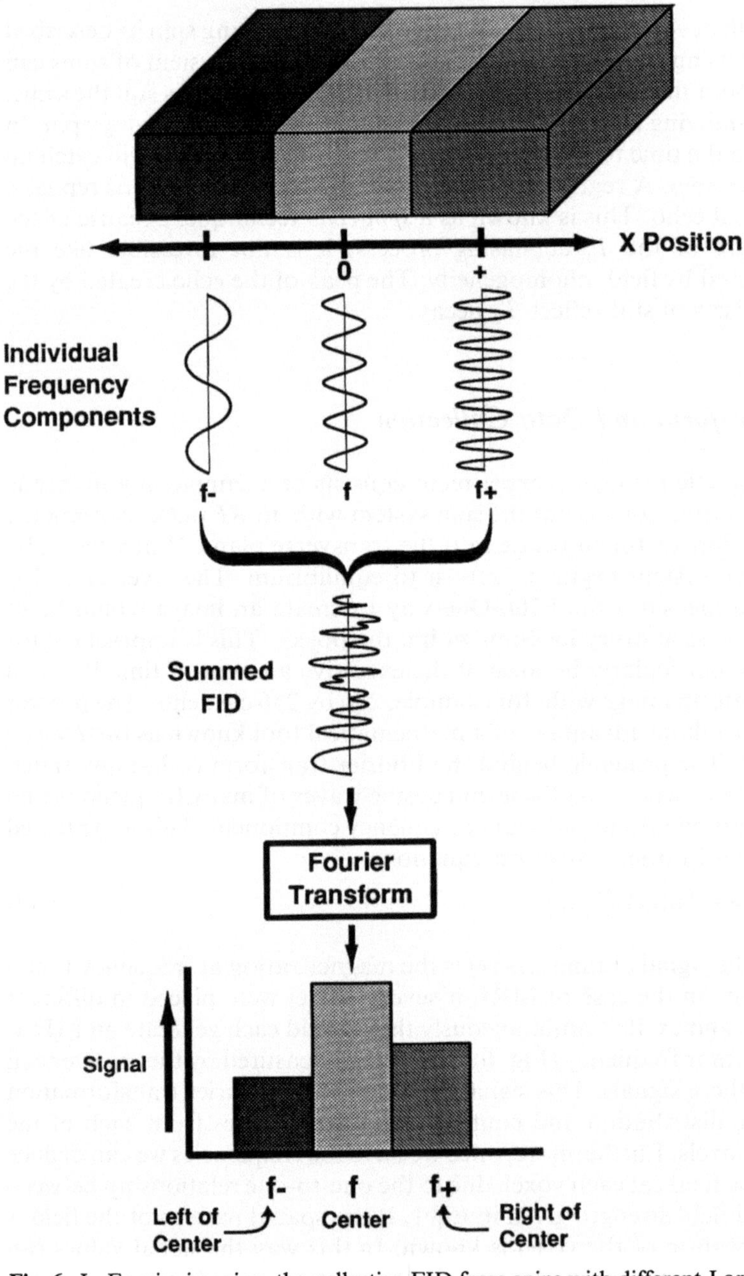

Fig. 6. In Fourier imaging, the collective FID from spins with different Larmor frequencies is measured and processed by a Fourier transformation to regain the individual frequency components of the input signal. By using magnetic field gradients to spatially vary the main field, frequency can be interpreted to give the location of the signal

and columns into the phase and frequency of the NMR signal. Spatial information is retrieved by performing a Fourier transform first along the columns of the image and then along the rows. This procedure is usually referred to as a *2DFT*, or two-dimensional Fourier transform.

The receiver coil is oriented along one axis in the transverse plane and samples one component of the oscillating signal. A second coil can be placed orthogonal to the first coil to sample the complementary component of the signal. By phase shifting the second signal by 90° and adding it to the first, an improvement of $\sqrt{2}$ in the signal-to-noise value can be gained. This is known as quadrature reception.

The MR signal is a complex valued signal. Ordinarily, it is the magnitude of the signal that is displayed in the image. Alternatively, the phase of the signal or the real or imaginary components of the signal can be displayed. Under ideal conditions for tissues with the same magnetic susceptibility values, the phase angle at every position in the image will be the same and displaying phase values will be uninformative. However, field inhomogeneity, an improperly tuned imaging sequence, or disturbances such as flow can result in variations in the phase. Another cause of phase disturbance is differences in the magnetic susceptibilities of different tissue. This is seen particularly at boundaries between air and tissue or between tissue and blood. The phase image is frequently used to detect these occurrences. If desired, just the real or imaginary portion of the signal can be examined. These correspond to the signals measured by the two orthogonal receiver demodulators.

In standard MRI data acquisition, Fourier transformation is an essential step of digital image processing before a recognizable image is obtained. Additional levels of digital signal processing are often used, such as smoothing and edge enhancement, to improve the appearance of the image. The image data can be manipulated to present a different form of the data, for example, an image composed of the T_1 or T_2 values at every pixel. There are an inexhaustible number of ways in which the data can be processed to provide different forms for display.

Spatial Localization

Throughout the development of MR imaging techniques, many methods have been used for data acquisition and image reconstruction, including the sensitive point and line methods and a number of variants of Fourier imaging. A number of texts provide details about the development and historical significance of each of these [17, 18]. We describe here the two-dimensional Fourier transform method that has become the standard for the majority of commercial systems today. In this method, the signal measured by the receiver has spatial information encoded into its frequency and phase. The 'raw data' as collected (the summed FIDs) do not make a recognizable image until they are decoded by 2DFT. In a typical data acquisition, position along the x direction is encoded into the frequency and position along the y direction into the phase.

Three steps are used to spatially localize points along the three axes in space. In a 2D data acquisition, the first of these is the slice selection. Consider the following example of a transaxial slice acquisition. Slice selection is performed by turning on a magnetic field gradient in the slice direction (shown along the z axis in Fig. 7a). A linear gradient G_z is used, such that the field in the z direction has the form

$$B(z) = B_0 + G_z \cdot z \tag{4}$$

With the z gradient on, every position along z has a different field value. The *RF* pulse used for excitation is tailored to contain frequencies corresponding only to the Larmor frequency of spins contained within the slice Δz. Only these spins will be affected and experience rotation of their magnetization vectors.

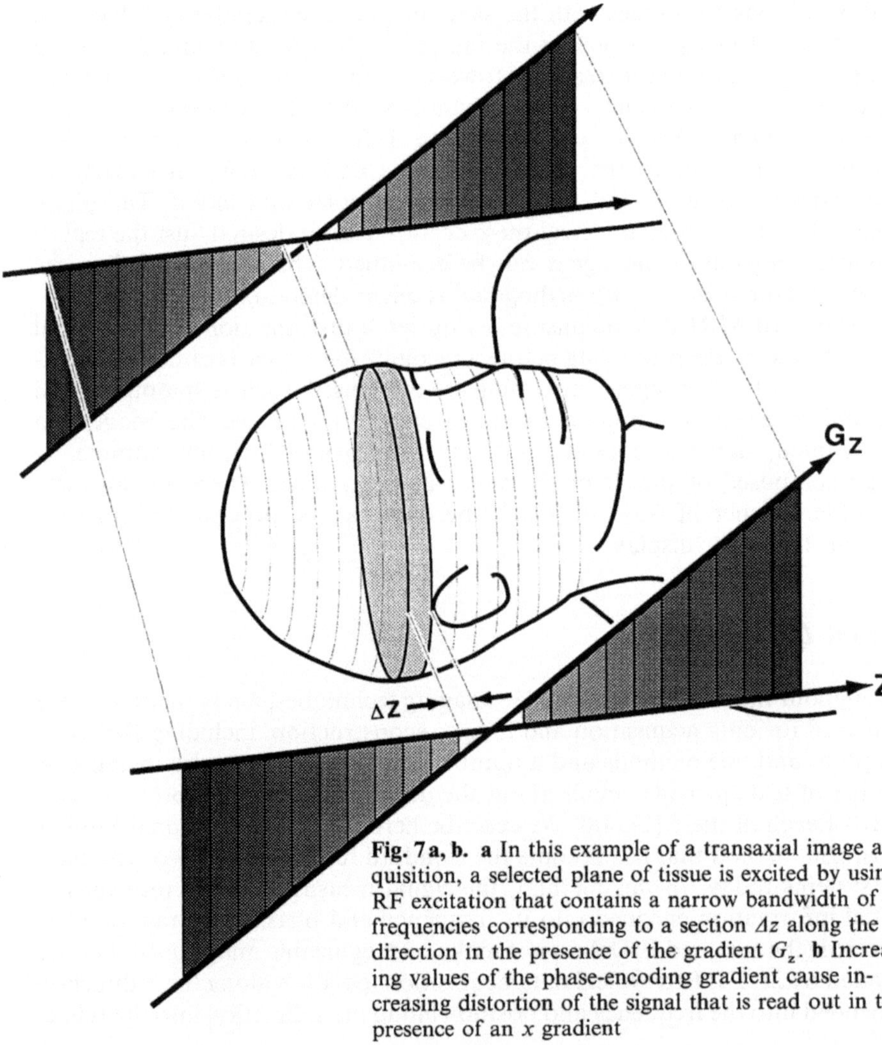

Fig. 7a, b. a In this example of a transaxial image acquisition, a selected plane of tissue is excited by using RF excitation that contains a narrow bandwidth of frequencies corresponding to a section Δz along the z direction in the presence of the gradient G_z. **b** Increasing values of the phase-encoding gradient cause increasing distortion of the signal that is read out in the presence of an x gradient

This is called *selective excitation*. These spins now have a transverse component to their magnetization and can generate signal.

Once a slice has been selected, phase encoding is used to provide differentiation along the y direction. It is convenient to first skip to the third axis of spatial localization and discuss frequency encoding. This is performed along the readout direction, shown as x in Fig. 7b. The readout gradient remains on while the FID is being sampled. The readout gradient is given by

$$B(x) = B_0 + G_x \cdot x \tag{5}$$

Spins at different positions along x will generate signal at different frequencies. The combined FID will be decomposed by Fourier transformation to give the individual frequency components. These will correspond to x position.

Between the time of the *RF* excitation with slice selection and the time of the signal readout, the phase-encoding gradient is turned on such that the field in the y direction has a linear dependence of the form

$$B(y) = B_0 + G_y \cdot y \tag{6}$$

The phase-encoding gradient is turned on for a short period of time and then removed. Slice excitation and signal readout are repeated for N_{pe} different values of the phase-encoding gradient. During the time that the phase encoding gradient is on, spins in different rows along the y direction precess at different frequencies. When the phase-encoding gradient is turned off, the

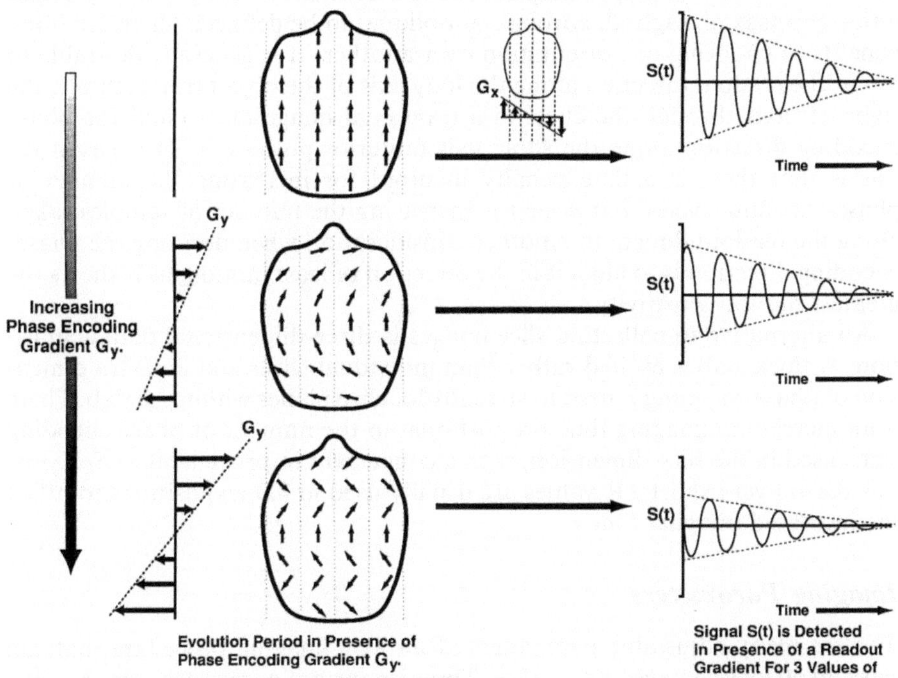

Figure 7b

spins are again in the same field B_0 and all return to the same precession frequency f_L. However, they will have gained or lost a phase angle as a function of the y direction. In subsequent phase-encoding cycles this will be repeated, but at increasing strengths of the phase-encoding gradient. Looking along each column, the combined FID will be increasingly distorted with each phase-encoding step. The effect on the FID of a changing gradient over N_{pe} of steps will be the same as the effect of sampling the FID continuously in the presence of a constant gradient, as was done in the readout direction. Figure 7b demonstrates the effect of three increasing values of the phase-encoding gradient on the FID signal. Fourier transformation along the y direction will give the frequency components along y and thus position. The two dimensions of Fourier transformation done in x and y are identical.

The acquired data are in the form of rows of sampled FIDs with increasing phase encoding gradient values. Because of the unique relationship between position and precession frequency established by the use of the magnetic field gradients, the acquired data represent the spatial frequency contributions of the image and so are offen referred to as k-space data, where k is the wave number or value of the spatial frequency. In the majority of images encountered when imaging human beings and animals with MRI, the largest contributions to the image are made by the low spatial frequencies, with decreasing contributions at higher k values. The highest k values contribute details such as edges and features of small dimension.

The magnet's three gradients can be used interchangeably to perform slice selection, phase encoding, or signal readout. This allows images of any orientation (transaxial, sagittal, coronal, or oblique) to be defined. There are often benefits to choosing one orientation over another. It is generally desirable to orient the readout direction along the long axis of the object (for example, the right-left axis through the chest in a transaxial image plane) and the phase encoding direction along the short axis (anterior-posterior). The reason for this is that there is a time penalty incurred for increasing the number of phase-encoding cycles, but none for increasing the number of samples taken along the readout direction. Another consideration when deciding the phase-encoding direction is to align it in the direction of least motion, as it shows the greatest motion sensitivity.

An alternative to collecting slice images is three-dimensional data acquisition. A thick slab is excited rather than individual slices and an extra dimension of phase encoding is used to spatially localize planes within the slab. There is an increase in imaging time proportional to the number of phase encoding steps used in the slice dimension, with a concurrent improvement in S/N. For this reason very short TR values are usually used in 3D acquisitions to offset the increased imaging time.

Imaging Parameters

The are many adjustable parameters of an MR imaging procedure that can have an effect on image appearance. Three important parameters are the *pulse repetition interval TR,* the *echo delay time TE,* and the *flip angle θ.* TR is the time between successive RF excitations of a given slice. The intrinsic T_1 value of the tissue determines the rate at which longitudinal magnetization recovers. By choosing TR appropriately, one can control whether a portion or all of that magnetization is allowed to recover. A long TR means a higher signal level but also lengthens the total imaging time. Since in most cases a heterogeneous group of tissues with a range of T_1 values will be imaged simultaneously, TR must be chosen judiciously to best portray all the tissues of interest [19].

TE is the time delay between the RF excitation and the measurement of the signal. Transverse magnetization created by the RF pulse begins to dephase immediately after the pulse due to T_2 and other effects such as field inhomogeneity. A 180° refocusing pulse at time TE/2 creates a spin echo at time TE that is corrected for field inhomogeneity effects. If no spin echo is performed, the rate of signal decay will be faster, as characterized by T_2^*. A short TE will measure maximal signal early in the decay, while a long TE will measure less signal later in the decay.

The flip angle is the angle through which the magnetization vector M is rotated away from the longitudinal direction by the RF pulse. A 90° pulse will maximize the transverse component but will also require the longest recovery time before the next excitation. Since the recovery rate varies for different tissues depending on T_1, a single choice of flip angle will have a different effect on each tissue. Choice of TR, TE, and flip angle will have an impact on image, contrast and signal-to-noise, as well as on the total imaging time and the ability to obtain multiple sections in a single experiment [20].

Changing the above parameters gives MRI a powerful control of image contrast. Contrast is defined by the expression

$$C = (I_a - I_b)/I_b \tag{7}$$

where I_a is the feature signal and I_b is the background signal. The MR image intensity depends not only on the hydrogen density distribution but also on the relaxation characteristics of the nuclei. The intensity dependence in a spin-echo experiment is approximated by the following expression

$$I = \frac{N(H)\sin\theta\,\{1 - \exp(-\mathrm{TR}/T_1)\}\exp(-\mathrm{TE}/T_2)}{1 - \cos\theta\exp(-\mathrm{TR}/T_1)} \tag{8}$$

Since water content in human soft tissue varies only by about 20%, hydrogen density $N(H)$ accounts for only a small portion of the contrast seen with MRI. Certain pathological conditions, such as edema of fluid-containing cysts, will have a more marked increase in water content und may be detectable on that basis alone.

Most of the contrast is generated by T_1 and T_2 differences among tissues. The T_1 value of a tissue determines the rate at which its longitudinal magnetization M_z recovers after RF excitation. The TR value chosen will determine how much recovered M_z is available to become transverse magnetization M_{xy} with the next RF excitation. Since signal is determined by M_{xy}, we can control signal intensity by adjusting TR. A tissue with a long T_1 will appear dark on a short TR image. If the goal is simply to maximize signal, it stands to reason that one should allow a TR long enough for M_z to recover completely. However, often considerations of contrast and imaging time make it advantageous to shorten TR at the expense of signal. Adjusting TR is one strategy for improving the contrast between a feature and its surrounding tissue. The contrast between tissues is shown above to be proportional to the difference of signal intensity and therefore to the difference in M_{xy} values. Two tissues with different T_1 values will recover at different rates, and a prudent choice of TR can maximize the difference in the amount of recovered M_z between the two.

TE is used to control contrast in a similar way. After M_{xy} is produced by RF excitation, it begins to dephase because of T_2 effects and field inhomogeneity. The size of the signal measured by the receiver decreases as the time delay TE increases. The rate of the signal decay is characterized by the T_2 value of the tissue. Signal can be maximized by using as short a TE value as possible. A long TE value will make tissues with a short T_2 appear dark. For two tissues with different T_2 values, the difference in the amount of measurable M_{xy} can be improved by a longer TE value, improving contrast by sacrificing signal.

An additional adjustable parameter, flip angle θ, can be used to optimize T_1-related contrast. The optimal flip angle will depend on the T_1 value of the tissue as well as on the TR value used. At short TR values (TR $< T_1$), reduced flip angles can increase signal-to-noise ratio as well.

Figure 8 illustrates the behavior of the signal difference as a function of choice of TR and TE for tissues with different T_1 and T_2 values. Two tissues A and B have relative hydrogen density values reflected by M_{0A} and M_{0B}. After a 90° RF excitation, T1 relaxation occurs at different rates, T_{1A} and T_{1B}, for the two tissues. The choice of the repetition time TR determines the relative amounts of magnetization $M_{xy\,0A}$ and $M_{xy\,0B}$ that become transverse magnetization. $M_{xy\,0A}$ and $M_{xy\,0B}$ decay with relaxation rates T_{2A} and T_{2B} respectively. Choice of echo delay time TE determine the relative signal intensities I_A and I_B at the time of readout. It is clear from this example that the final intensity difference is affected by both TR and TE as well as by the hydrogen density, T_1 and T_2 values of the tissues.

When flowing blood is being imaged, several other considerations become important. Minimizing the voxel size is particularly important in flow imaging, as intravoxel dephasing is the primary cause of signal loss from moving spins. Any velocity dispersion that exists among the spins in a voxel will result in signal drop. Efforts to correct for velocity dispersions using flow-compensating gradient waveforms are partial at best for most cases of physiological flow and motion. Often the decrease in S/N is outweighed by the benefit of increased contrast due to the regained flow signal when resolution is reduced. Another important way to reduce the dephasing that occurs between excitation and signal readout is to minimize the TE value, although this may affect T_2 contrast.

Many of the procedures used for imaging blood rely on the inflow of spins that have received not previous RF excitation into a volume of stationary

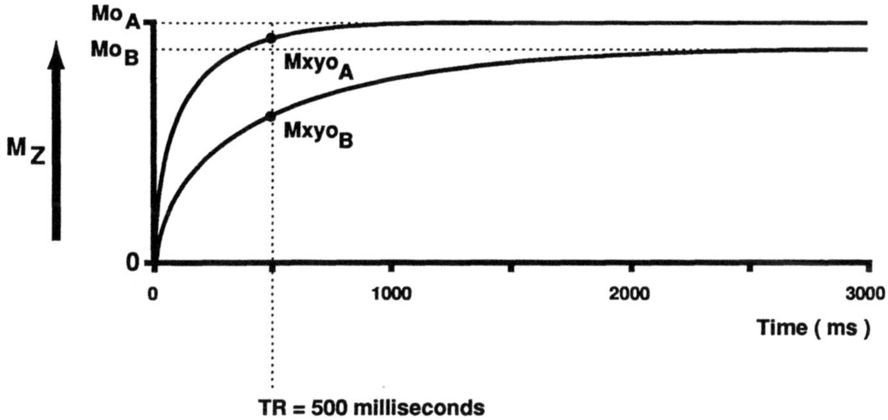

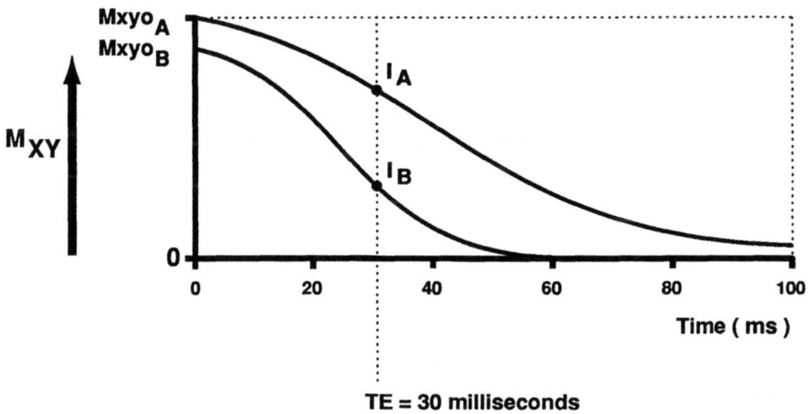

Fig. 8. The effect of choice of TR and TE on the relative signal intensities I_A and I_B of tissues *A* and *B* with hydrogen density T_1, and T_2 values M_{0A}, T_{1A} and T_{2A} for tissue *A* and M_{0B}, T_{1B} and T_{2B} for tissue *B*

tissue that is highly saturated by frequent RF excitation. This gives a bright signal from blood while suppressing background signal [21, 22]. In these techniques, TR must be selected to give sufficient background suppression while allowing enough time for fresh spins to enter the volume. By generating contrast in this way, T_1 effects become less important.

Instrumentation

The whole-body MRI scanner has four major subsystems: the magnet, the gradient system, the RF transmitter/receiver system, and the computer system for data acquisition and display (Fig. 9) [17, 18, 23].

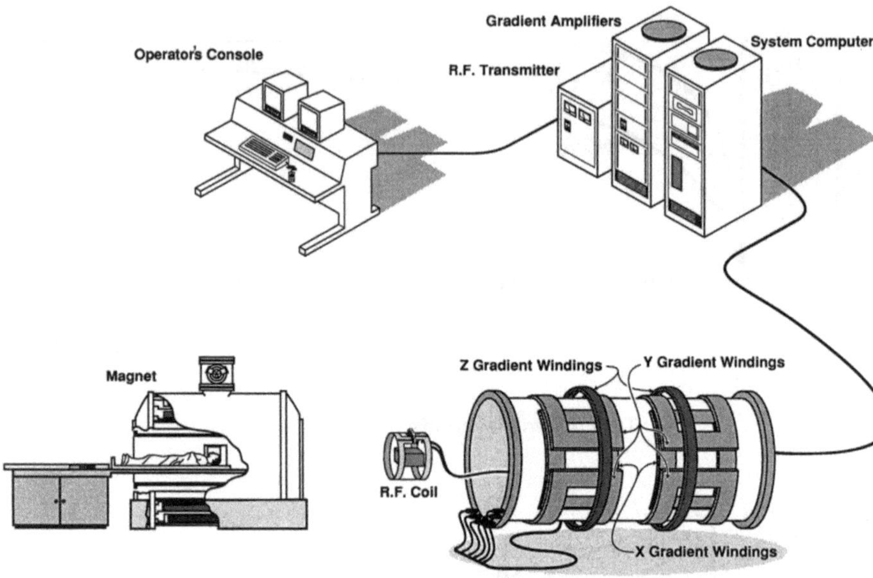

Fig. 9. Major components of MRI apparatus are the magnet, gradient subsystem, RF coils and associated electronics and the computer system for data acquisition, processing, and display

Magnet Subsystem

The magnet is the central component of the imaging system and can be permanent, resistive, or superconducting. The requirements for operating field strength and field homogeneity usually determine which magnet can be used. In general, for clinical imaging of hydrogen at a field strength greater than about 0.3 Tesla, a superconducting magnet is necessary. Resistive or permanent magnets perform adequately below this value. Some other considerations for magnet selection are the ease of access, cost, and siting.

Magnetic Field Gradients

The gradient subsystem provides the spatial dependence to the magnetic field needed for spatial localization. The simplest and most common design uses linear magnetic field gradients along each of three mutually orthogonal directions. While the actual physical design of each may differ to allow for patient access to the magnet, they perform identical functions and can be interchanged to acquire transaxial, sagittal, coronal, or oblique imaging planes. Some of the requirements on gradient design are strength, linearity, and the ability to switch rapidly. An increase in gradient strength is the easiest method of increasing spatial resolution. Gradient linearity assures that the pixel dimension is the same at every point in the image. Many imaging systems require the gradients to change values very rapidly. Fast switching allows a short TE and other factors needed for high speed and angiographic imaging. The magnetic field exerts a force on the current-carrying wires of the gradient coil. When the gradients are switched, an audible knocking is heard as the coil reacts to the forces upon it. This is the source of the rhythmic pattern heard during an imaging procedure.

The gradient subsystem is the one most critical to the performance of angiographic techniques. In flow imaging, specialized gradient waveforms are used that compensate for phase differences accumulated by moving spins. These waveforms are more demanding in terms of accuracy and speed than those used for conventional imaging. Another means of minimizing the phase accumulation is by reducing TE. Both of these procedures are limited by gradient performance. In general, stronger gradients and faster rise times improve the capability for angiographic imaging.

Transmitter/Receiver

The transmitter coil provides the magnetic field B_1 in the form of a radio-frequency magnetic field oscillating in the transverse plane. The receiver coil detects the induced voltage during relaxation. Because of the intermittent nature of transmission and reception, these two functions can often be served by the same coil. The performance of the receiver coil is strongly affected by how well the object fills the coil. For this reason, special coils are built for almost every body part. Most coils surround the body in order to produce a uniform B_1 field. Surface coils that span only a limited area produce better S/N images for anatomy close to the body surface. The sensitivity of these surface coils falls off quickly with depth and suffers from nonuniform intensity shading.

Computer System

The computer system is responsible for coordinating the entire system performance. Its function can be divided into two major areas. It is the controller for all hardware operation and maintains the timing of all events in the MRI pulse sequence. Once the data have been acquired, it also carries out all image manipulations including processing by 2DFT, disk storage, display, and archiving.

Resolution, Signal-to-Noise Ratio and Imaging Time

Resolution and Field of View

Image quality will be affected by its resolution, signal-to-noise ratio, and imaging time [24, 25]. The resolution in MRI is measured by the voxel dimension used in the image. It is determined by the frequency bandwidth of the RF excitation and the strength of the magnetic field gradients. Achievable resolution in a whole-body image is $0.5 \times 0.5 \times 2$ mm. Despite increasing capabilities of the hardware, there are other limitations to the minimum voxel dimension. As the voxel size decreases, signal strength decreases in proportion to the reduction in the number of nuclei it contains. Signal-to-noise ratio becomes a practical limitation. Even with improvements in noise reduction techniques, thermal motion and diffusion within a voxel may become a more fundamental constraint. Another consideration is the relationship between resolution, the size of the field of view (FOV), and the imaging time. For a fixed matrix size and imaging time, the FOV will be reduced in proportion to the reduction in voxel size. To maintain FOV at the higher resolution, additional measurements are necessary. If they are taken in the phase-encoding direction, the imaging time increases proportionally.

When the object size exceeds the imaging FOV, a condition known as *aliasing* results, whereby signal from tissue outside of the FOV is interpreted to be overlapping with signal from tissue within the FOV. This occurs when the object contain frequencies above the maximum frequency $f_{\max}$ that can be unambiguously assigned; $f_{\max} = 102\ T_s$, where T_s is the sampling interval. The result is image wraparound. For example, in a head image for which the FOV in the anterior/posterior direction is smaller than the size of the head, signal from the anterior of the head that is above the FOV will appear at the bottom of the image, overlapping the posterior segment of the image. This can be overcome by increasing the FOV, by increasing the number of samples at the same resolution, or by decreasing the resolution for the same number of samples in that direction. An alternative is to eliminate outside tissue by selective excitation or presaturation.

Signal-to-Noise Ratio

Signal-to-noise ratio (S/N) is another important contributor to image quality. Parameters of the imaging sequence such as TR, TE, and flip angle will all affect the total signal. Increasing TR, decreasing TE, and using a flip angle of 90° with a long TR will all increase the total signal but may be undesirable from the standpoints of contrast and imaging time. Reduction of the voxel size provides an improvement in resolution but proportionally reduces the number of signal generating spins while noise remains the same. If more measurements are taken to provide the resolution, the extra measurements bring back some

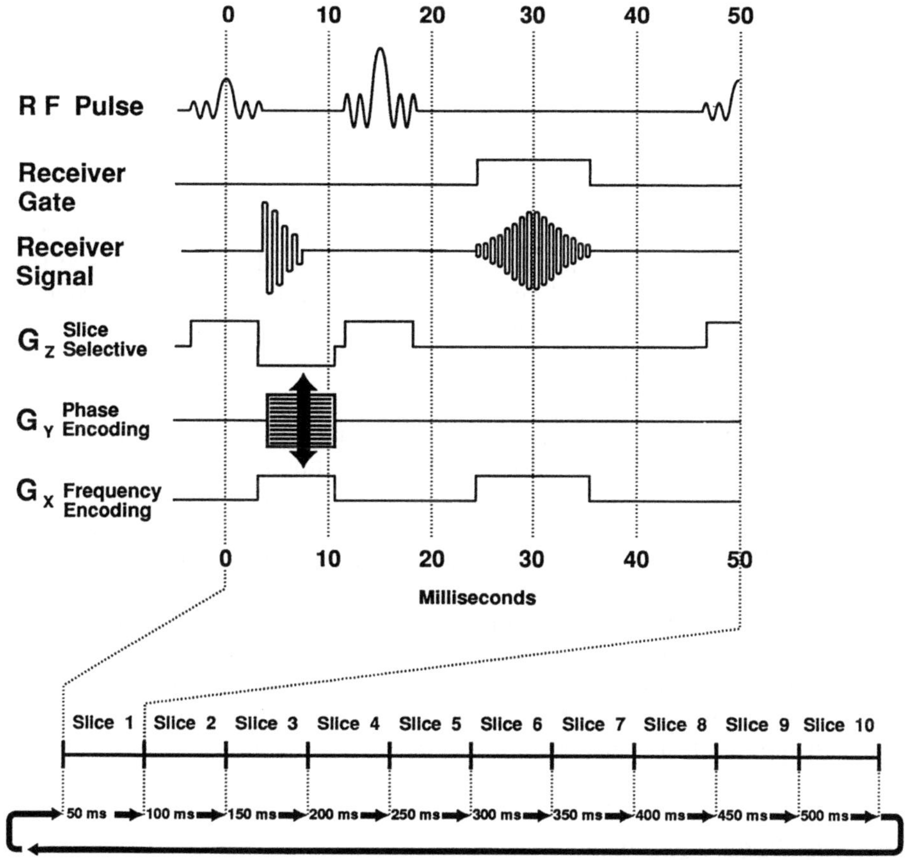

Fig. 10. A pulse timing diagram showing the strengths and timing of transmitted RF excitations, gradients G_x, G_y, and G_z, and signal reception for a typical spin echo data acquisition. In this example, data collection is complete in less than 50 ms. This allows excitation and data acquisition to be performed at nine other slice locations before the interval between excitations, TR = 500 ms, elapses for slice 1

S/*N*. The total signal will also be affected by the tuning of rephasing gradients for the slice selection and readout gradients. Incomplete rephasing will result in reduced signal. Another important determinant of signal is the frequency bandwidth of the voxel's spectrum. Distributing the total signal over a larger bandwidth reduces *S*/*N* by increasing the allowable noise bandwidth.

For a fixed imaging sequence, the signal intensity will also be affected by the operating frequency of the magnet. Signal increases as the square of frequency, ignoring variations in relaxation times. The noise dependence is not as easily quantified. Theory shows that at high frequency noise increases in proportion to frequency, while at low frequency it is proportional to the square root of frequency. The dependence of *S*/*N* on operating frequency has been reported experimentally to vary to the first power or less. For a well-tuned system, the body itself is the biggest noise source because random currents in the body are picked up as noise by the receiver coil.

Imaging Time

To acquire a full data set, excitation and readout are repeated after time TR for each value of the phase-encoding gradient. Repeated measurements at the same value of the phase-encoding gradient may also be obtained in order to improve the S/N using signal averaging.

$$\text{Total time} = TR \cdot N_{\text{pe}} \cdot N_{\text{acq}} \tag{9}$$

where N_{acq} is the number of repetitions at each phase-encoding step.

In general, the acquisition of multiple slices does not affect the imaging time. *Multislicing* is achieved by using the time between the completion of data acquisition and the start of the next excitation at TR to acquire data at additional slice locations. For example, for a double spin echo sequence with TE values of 30 and 60 ms and a TR value of 2000 ms, generously allowing 100 ms to complete data readout, an interval of 1900 ms is available before the next excitation occurs. In this time 19 additional slices can be acquired by moving the slice selection to a new location that is parallel and usually contiguous to the previous slices. In this way, multislicing improves the duty cycle of the acquisition without increasing the total imaging time.

Imaging Pulse Sequence

All of the necessary components are now in place, and we can talk about the *pulse sequence*. All of the parameters of the imaging technique are specified with the pulse sequence, and this is often shown schematically in the pulse timing diagram. The number, strength, timing, and waveform of the RF excitations, the duration and strength of the three magnetic field gradients, and the length and sampling rate of the readout period are all selected to determine the TR, TE, resolution, FOV, S/N, and contrast for the tissues being imaged. A priori knowledge of the T_1 and T_2 values helps with the selection of these parameters. Figure 10 is an example of a pulse timing diagram for a spin-echo pulse sequence with TR = 500 ms and TE = 30 ms. At the start of the experiment at $t = 0$, a 90° RF pulse is transmitted in the presence of a slice selection gradient along the z axis. After the pulse is completed, the slice select gradient is reversed to correct the phase dispersion caused by the slice selection. At the same time, the phase-encoding gradient is turned on at its first value. The readout gradient also appears. This is a dephasing lobe used to dephase the spins so that a full gradient echo can be measured during readout. At time $t = 15$ ms, a 180° RF pulse is transmitted in the presence of the slice select gradient to create a spin echo. Because the 180° pulse creates a mirror image of the spin system, the slice select gradient acts in a positive sense prior to the 180° pulse and in a negative afterwards. This is also why the dephasing lobe of the readout gradient appears to be in the same direction as during readout. When the 180° pulse is completed, the readout gradient is turned on and the signal is sampled symmetrically about the echo time $t = 30$ ms. After time TR, the experiment is repeated for each new value of the phase-encoding gradient. In practice, because data acquisition is over in less than 50 ms, the waiting time can be spent acquiring other slices until it is time to return to the first slice at TR = 500 ms.

Many special function pulse sequence have been developed to address specific imaging needs. The spin-echo pulse sequence has been a standard over the years, but it has some drawbacks. The time spent issuing 180° pulses in order to correct for field inhomogeneities also serves to prolong the TE value. If the 180° pulse is removed and a shorter TE is used, the loss of signal due to field inhomogeneity can be offset to some extent by the reduced T_2 decay. Sequences that create the signal echo by dephasing and rephasing the spins using the readout gradient rather than 180° pulses are referred to as *gradient-reversal* or *gradient-echo* sequences. Through the development of MRI technology, as methods for signal acquisition and noise reduction have improved, it has became possible to maintain more than adequate S/N ratios while increasing resolution, reducing scan times, and improving contrast. In addition to manipulating TR and TE for contrast, the angle of rotation of the magnetization vector M by RF excitation can also be adjusted. *Partial-flip* sequences, as they are called, create less transvese magnetization and therefore less signal but generate another degree of T_1 contrast. More importantly, by rotating the longitudinal magnetization vector only partway into the transverse plane, less time is required for it to recover. This allows TR to be reduced substantially and lowers the total imaging time. Many sequences combine these features, most notably the FLASH [26] or GRASS [27] sequences. These partial-flip, gradient-echo sequences have a unique image appearance due to their reduced flip angle, TR and TE. Two-dimensional multislice and 3D sequences of this type are commonly used to perform MR angiography.

Summary

It is hoped that the reader is left considering the many parameters that are involved in defining an MR imaging procedure and how these interact with the physical properties of the tissues. Therein lies the flexibility of MRI as well as the frustration when one is faced with an overwhelming number of techniques from which to select. The considerations become greater with the imaging of blood flow.

This chapter was intended to familiarize the reader with the essential components of magnetic resonance imaging. It has not been possible to be exhaustive in the coverage; many important and interesting topics have been only briefly mentioned or omitted altogether. Topics such as chemical shift imaging, fast imaging, use of contrast agents, and cardiac gating are relevent to the study of flow and can be studied in the texts listed in the references.

References

1. Axel L et al. (1986) ACR glossary of MR terms. American College of Radiology
2. Castleman KR (1979) Digital image processing. Prentice-Hall, Englewood Cliffs
3. Rosenfeld A, Kak AC (1982) Digital picture processing. Academic, Orlando
4. Gonzalez RC, Wintz P (1977) Digital image processing. Addison-Wesley, Reading, MA
5. Abragham A (1961) The principles of nuclear magnetism. Clarendon, Oxford
6. Curry TS, Dowdey JE, Murry RC (1978) Nuclear magnetic resonance. In: Christensen EE, Curry TS, Dowdey JE (eds) Introduction to the physics of diagnostic radiology. Lea and Febiger, Philadelphia
7. Farrar TC, Becker ED (1971) Pulse and Fourier transform NMR. Academic, New York
8. Fukushima E, Roeder SBW (1981) Experimental pulse NMR: a nuts and bolts approach. Addison-Wesley, Reading, MA
9. Mansfield P, Morris PG (1982) NMR imaging in biomedicine. Academic, New York
10. Crooks LE, Mills CM, Davis PL, Brant-Zawadzki M, Hoenninger JC, Arakawa M, Watts JC, Kaufman L (1982) Visualization of cerebral abnormalities by NMR imaging: The effects of imaging parameters on contrast. Radiology 144: 843–852
11. Damadian R (1971) Tumor detection by nuclear magnetic resonance. Science 171: 1151–1153
12. Davis PL, Kaufman L, Crooks LE, Margulis AR (1981) NMR characteristics of normal and abnormal rat tissues. In: Kaufman L, Crooks LE, Margulis AR (eds) Nuclear magnetic resonance imaging in medicine. Igaku-Shoin, New York
13. Koenig SH, Brown RD (1986) Relaxometry of tissue. In: CRC handbook on NMR in cells. CRC Press, Boca Raton
14. Bottomly PA, Foster TH, Argersinger RE, Pfeifer LM (1984) A review of normal tissue hydrogen NMR relaxation times and relaxation mechanisms from 1–100 MHz: dependence on tissue type, NMR frequence, temperature, species, excision and age. Med Phys 11: 425–448
15. Bracewell RN (1978) The Fourier transform and its applications. McGraw-Hill, New York
16. Morris PG (1986) Nuclear magnetic resonance imaging in medicine and biology. Clarendon, Oxford
17. Kaufman L, Crooks LE, Margulis AR (1981) Nuclear magnetic resonance imaging in medicine. Igaku Shoin, Tokyo
18. Partain CL, Price PR, Patton JA, Kulkarni MV, James AE (1988) Magnetic resonance imaging, vol II. Saunders, Philadelphia
19. Ortendahl DA, Hylton NM, Kaufman L, Watts JC, Crooks LE, Mills CM, Stark DD (1984) Analytical tools for magnetic resonance imaging. Radiology 153: 479–488
20. Edelstein WA, Bottomly PA, Hart HR et al. (1983) Signal-to-noise and contrast in nuclear magnetic resonance. J Comput Assist Tomogr 7: 391
21. Kaufman L, Crooks LE, Sheldon PE, Rowan W, Miller T (1982) Evaluation of NMR imaging for detection and quantification of obstructions in vessels. Invest Radiol 17: 554–560
22. von Schulthess GK, Fisher M, Crooks LE, Higgins CB (1985). Gated MR imaging of the heart: intracardiac signals in patients and healthy subjects. Radiology 156: 125–132
23. Maudsley AA (1984) Electronics and instrumentation for NMR imaging. IEEE Trans Nucl Sci NS 31: 990
24. Hoult DI, Lauterbur PC (1979) The sensitivity of the zeugmatographic experiment involving human samples. J Magn Reson 34: 425

25. Brunner P, Ernst RR (1979) Sensitivity and performance time in NMR imaging. J Magn Reson 33:83
26. Haase A, Frahm J, Matthaei D, Hanicke W, Merboldt KD (1986) FLASH imaging. Rapid NMR imaging using low flip-angle pulses. J Magn Reson 67:258–266
27. Glover GH, Pelc NJ, Shimakawa A (1987) GRASS movie techniques for gated studies. Proceedings 1987 Topical Conference on Fast Magnetic Resonance Imaging Techniques. Case Western Reserve University, Cleveland, Ohio

Chapter 6

Basics of MR Angiography

D. E. BOHNING

Introduction

In any angiographic technique, the different interaction of the technique with blood flowing in the vessels and stationary tissue determines the contrast and ultimate image quality. In MR angiography, these interactions fall into two general categories:
a) phase-related and
b) time-of-flight.

Phase-related effects are due to the phase difference between spins moving along a magnetic field gradient and stationary spins. Such phase differences are an inherent result of the interaction between flowing blood and the imaging gradients but can also be induced or modified with special velocity-sensitive gradients added to the standard imaging sequence. Phase-based angiography exploits this interaction for phase-based contrast, i.e., modulation of image intensity, or for explicit use of phase in complex, i.e., real and imaginary, image manipulation.

Time-of-flight effects are due to the replacement of spins within the imaging volume by spins with a significantly different level of magnetization. The difference in magnetization creates contrast between blood in the vessels and the surrounding tissue, for example, partially saturated spins being carried out of the image volume by blood flow to be replaced by fully magnetized spins, or spins purposely presaturated upstream replacing signal-producing spins within the volume. Time-of-flight techniques can rely solely on this effect, or they can use the subtraction of images with different characteristics for contrast enhancement.

The basic parameters in both phase-related and time-of-flight effects are the magnitudes and timings of the MR pulse sequence's gradients, and its repetition time, TR, and echo time, TE, in relation to the relative T_1 and T_2 of blood and tissue and the velocity of blood flow into and through the imaging volume. Further control of the interaction is possible by preconditioning the spins with inversion or presaturation pulses at some time *TI*, prior to the imaging pulse sequence proper. Typical spin echo and gradient echo pulse sequences used for MR angiography are diagramed in Figs. 1 and 2 respectively.

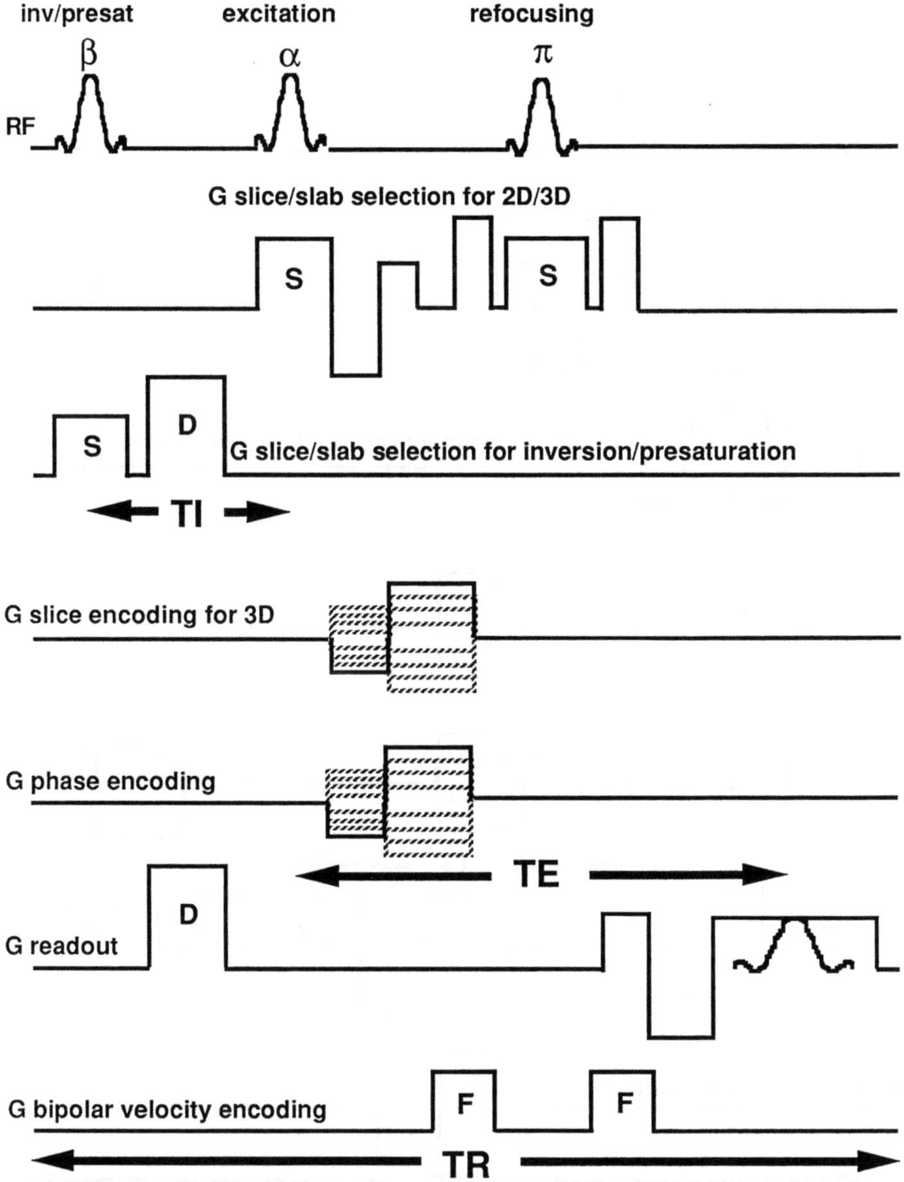

Fig. 1. Spin-echo pulse sequence for angiography showing slice/slab selection gradients (*S*), spin-dephasing gradients (*D*), and flow-encoding gradients (*F*) which can be applied in any direction

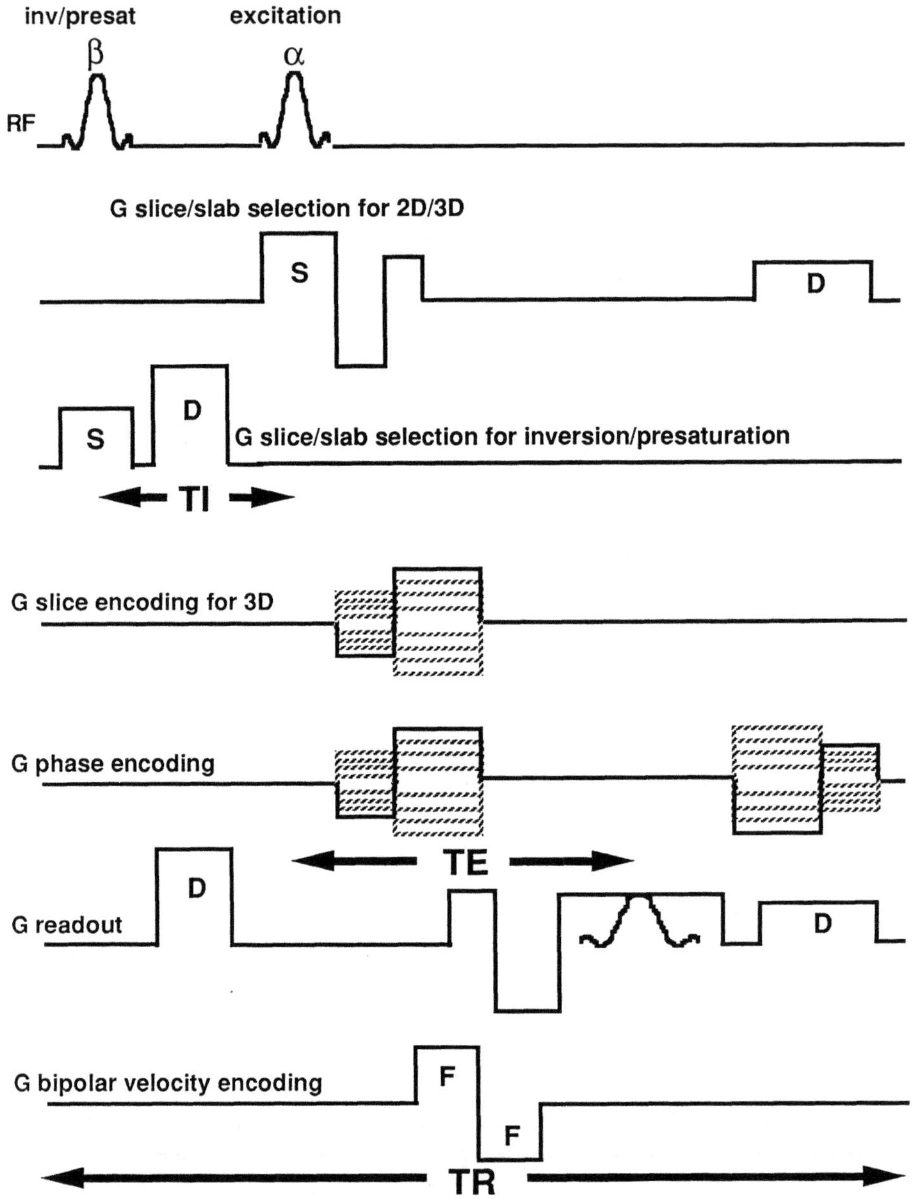

Fig. 2. Gradient-echo pulse sequence for angiography showing slice/slab selection gradients (*S*), spin-dephasing gradients (*D*), and flow-encoding gradients (*F*) which can be applied in any direction

Table 1. Comparative summary of angiographic methods by reference, relative acquisition/reconstruction/postprocessing time, and contrast-to-noise (*C/N*)

	Phase-based		Time-of-flight	
	Inherent	Induced	Passive	Active
2D projection	5 SE, 6 SE [a] 2/2/1 [b] 9 [c]	11 GE, 12 GE 2/2/1 18	– 1/1/0 0 SE, 0 GE	27 SE, 28 IR 2/2/1 9 SE, 3 GE
MS reformatted	– [a] 8/8/64 [b] 34 [c]	– 8/8/64 69	22 GE, 23 GE 4/4/64 12 SE, 52 GE	21 GE 8/8/64 22 SE, 61 GE
3D reformatted	– [a] 8/12/64 [b] 63 [c]	13 GE, 14 GE 8/12/64 125	25 GE, 26 GE 4/6/64 22 SE, 37 GE	24 GE 8/12/64 42 SE, 69 GE

[a] References; spin echo (SE), gradient echo (GE) and inversion recovery (IR)
[b] Relative acquisition/reconstruction/postprocessing time for 64 slices, 16 projections; GE given, SE single slice acquisition $\approx 20 \times$ GE single slice acquisition (note: between techniques [columns], acquisition, etc., times have different scale)
[c] Relative contrast-to-noise behavior, based on parameters in Appendix B

Classification of the various NMR angiographic methods is attempted in Table 1. To facilitate the comparison, the table includes estimates of relative acquisition/reconstruction/postprocessing times for a 16-view angiogram and relative contrast-to-noise for the techniques. They are offered only to assist the general overview; publications on the specific techniques should be consulted for more rigorously derived estimates.

A full consideration of the interactions between blood flow and the NMR angiographic process necessarily depends on an understanding of blood flow under both physiologic and pathologic conditions. However, under physiologic conditions, the blood flow is free of turbulence throughout the cardiac cycle, and useful estimates of the effect of flow on the NMR signal can be made by assuming simple plug flow, that is, a constant velocity across the vessel lumen [1–4].

Phase-based Methods

The present interest in MR angiography was initiated by the projection/subtraction angiograms produced by Wedeen et al. [5, 6] according to a phase-based method described by Macovski [7]. The measurement direction was aligned with the vessels of interest, and two thick-slice cardiac-triggered images were acquired, one at diastole and the other at systole. Due to the different blood velocities, the phase shifts of the blood in the two images were inherently different. Since the signal from stationary tissue remained the same in both images, subtracting the two images created a difference image in which only flowing blood remained.

Wedeen's demonstration that inherent velocity-induced phase shifts could be used to produce angiograms focused attention on Moran's [8] method for flow-encoding images with phase shifts induced by special bipolar gradient pulses like those in Figs. 1 and 2 labeled with an "F", opening the way to the design of pulse sequences and protocols specifically aimed at phase-based imaging and quantification of flow. The basis for active phase control with switched magnetic field gradients is described below.

A spin which moves a distance Δx along a positive magnetic field gradient G will be subjected to a slightly higher field than the stationary spins it leaves behind, by an amount $G\,\Delta x$. By the Larmor equation,

$$\omega = \gamma B_0 \tag{1}$$

where ω = the frequency of precession,
γ = the gyromagnetic ratio, and
B_0 = the magnetic field, it will then precess at a slightly higher frequency,

$$\omega = \gamma B_0 + \gamma G \Delta x.$$

Hence, if both spins are excited together at time $t = 0$, at time $t = T$ the spin at Δx will have a phase angle of

$$\Phi_{\Delta x} = T(\gamma B_0 + \gamma G \Delta x)$$

as compared with

$$\Phi_0 = T(\gamma B_0)$$

for the spins that have not moved, the difference

$$\Phi_{\Delta x} - \Phi_0 = \gamma T G \Delta x$$

being proportional to the distance moved, Δx, and the area (duration-strength product) of the gradient, TG. In general, the phase acquired by a "spin" at the position x moving along a gradient G is given by:

$$\Phi = \gamma \int_0^T G(t)\, x(t)\, \mathrm{d}t \tag{2}$$

where γ is the gyromagnetic ratio, $G(t)$ is the gradient waveform, and $x(t)$ is the position of the "spin" at time t. The integral is taken over the entire period, T, during which the "spin" is exposed to the gradient.

For a spin moving at a constant velocity, V_0, in a constant gradient, $G(t) = G$,

$$x(t) = x_0 + V_0 t, \tag{3}$$

and

$$\Phi = \gamma \int_0^T G\,(x_0 + V_0\, t)\, dt,$$
$$= \gamma \int_0^T G\, x_0\, dt + \gamma \int_0^T G\, V_0\, t\, dt.$$

Hence, the phase difference for the moving spins is

$$\Delta\Phi = \gamma \int_0^T G\, V_0\, t\, dt$$
$$= \gamma\, G\, V_0 \int_0^T t\, dt$$
$$= \gamma\, G\, V_0\, T^2/2. \qquad (4)$$

Although, in general, the exact motion of the "spin", $x(t)$, is not known, any continuous, differentiable function can be expanded in a Taylor series:

$$x(t) = x|_{t=0} + [dx/dt|_{t=0}]\, t + [d^2x/dt^2|_{t=0}]\, t^2/2 + \ldots$$
$$= X_0 \quad + \quad V_0\, t \quad + \quad A_0\, t^2/2 \quad + \ldots \qquad (5)$$

X_0 being the position of the "spin" at time 0, V_0 its velocity, and A_0 its acceleration. Hence, from Eq. (2), the phase can be written:

$$\Phi = \gamma \int G(t)\,(X_0 + V_0\, t + A_0\, t^2/2 + \ldots)\, dt;$$
$$= \gamma\, X_0 \int G(t)\, dt + \gamma\, V_0 \int G(t)\, t\, dt + \gamma\, A_0 \int G(t)\, t^2\, dt/2 + \ldots$$

In this form, it can be seen that the integrals in the individual terms of the sum correspond to the increasing moments of the gradient waveform,

$$M_i = \int G(t)\, t^i\, dt, \qquad (6)$$

and

$$\Phi = \gamma\, X_0\, M_0 + \gamma\, V_0\, M_1 + \gamma\, A_0\, M_2/2 + \ldots \qquad (7)$$

By using a gradient waveform which makes M_1 equal to zero ("gradient moment nulling"), the velocity-induced phase shift will be removed, or "velocity compensated." If a gradient waveform for which M_0 is identically zero is added, for example, bipolar gradient pulses like those labeled with an "*F*" in the spin echo sequence of Fig. 1 and the gradient echo sequence of Fig. 2, then stationary spins will not be affected, and the gradient can be used for flow encoding as described by Moran [8]. This gradient can then be scaled to give M_1 any desired value, such as $\pi/\gamma\, V_s$. In this example spins moving with velocity V_s will experience a phase shift $\Phi = \pi$ and will be 180° out of phase with the signal from stationary tissue; thus, maximum contrast will be created.

Note that in the spin echo sequence (Fig. 1), the 180° pulse inverts the spins, so the second lobe of the bipolar gradient, placed after the 180° pulse, has the same sign as the first lobe. In the gradient echo sequence (Fig. 2), the second lobe must have a sign opposite to the first to achieve the proper balance.

Although van Dijk [9] and Bryant et al. [10] had earlier applied Moran's ideas to phase display of cardiac wall motion and the measurement of flow, respectively, Dumoulin et al. [11, 12] were the first to flow-encode phase shifts for angiography. Adding bipolar gradients to a flow-compensated gradient echo pulse sequence and alternating their sign on alternate acquisitions, they created two thick-slice projection images with opposite flow sensitivities. The complex images were then subtracted to create the final projection angiogram. This process is described in detail in Appendix A and Fig. 3.

A measure of the vessel contrast-to-noise (C/N) for such an angiogram is given by

$$C/N \approx [I_b - I_t^A]/[N_t^A]$$

where I_b and I_t^A are the blood (vessel) and angiogram residual tissue signal respectively, and N_t^A is the angiogram residual tissue noise level. From Eq. A10 in Appendix A, a vessel diameter of d_b, a tissue thickness of d_t, and assuming a uniform level of uncorrelated noise per unit thickness, N, in the two images to be subtracted, the C/N would be approximately given by:

$$\begin{aligned} C/N &\approx [I_v - \sqrt{d_t 2 N^2}]/[\sqrt{d_t 2 \mathrm{N}^2}] \\ &\approx [\sqrt{2}\, d_b \sin(\pi v/2 V_s) - \sqrt{2 d_t} N]/[\sqrt{2 d_t} N] \\ &\approx [d_b \sin(\pi v/2 V_s) - \sqrt{d_t} N]/\sqrt{d_t} N] \end{aligned}$$

for 2D or multislice acquisitions. It would be a factor of $\sqrt{n_{3D}}$ higher for 3D, where n_{3D} is the number of 3D slices. Of course, the stationary tissue subtrac-

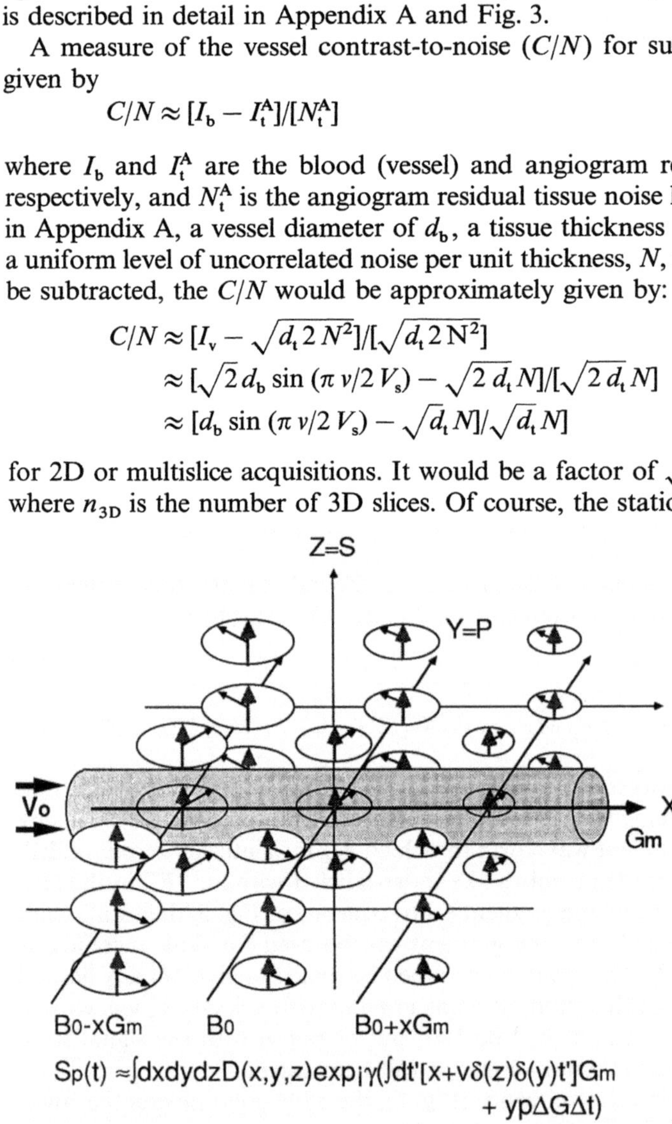

Fig. 3. Projective complex subtraction angiography (G_m = measurement gradient, G_v = flow encoding gradient). **a** Spin behavior for flow-compensated sequence. **b** Spin behavior for velocity-sensitive sequence. **c** Phase-based substraction angiogram

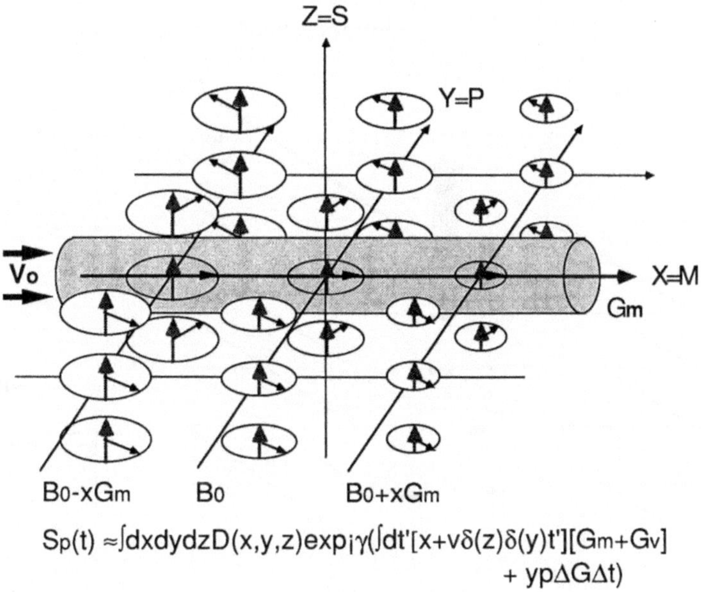

Fig. 3b

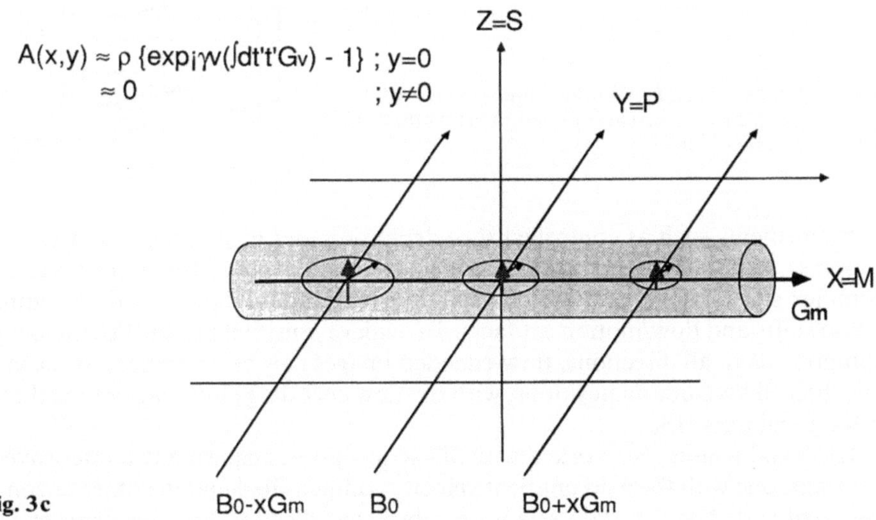

Fig. 3c

tion is never as perfect as indicated in Appendix A, due to flow and motion artifacts as well as to misregistrations from patient movements. In Table 1, relative C/N values have been computed assuming the parameter values given in Appendix B.

Although phase-based subtraction angiography is more frequently performed with gradient echo acquisitions due to their shorter acquisition times and greater phase sensitivity, it can be performed with virtually any pulse sequence, provided a phase difference between flowing blood and stationary tissue is produced.

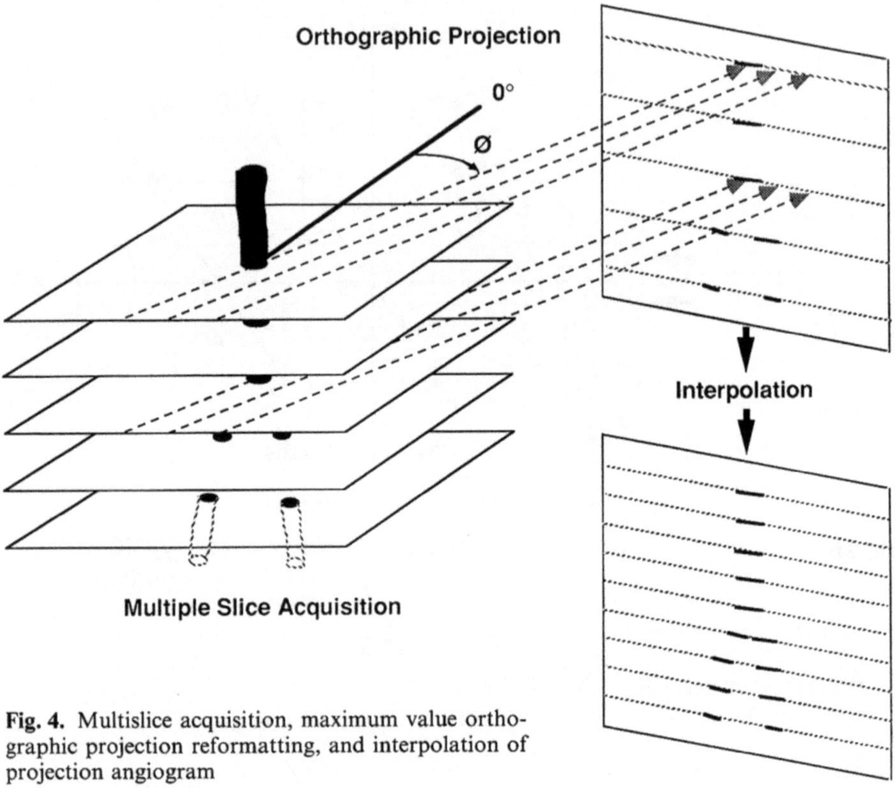

Fig. 4. Multislice acquisition, maximum value orthographic projection reformatting, and interpolation of projection angiogram

Refinements such as controlled slice dephasing and high repetition rates [12] suppress stationary tissue signal, reducing dynamic range for more accurate complex subtractions, and are effective in averaging out spurious flow-coding phase shifts and flow/motion artifacts. To achieve sensitivity to the flow velocity components in all directions, flow-encoded images can be combined, for example, three 3D volume acquisitions, with the flow encoding along each of the three orthogonal axes [13].

Laub and Kaiser also worked with 3D acquisitions, acquiring two interleaved data sets, one with the read gradient velocity compensated and the other uncompensated [14]. The two data sets were subtracted slice by slice, and then maximum intensity orthographic projections [15, 16] were made to create projection angiograms (Fig. 4). The authors emphasized the importance of short echo times for reducing artifacts.

Time-of-Flight Methods

The "refreshment" effect of flowing blood [17], described as "paradoxial enhancement" [18] and "flow-related enhancement" [19], was noted at the outset of medical MR tomography. Stationary tissue cancellation was produced by Wehrli [20], who subtracted images alternately preceded by selective and non-selective presaturation pulses. Time-of-flight angiography, however, began with Frahm et al. [21], who acquired a projective angiogram line by line.

The present form [22, 23] uses a contiguous or overcontinuous stack of thin transverse 2DFT gradient-echo images. The refreshment effect maintains signal in the vessels, while the rapid repetition pulse sequence saturates stationary tissue. Then, a maximum value orthographic projection [15, 16] is used to produce an angiogram, as shown in Fig. 4. Though this method requires a postacquisition projection, it avoids a second acquisition and image subtraction and the attendant problems from time-dependent artifacts, e. g., patient movement. It also allows postacquisition projection at any desired angle.

Dumoulin et al. [24] explored the time-of-flight method with 3D acquisitions. In this protocol, to increase contrast, a second data set in which incoming blood had been presaturated was acquired and subtracted from the first. Despite the subtraction, this 3D version of the "inflow" technique appears less effective than multiple thin slice acquisition due to excitation of the entire sample volume, and the consequent reduction of spin refreshment and vessels with slow flow. Masaryk et al. [25, 26] successfully imaged the intra- and extracranial vasculature relying only on spin refreshment by simultaneously acquiring two separate sagittal 3D slabs rather than the much larger transverse slice volume used by Dumoulin et al. [24].

Spin-Echo Time-of-Flight

In a spin-echo sequence (Fig. 1), both a 90° RF excitatin pulse and a 180° RF refocusing pulse are required to produce a signal. Assume plug flow in a vessel transected by an imaging volume of thickness d as shown in Fig. 5. By the time the 180° pulse is given at $t = \mathrm{TE}/2$, spins just entering the slice and excited by the 90° pulse will have moved a distance $V \cdot \mathrm{TE}/2$ across the slice, to be replaced by fully magnetized spins. By the time of the next 90° excitation pulse at $t = \mathrm{TR}$, a time interval $\mathrm{TR} - \mathrm{TE}/2$ later, these spins have moved a distance of $V \cdot (\mathrm{TR} - \mathrm{TE}/2)$ across the slice and been replaced by fully magnetized spins. Hence, for $0 \leq V \leq d/\mathrm{TR}$, three different spin fractions can be identified within the slice: (a) fully magnetized spins which have penetrated the slice to a depth of $\mathrm{V} \cdot (\mathrm{TR} - \mathrm{TE}/2)$ contributing the fraction

$$V \cdot (\mathrm{TR} - \mathrm{TE}/2)/d, \tag{8A}$$

(b) spins that experienced only a 180° pulse contributing the fraction

$$V \cdot (\mathrm{TE}/2)/d, \tag{8B}$$

and (c) spins that experienced both 90° and 180° pulses with a relative population

$$(d - V \cdot \mathrm{TR})/d. \tag{8C}$$

The signal from each of these three different fractions is then

$$I_A \approx V \cdot [\mathrm{TR} - \mathrm{TE}/2]/d \qquad (9\,\mathrm{A})$$

$$I_B \approx [V \cdot \mathrm{TE}/2]\,[1-2\exp(-(\mathrm{TR}-\mathrm{TE}/2)/T_{1b})]/d \qquad (9\,\mathrm{B})$$

$$I_C \approx [d-V \cdot \mathrm{TR}]\,[1-2\exp(-(\mathrm{TR}-\mathrm{TE}/2)/T_{1b})+\exp(-\mathrm{TR}/T_{1b})]/d \qquad (9\mathrm{C})$$

Adding these three contributions, including the T_{2b} decay factor for blood

$$E_{2b} = \exp(-\mathrm{TE}/T_{2b}),$$

and accounting for the third fraction loss of $V \cdot \mathrm{TE}/2$ by the time of the 180° pulse, the total spin-echo signal from the blood with the vessel is given by

$$\begin{aligned} I_b^{se} \approx E_{2b} \cdot \{ & V \cdot (\mathrm{TR}-\mathrm{TE}/2) \\ & + [V \cdot \mathrm{TE}/2]\,[1-2\exp(-(\mathrm{TR}-\mathrm{TE}/2)/T_{1b})] \\ & + [d-V \cdot \mathrm{TR}-V \cdot \mathrm{TE}/2]\,[1-2\exp(-(\mathrm{TR}-\mathrm{TE}/2)/T_{1b}) \\ & + \exp(-\mathrm{TR}/T_{1b})]\}/\mathrm{d} \end{aligned} \qquad (10)$$

which holds as long as $0<(d-V\mathrm{TR})/d \leq V\,\mathrm{TE}/2\,\mathrm{d}$. For $V=d/\mathrm{TR}$, all spins are replaced by fully magnetized spins and Eq. 10 takes its maximum value

$$I_b^{se} \approx E_{2b} \cdot [(\mathrm{TR}-\mathrm{TE}/2) - (\mathrm{TE}/2)\exp(-\mathrm{TR}/T_{1b})]/\mathrm{TR} \qquad (11)$$

The signal for stationary tissue is just

$$I_t^{se} \approx E_{2t} \cdot [1-2\exp(-(\mathrm{TR}-\mathrm{TR}/2)T_{1t}) + \exp(-\mathrm{TR}/T_{1t})] \qquad (12)$$

Estimates of the C/N for viewing the vessel are

$$C/N\,(\mathrm{tof, se, tr}) \approx \{I_b^{se} - I_t^{se}\}/N \qquad (13)$$

for vessels perpendicular to the imaging slab, and

$$C/N\,(\mathrm{tof, se, ip}) \approx \{d_b \cdot I_b^{se} - d_t \cdot I_t^{se}\}/\sqrt{d_t}N \qquad (14)$$

for in plane vessels.

Note that in the carotids, with an average velocity of 0.2 m/s, a spin-echo TR of 400 ms would allow complete refreshment about 80 mm into the slab, while for a typical gradient-echo sequence, TR = 20 ms, only 4 mm would be refreshed. Hence, in-plane time-of-flight entails large TR values and long imaging times, though only one or two (for subtraction) thick-slice acquisitions may be required.

Gradient-Echo Time-of-Flight

The gradient-echo sequence uses gradient refocusing to form an echo rather than relying on a 180° RF pulse. Thus, the signal is the sum of contributions from spins successively excited (and saturated!) as they flow across the slice, those a distance VTR into the slice having received one excitation, those spins from VTR to 2VTR having received two excitations, those from 2VTR to 3VTR having received three excitations, etc., up to a maximum of $n = d/V \cdot \mathrm{TR}$ excitations when the blood has flown on out of the slice.

If $E_{1b} = \exp(-\mathrm{TR}/T_{1b})$ and $E_{2b} = \exp(-\mathrm{TR}/T_{2b})$ are the blood spin-lattice and spin-spin relaxation factors respectively, the general expression for the longitudinal magnetization in blood which has received n excitations with tip angle α is

$$\begin{aligned} M_b^n &\approx 1 - (1 - M_b^{n-1} - \cos\alpha)\, E_{1b} \\ &\approx (1 - E_{1b}) + E_{1b}\cos\alpha\, M_b^{n-1} \\ &\approx (1 - E_{1b}) + E_{1b}\cos\alpha\,\{(1 - E_{1b}) + E_{1b}\cos\alpha\, M_b^{n-2}\} \\ &\approx (1 - E_{1b})(1 + E_{1b}\cos\alpha) + E_{1b}^2\cos^2\alpha\, M_b^{n-2} \\ &\approx (1 - E_{1b})(1 + E_{1b}\cos\alpha) \\ &\quad + E_{1b}^2\cos^2\alpha\,\{(1 - E_{1b}) + E_{1b}\cos\alpha\, M_b^{n-3}\} \\ &\approx (1 - E_{1b})(1 + E_{1b}\cos\alpha + E_{1b}^2\cos^2\alpha) + E_{1b}^3\cos^3\alpha\, M_b^{n-3}. \end{aligned}$$

In general,

$$\begin{aligned} M_b^n &\approx (1 - E_{1b}) \sum_{i=0}^{n-2} \{E_{1b}\cos\alpha\}^i + E_{1b}^{n-1}\cos^{n-1}\alpha \\ &\approx (1 - E_{1b})(1 - E_{1b}^{n-1}\cos^{n-1}\alpha)/(1 - E_{1b}\cos\alpha) \\ &\quad + E_{1b}^{n-1}\cos^{n-1}\alpha, \end{aligned}$$

or

$$M_b^n \approx M_b^{ss}(1 - E_{1b}^{n-1}\cos^{n-1}\alpha) + E_{1b}^{n-1}\cos^{n-1}\alpha, \tag{15}$$

where

$$M_b^{ss} \approx (1 - E_{1b})/(1 - E_{1b}\cos\alpha) \tag{16}$$

is the well-known expression for steady-state magnetization.

The total blood signal, I_b^{ge}, is then

$$\begin{aligned} I_b^{ge}\,(1 \le n \le d/V \cdot \mathrm{TR}/ &\approx \sin\alpha\, E_{2b} \\ &\quad \cdot [\mathrm{VTR} \sum_{i=1}^{n-1} M_b^i + (d - \{n-1\}\mathrm{VTR})\, M_b^n]/d \\ &\approx \sin\alpha\, E_{2b} \\ &\quad \cdot [\mathrm{VTR} \sum_{i=1}^{n} M_b^i + (d - n\mathrm{VTR})\, M_b^n]/d. \end{aligned} \tag{17}$$

$$\approx \sin\alpha\, E_{2b}$$
$$\cdot [\mathrm{VTR} \sum_{i=1}^{n} \{M_b^{ss}(1 - E_{1b}^{i-1}\cos^{i-1}\alpha) + E_{1b}^{i-1}\cos^{i-1}\alpha\}$$
$$+ (d - n\,\mathrm{VTR})\,\{M_b^{ss}(1 - E_{1b}^{n-1}\cos^{n-1}\alpha)$$
$$+ E_{1b}^{n-1}\cos^{n-1}\alpha\}]/d. \tag{18}$$

$$\approx \sin\alpha\, E_{2b}$$
$$\cdot [\mathrm{VTR}\left\{\sum_{i=1}^{n} M_b^{ss} + \sum_{i=1}^{n}(1 - M_b^{ss})\,E_{1b}^{i-1}\cos^{i-1}\alpha\right\}$$
$$+ (d - n\mathrm{VTR})\,\{M_b^{ss} + (1 - M_b^{ss})\,E_{1b}^{n-1}\cos^{n-1}\alpha\}]/d. \tag{19}$$

$$\approx \sin\alpha\, E_{2b}$$
$$\cdot [\mathrm{VTR}\,\{\mathrm{n}M_b^{ss} + (1 - M_b^{ss})\,(E_{1b}^{n}\cos^{n}\alpha - 1)/(E_{1b}\cos\alpha - 1)\}$$
$$+ (\mathrm{d} - n\mathrm{VTR})\,\{M_b^{ss} + (1 - M_b^{ss})\,E_{1b}^{n-1}\cos^{n-1}\alpha\}]/d. \tag{20}$$

For stationary tissue $V = 0$, and n will be equal to the number of phase-encoding steps, 256 for a 256×256 image, giving a signal of

$$I_t^{ge}(n = 256) = \sin\alpha\, E_{2t}\cdot(1 - \mathrm{E}_{1t})\,(1 - E_{1t}^{255}\cos^{255}\alpha)/(1 - E_{1t}\cos\alpha)$$
$$+ E_{1t}^{255}\cos^{255}\alpha$$
$$\approx \sin\alpha\, E_{2t}\cdot(1 - E_{1t})/(1 - E_{1t}\cos\alpha), \tag{21}$$

the standard expression for the steady-state signal for a short repetition time gradient-echo sequence.

For $V \geq \mathrm{d/TR}$, the spins are always fully refreshed, receiving only one excitation, and the signal is simply

$$I_b^{ge}(n = 1) \approx \sin\alpha\, E_{2b} \tag{22}$$

Of course, should the flow rate be so high or the slice so thin that there is insufficient time for the echo to be refocused, i.e., $V > d/\mathrm{TE}$, then no signal will be generated!

The vessel C/N for slabs transverse to the vessel has the general behavior

$$C/N\,(\mathrm{tof, ge, tr}) \approx \{I_b^{ge}(1 \leq n \leq d/V\cdot\mathrm{TR}) - I_t^{ge}(n = 256)\}/N \tag{23}$$

Similarly, the C/N for inplane vessels will be given by

$$C/N\,(\mathrm{tof, ge, ip}) \approx \{\sqrt{d_b}\cdot I_b^{ge}(1 \leq n \leq \mathrm{FOV}/V\cdot\mathrm{TR})$$
$$- \sqrt{d_t}\cdot I_b^{ge}(n = 256)\}/\sqrt{d_t}\,N \tag{24}$$

Inversion/Presaturation Pulse Preconditioning – Spin Labeling

In both the spin-echo (Fig. 1) and gradient-echo (Fig. 2) implementations of the time-of-flight technique, the incoming spins can be preconditioned with an inversion or presaturation pulse. The preconditioning pulse can be applied to a slab adjacent to (as shown in Figs. 5 and 6) or a slab at some distance from the imaged volume to maximize the effect on vessel contrast. These so-called spin labeling techniques, i.e., techniques in which spins are actively preconditioned by saturation or inversion to create vessel contrast, rather than relying on refreshment or relaxation effects, were applied to angiography by Dixon et al. [27] and Nishimura et al. [28]. Dixon et al. [27] used a surface coil placed on the neck and a Gaussian RF pulse applied while a gradient ramped down to invert flowing spins by adiabatic fast passage. Subtracting nonselective SE projection image with and without blood "labeling" allowed the separation of blood and stationary tissue images of the head. Nishimura et al. [28] actively labeled spins using a 180° inversion pulse. They first used a selective 180° pulse to invert spins in the region of interest, then acquired a standard thick-slice projection spin-echo image. In the second acquisition, the inversion pulse was made nonselective. Complex subtraction of the two images removed the static component and left a blood signal with amplitude proportional to exp $(-\mathrm{TI}/T_{1b})$, where TI is the inversion delay time and T_{1b} is the spin-lattice relaxation

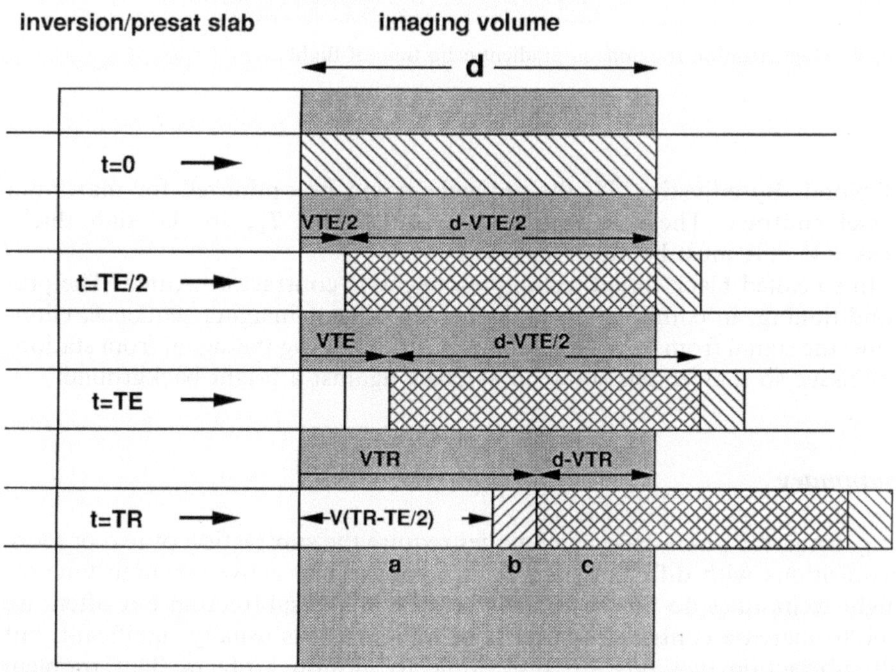

Fig. 5. Magnetization fractions in spin-echo time-of-flight. **a** Fully magnetized spins; **b**, 180° only; **c** 90° and 180°

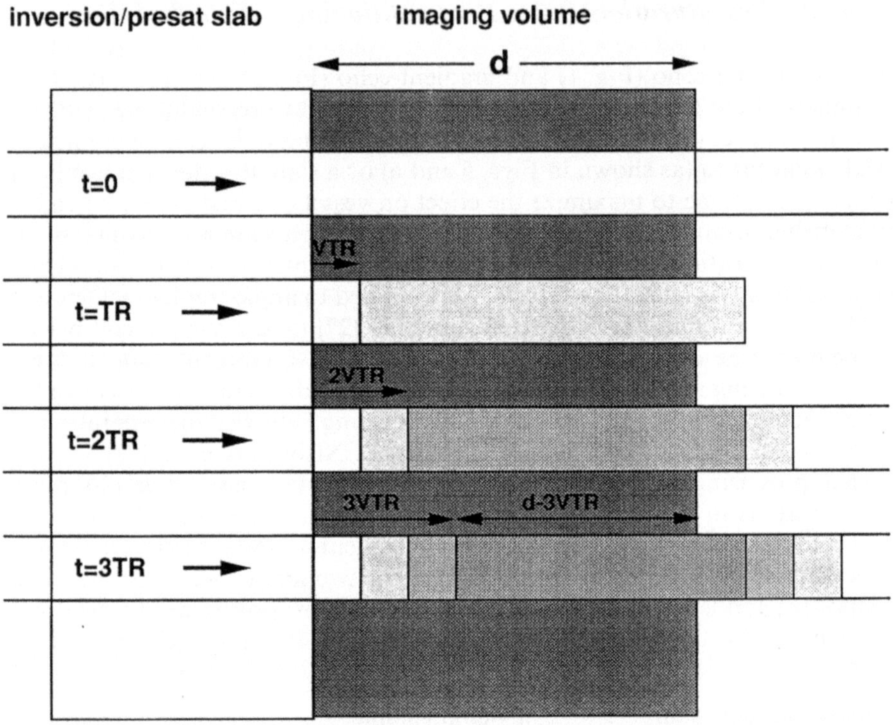

Fig. 6. Magnetization fractions in gradient-echo time-of-flight

of blood. By adjusting TI, the sequence could be optimized for maximum vessel contrast. These techniques are limited by T_1, so the slab thickness = $V \cdot \mathrm{TR}$ must be much less than $V \cdot T_1$.

In so-called black blood techniques, negative contrast is created. The preconditioning, in combination with the subsequent imaging sequence, eliminates the signal from flowing blood while maintaining the signal from stationary tissue so that vessels show up as black against a bright background.

Summary

In general, the phase-based techniques require the subtraction of two or more acquisitions with different parameter settings. The active-labelling time-of-flight techniques do not necessarily require image subtraction but often use one to increase contrast. The increase in contrast is usually significant, but all subtraction methods are susceptible to motion artifacts. This problem can be reduced by interleaved acquisition of the projections to be subtracted [29, 30].

Typically, only the spin-refreshment (passive) time-of-flight techniques using multiple thin slices and short TR gradient-echo sequences for static tissue suppression provide adequate contrast without subtraction. However, many slices are needed to achieve adequate resolution and a time-consuming postacquisition orthographic projection is required. Compared with phase-based angiography, these angiographic techniques are much less dependent on scanner performance, eddy currents, and B_0 and RF homogeneity.

Though the C/N values in Table 1 have not been derived in a completely rigorous manner, they do allow some general observations. Phase-based angiography using special gradients to induce phase difference rather than depending on the inherent interaction between blood flow and the imaging gradients doubles C/N, so, since both require the same time, it is easy to understand why inherent time-of-flight angiography has not been pursued beyond the original work [5, 6]. Passive 2D projection time-of-flight angiography is impossible because the vessels are lost in the thick slab of tissue, giving no contrast. However, active 2D projection time-of-flight is being done, mostly with a subtraction to boost C/N, and mostly with spin-echo acquisitions, because the TR is then large enough for refreshment of a reasonably large field of view.

The large signal-to-noise (S/N) advantage of 3D over multislice is offset partially by inhomogeneities over the extremes of the slice in phase-based methods, and partially by the limited refreshment of the entire slice set in time-of-flight techniques. Hence, until system tolerances improve, multislice reformatted methods have practical advantages over their 3D counterparts with regard to phase-based techniques, as well as inherent advantages with regard to both passive and active time-of-flight techniques.

In the future, angiograms from all of the techniques will improve as shielded gradients, short echotimes (TE $\leq$ 3 ms), and pulse crafting for better slice selection become widespread. However, multiple thin-slice and 3D chunk angiographic sequences which take advantage of both phase-based and time-of-flight effects as well as preconditioning will likely predominate, the phase information giving sensitivity to slow flow and the potential for quantitation, and the preconditioning suppressing tissue signal or enhancing the refreshment effect for in-plane sensitivity as well as higher overall blood signal for better phase determination. Increased computing power will eliminate post-processing delays and give the capability of handling multiple volume datasets for retrospectively gated multiphase angiograms.

Appendix A. Reconstruction of complex subtraction projection angiogram

To see how the subtraction of two images with different velocity sensitivities produces an angiogram in which only the vessels appear, the makeup of the signal from each must be examined.

The signal for the p-th profile is given by

$$S_p(t) = \int dx\, dy\, dz\, D(x, y, z) \cdot \exp i\gamma \left(\int_0^t dt' [x + v(x, y)\, t'] [G_m + G_v] + y p\, \Delta G_p \Delta t_p \right) \quad (A\,1)$$

$$= \int dx\, dy\, dz\, D(x, y, z) \cdot \exp i\gamma \left(\int_0^t dt' [x + v(x, y)\, t'] [G_m + y p\, \Delta G_p \Delta t_p + \int_0^t dt'\, x G_v + \int_0^t dt'\, v(x, y)\, t'\, G_v \right)$$

Since x and v are assumed constant in time, this becomes

$$= \int dx\, dy\, dz\, D(x, y, z) \cdot \exp i\gamma \left(x \int_0^t dt'\, G_m + v(x, y) \int_0^t dt'\, t'\, G_m + y p\, \Delta G_p \Delta t_p + x \int_0^t dt'\, G_v + v(x, y) \int_0^t dt'\, t'\, G_v \right) \quad (A\,2)$$

Now, both the first moment of the measurement gradient G_m and the zeroth moment of the bipolar velocity encoding gradient G_v are zero; hence, in general,

$$S_p(t) = \int dx\, dy\, dz\, D(x, y, z) \cdot \exp i\gamma \left(x \int_0^t dt'\, G_m + y p \Delta G_p \Delta t_p + v(x, y) \int_0^t dt'\, t'\, G_v \right)$$

For a resolution of r, each measurement profile consists of $2r$ samples of this signal taken at intervals of Δt_m, the n-th sample being given by

$$S_p(t_n) = \int dx\, dy\, dz\, D(x, y, z) \cdot \exp i\gamma \left(n \Delta t_m x G_m + y p\, \Delta G_p \Delta t_p + v(x, y) \int_0^T dt'\, t'\, G_v \right) \quad (A\,3)$$

where T is the time at which the bipolar velocity-encoding gradient is turned off. Measuring a complete set of p profiles in this way would make up the complete acquisition; typically 256 profiles, each consisting of 512 samples.

Letting $k_x = \gamma G_m \Delta t_m$ and $k_y = \gamma \Delta G_p \Delta t_p$, this becomes:

$$S_p(t_n) = \int dx\, dy\, dz\, D(x, y, z) \exp(i n k_x x) \exp(i p k_y y) \cdot \exp \left(i \gamma v(x, y) \int_0^T dt'\, t'\, G_v \right), \quad (A\,4)$$

We know that a 2D image can be recovered from this set of signal samples by use of the inverse discrete Fourier transform,

$$\mathfrak{D}(x,y) = \sum_{p}\sum_{n} S_p(t_n) \exp(-i n k_x x) \exp(-i p k_y y), \tag{A5}$$

so we have

$$\mathfrak{D}(x,y) = \sum_{p}\sum_{n}\Bigg[\int dx'\,dy'\,dz'\,D(x',y',z') \exp(i n k_x x') \exp(i p k_y y') \cdot \exp\left(i\gamma v(x',y',z') \int_0^T dt'\,t'\,G_v\right)\Bigg] \cdot \exp(-i n k_x x) \exp(-i p k_y y) \tag{A6}$$

and the two different acquisitions used in the phase-based subtraction are then:

$$\mathfrak{D}^1(x,y) = \sum_{p}\sum_{n}\Bigg[\int dx'\,dy'\,dz'\,D(x',y',z') \exp(i n k_x x') \exp(i p k_y y') \cdot \exp(i\gamma v(x',y',z') \int_0^T dt'\,t'\,G_{v1}\Bigg] \cdot \exp(-i n k_x x) \exp(-i p k_y y)$$

for the velocity-compensated acquisition, and

$$\mathfrak{D}^2(x,y) = \sum_{p}\sum_{n}\Bigg[\int dx'\,dy'\,dz'\,D(x',y',z') \exp(i n k_x x') \exp i p k_y y') \cdot \exp(i\gamma v(x',y',z') \int_0^T dt'\,t'\,G_{v2}\Bigg] \cdot \exp(-i n k_x x) \exp(-i p k_y y)$$

for the velocity-senitive acquisition, with the angiogram given by a complex subtraction of the two

$$A(x,y) = \mathfrak{D}^2(x,y) - \mathfrak{D}^1(x,y). \tag{A7}$$

$$= \sum_{p}\sum_{n}\Bigg[\int dx'\,dy'\,dz'\,D(x',y',z') \exp(i n k_x x') \exp(i p k_y y') \cdot \exp(i\gamma v(x',y',z') \int_0^T dt'\,t'\,G_{v2}\Bigg] \exp(-i n k_x x) \exp(-i p k_y y)$$

$$- \sum_{p}\sum_{n}\Bigg[\int dx'\,dy'\,dz'\,D(x',y',z') \exp(i n k_x x') \exp(i p k_y y') \cdot \exp\left(i\gamma v(x',y',z') \int_0^T dt'\,t'\,G_{v1}\right)\Bigg] \exp(-i n k_x x) \exp(-i p k_y y)$$

$$= \sum_{p}\sum_{n}\Bigg[\int dx'\,dy'\,dz'\,D(x',y',z') \exp(i n k_x x') \exp(i p k_y y') \cdot \left\{\exp\left(i\gamma v(x',y',z') \int_0^T dt'\,t'\,G_{v2}\right) - \exp\left(i\gamma v(x',y',z') \int_0^T dt'\,t'\,G_{v1}\right)\right\}\Bigg] \cdot \exp(-i n k_x x) \exp(-i p k_y y) \tag{A8}$$

Since the expression within the braces,

$$\left\{\exp\left(i\gamma v(x',y',z')\int_0^T dt'\,t'\,G_{v2}\right) - \exp\left(i\gamma v(x',y',z')\int_0^T dt'\,t'\,G_{v1}\right)\right\},$$

is exactly zero for $v(x',y',z') = 0$, that is, for stationary tissue $A(x,y)$ is, indeed, an image of the isolated vessels.

We can follow this through with our simple illustration shown in Fig. 3, where $v(x',y',z') = v\,\delta(y-0)\,\delta(z-0)$ and $D(x',y',z') = \varrho$, to see how the intensity of the vessels in the angiogram depends on the size of the bipolar velocity-encoding gradients.

$$\begin{aligned} A(x,y) = \sum_p \sum_n \Bigg[& \int dx'\,dy'\,dz'\,\varrho \exp(i n k_x x') \exp(i p k_y y') \\ & \cdot \left\{\exp\left(i\gamma v\,\delta(y-0)\,\delta(z-0)\int_0^T dt'\,t'\,G_{v2}\right)\right. \\ & \left. - \exp\left(i\gamma v\,\delta(y-0)\,\delta(z-0)\int_0^T dt'\,t'\,G_{v1}\right)\right\}\Bigg] \\ & \cdot \exp(-i n k_x x) \exp(-i p k_y y) \end{aligned}$$

$$\begin{aligned} = \sum_p \sum_n \Bigg[& \int dx'\,dy'\,\varrho \exp(i n k_x x') \exp(i p k_y y') \\ & \cdot \left\{\exp\left(i\gamma v\,\delta(y-0)\int_0^T dt'\,t'\,G_{v2}\right)\right. \\ & \left. - \exp\left(i\gamma v\,\delta(y-0)\int_0^T dt'\,t'\,G_{v1}\right)\right\}\Bigg] \\ & \cdot \exp(-i n k_x x) \exp(-i p k_y y) \end{aligned}$$

since the vessel is one unit thick ($\Delta z = 1$) and at $z = 0$.

$$\begin{aligned} = \sum_p \sum_n \Bigg[& \int dx'\,\varrho \exp(i n k_x x') \left\{\exp\left(i\gamma v \int_0^T dt'\,t'\,G_{v2}\right)\right. \\ & \left. - \exp\left(i\gamma v \int_0^T dt'\,t'\,G_{v1}\right)\right\}\Bigg] \\ & \cdot \exp(-i n k_x x) \exp(-i p k_y y) \end{aligned}$$

since the vessel is one unit wide ($\Delta y = 1$) and at $y = 0$. Now, integrating x' from -1 to 1

$$\begin{aligned} = \sum_p \sum_n \Bigg[& \varrho \{\exp(i n k_x) - \exp(-i n k_x)\}/(i n k_x) \\ & \cdot \left\{\exp\left(i\gamma v \int_0^T dt'\,t'\,G_{v2}\right) - \exp\left(i\gamma v \int_0^T dt'\,t'\,G_{v1}\right)\right\}\Bigg] \\ & \cdot \exp(-i n k_x x) \exp(-i p k_y y) \end{aligned}$$

$$= \left\{ \exp\left(i\gamma v \int_0^T dt'\, t'\, G_{v2} \right) - \exp\left(i\gamma v \int_0^T dt'\, t'\, G_{v1} \right) \right\}$$
$$\cdot \exp(-i n k_x x) \exp(-i p k_y y)$$
$$\cdot \sum_p \sum_n [\varrho\, 2 \sin(n k_x)/(n k_x)] \cdot \exp(-i n k_x x) \exp(-i p k_y y)$$

But the inverse Fourier transform of a sinc-gauss function is just a step function, hence

$$A(x,y) = \left\{ \exp\left(i\gamma v \int_0^T dt'\, t'\, G_{v2} \right) - \exp\left(i\gamma v \int_0^T dt'\, t'\, G_{v1} \right) \right\} \varrho \quad ; \; y=0$$
$$= 0 \quad ; \; y \neq 0 \qquad \text{(A 9)}$$

Just our vessel at $y=0$ with constant density along x, modified by

$$\left\{ \exp\left(i\gamma v \int_0^T dt'\, t'\, G_{v2} \right) - \exp\left(i\gamma v \int_0^T dt'\, t'\, G_{v1} \right) \right\}$$

If the bipolar gradient in the first acquisition (Fig. 3 a) is zero (fully velocity compensated) and that for the second (Figure 3 b) has a first moment of $\pi/2\gamma V_s$, then

$$A(x,y) = \{\exp(i v \pi/2 V_s) - 1\}\, \varrho$$

or

$$|A(x,y)|$$
$$= \varrho \sqrt{\{\cos(v\pi/2V_s) + i \sin(v\pi/2V_s) - 1\}\{\cos(v\pi/2V_s) - i \sin(v\pi/2V_s) - 1\}}$$
$$= \varrho \sqrt{\{\cos(v\pi/2V_s) - 1\}^2 + i \sin^2(v\pi/2V_s)}$$
$$= \varrho \sqrt{\{\cos^2(v\pi/2V_s) - 2\cos(v\pi/2V_s) + 1\} + \sin^2(v\pi/2V_s)}$$
$$= \varrho \sqrt{2\{1 - \cos(v\pi/2V_s)\}} \qquad \text{(A 10)}$$

so that when $v = V_s$ the magnitude of the vessels in the phase-based subtraction angiogram will be greatest.

Appendix B. Parameter values for Table 1 contrast-to-noise estimates

T_1 (1.5 Tesla) for blood = 200 ms
T_2 (1.5 Tesla) for blood = 60 ms
T_1 (1.5 Tesla) for tissue = 700 ms
T_2 (1.5 Tesla) for tissue = 70 ms

TR for spin echo = 400 ms
TE for spin echo = 20 ms

TR for gradient echo = 20 ms
TE for gradient echo = 10 ms
α = gradient echo excitation pulse angle = 20°

V_s = maximum phase sensitive velocity = 0.4 m/s
v = mean vessel (carotid) blood velocity = 0.2 m/s

n = number of slices for multislice or 3D = 64
d = slice thickness = 2.0 mm
d_t = projection slab thickness = 100.0 mm
d_b = blood vessel (carotid) diameter = 5.0 mm
FOV = 128 mm

N = fraction of noise per unit signal = 0.1

References

1. Axel L (1984) Blood flow effects in magnetic resonance imaging. AJR 143:1157–1166
2. Wehrli FW, Bradley WG Jr (1988) Magnetic resonance flow phenomenon and flow imaging. In: Wehrli FW, Shaw D, Kneeland HB (eds) Magnetic resonance imaging: principles, methodology and applications. VCH, New York, pp 469–519
3. Moran PR, Saloner D, Tsui BMW (1987) Flow imaging by suppression of signal from stationary matter. IEEE Trans Med Imaging 6:41
4. Gao J-H, Holland SK, Gore JC (1988) Nuclear magnetic resonance signal from flowing nuclei in rapid imaging using gradient echoes. Med Phys 15(6):809–814
5. Wedeen VJ, Meuli RA, Edelman RR et al. (1985) Projective imaging of pulsatile flow with magnetic resonance. Science 230:946–948
6. Meuli RA, Wedeen VJ, Geller SC, Edelman RR, Frank LR, Brady TJ, Rosen BR (1986) MR gated subtraction angiography: evaluation of lower extremities. Radiology 159:411–418
7. Macovski A (1982) Selective projection imaging: applications to radiography and NMR. IEEE Trans Med Imaging 1:42–47
8. Moran PR (1982) A flow velocity zeugmatographic interlace for NMR imaging. Magn Reson Imaging 1(4):197–203
9. van Dijk P (1984) Direct cardiac NMR imaging of heart wall and blood flow velocity. J Comput Assist Tomogr 8:429–436
10. Bryant DJ, Payne JA, Firmin DN et al. (1984) Measurement of flow with NMR imaging using a gradient pulse and phase difference technique. J Comput Assist Tomogr 8:588–593
11. Dumoulin CL, Hart HR (1986) Magnetic resonance angiography. Radiology 161:717–720
12. Dumoulin CL, Souza SP, Hart HR (1987) Rapid scan magnetic resonance angiography. Magn Reson Med 5:238–245
13. Dumoulin CL, Souza SP, Walker MF, Wagle WA (1989) Three-dimensional phase-contrast angiography. Magn Reson Med 9:139–149
14. Laub GA, Kaiser WA (1988) MR angiography with gradient motion refocusing. J Comput Assist Tomogr 12(3):377–382
15. Valk PE, Hale JD, Kaufman L, Crooks LE, Higgins CB (1985) MR imaging of the aorta with three dimensional vessel reconstruction. Radiology 157:721–725

16. Hale JD, Valk PE, Watts JC et al. (1985) MR imaging of blood vessels using three-dimensional reconstruction: methodology. Radiology 157: 727–733
17. Hinshaw WS (1976) Image formation by nuclear resonance: the sensitive point method. J Appl Physiol 47: 3709–3721
18. Crooks LS, Mills CM, Davis PD, Brant-Zawadski M, Hoenninger J, Arakawa M, Watts J, Kaufman L (1982) Visualization of cerebral and vascular abnormalities by NMR imaging. The effect of imaging parameters on contrast. Radiology 144: 843–852
19. Bradley WG, Waluch V (1985) Blood flow: magnetic resonance imaging. Radiology 154: 443–450
20. Wehrli FW, MacFall JR, Axel L et al. (1984) Approaches to in-plane and out-of-plane flow imaging. Noninvasive Med Imaging 1: 127–136
21. Frahm J, Merboldt KD, Hanicke W, Gyngell ML, Bruhn H (1988) Rapid line scan NMR angiography. Magn Reson Med 7: 79–87
22. Gullberg F, Wehrli FW, Shimakawa A, Simons MA (1987) MR vascular imaging with a fast gradient refocusing pulse sequence and reformatted images from transaxial sections. Radiology 165: 241–246
23. Groen JP, de Graaf RG, van Dijk P (1988) MR angiography based on inflow (abstract). In: Proceedings, 7th annual meeting of the Society of Magnetic Resonance in Medicine, p 906
24. Dumoulin CL, Cline HE, Souza SP, Wagle WA, Walker MF (1989) Three-dimensional time-of-flight magnetic resonance angiography using spin saturation. Magn Reson Med 11: 35–46
25. Masaryk TJ, Modic MT, Ross JS, Ruggieri PM, Laub GA, Lenz GW, Haake EM, Selman WR, Wiznitzer M, Harik SI (1989) Intracranial circulation: Preliminary clinical results with three-dimensional (volume) MR angiography. Radiology 171: 793–799
26. Masaryk TJ, Modic MT, Ruggieri PM et al. (1989) Three-dimensional (volume) gradient-echo imaging of the carotid bifurcation: preliminary clinical experience. Radiology 171: 801–806
27. Dixon WT, Du LN, Faul DD, Gado M, Rossnick S (1986) Projection angiograms of blood labeled by adiabatic fast passage. Magn Reson Med 3: 454–462
28. Nishimura DG, Macovski A, Pauly JM, Conally SM (1987) MR angiography by selective inversion recovery. Magn Reson Med 4: 193–202
29. Groen JP, van Dijk P, In den Kleef JTE (1987) MR subtraction angography by rapid sequential excitation (abstract). In: Proceedings of the 6th annual meeting of the society of magnetic resonance in medicine, p 868
30. Lanzer P, Bohning D, Groen J, Gross G, Nanda N, Pohost G (1990) Aortoiliac and femoropopliteal phase-based NMR angiography: a comparison between FLAG and RSE. Magn Reson Med 15: 372–385

Chapter 7

Phase-sensitive Angiography

S. P. Souza and C. L. Dumoulin

Nuclear magnetic resonance (NMR) has long been a powerful tool for laboratory chemical analysis, and magnetic resonance imaging (MRI) is now well established as a powerful and flexible tool for clinical diagnosis. Much of the usefulness and richness of magnetic resonance (MR) in medical and nonmedical settings derives from the fact that the MR signal is sensitive to a wide range of physical and chemical parameters.

Since the early days of MR it has been known that motion and flow affect the MR signal, and that this offers an opportunity to use MR for flow measurement [1, 2]. With the advent of MRI and its medical applications, these flow effects were encountered as artifacts degrading conventional stationary tissue images. In recent years many efforts have been undertaken to exploit the motion sensitivity of the MR signal in order to image blood flow in vivo and provide both a research tool and a clinically practical alternative to established but relatively invasive X-ray contrast-medium methods.

Techniques for flow imaging using MR can be generally divided into time-of-flight methods, which rely on a macroscopic displacement of material during the MR procedure, and phase-sensitive methods. Each group has its respective strengths and weaknesses. In this chapter we will discuss techniques which exploit the effect of motion on the phase of the MR signal to image blood flow.

Phase and Motion in MRI

We begin by reviewing the properties of a sample of magnetized protons or, more generally, spins. In an MR procedure the sample is immersed in a uniform static magnetic field with a particular orientation in space. The sample has a net magnetization which may have components in three dimensions (Fig. 1). Customarily, we divide the magnetization of the sample into two components. The longitudinal magnetization is the component parallel to the static magnetic field B_0. It is described by a signed magnitude but, since its direction is fixed by definition, carries no directional information. The transverse magnetization is the component in the plane perpendicular (transverse) to the static field. It is thus a two-dimensional quantity, having both a magnitude and an instantaneous direction (angle, or phase) within the transverse

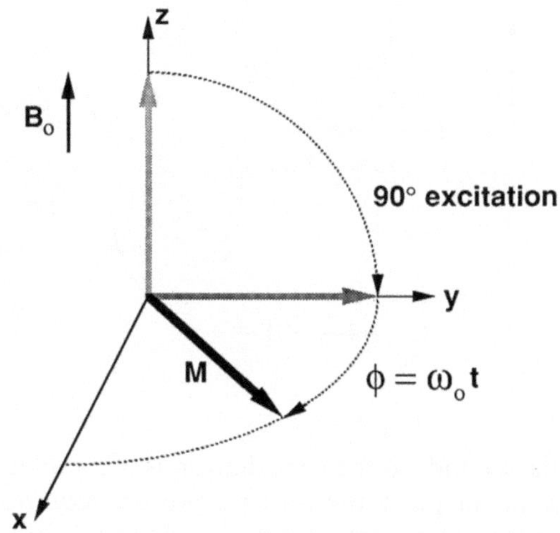

Fig. 1. Model of a sample of protons in a magnetic resonance procedure. Initially, the net magnetization M is aligned with the main magnetic field B_0 along the z axis. After a 90° radiofrequency excitation pulse is applied the magnetization lies in the x-y plane perpendicular to B_0, where it processes at the Larmor frequency ω_0. The phase ϕ is the instantaneous position of the magnetization in the x-y plane

plane. The zero or reference point for this angle is arbitrary, so only differences in phase are meaningful.

As shown in a previous chapter, it is the response of the longitudinal and transverse magnetization of a sample to applied radio frequency energy and linear magnetic field gradients, usually in the form of brief pulses, that makes magnetic resonance imaging possible. In particular, since the resonance frequency of a proton is directly proportional to the magnetic field it is immersed in, magnetic field gradients make the resonance frequency a function of position. The effects of motion on both longitudinal and transverse magnetization are frequently the sources of artifacts in conventional MR imaging, but they can be exploited to produce MR angiograms. Phase-sensitive methods rely on motional effects on transverse magnetization, time-of-flight methods on longitudinal magnetization.

In conventional MR imaging, simple unipolar (single-lobed) magnetic field gradient pulses are used to encode a proton's position by shifting its frequency (as on the readout axis) or phase by an amount dependent on its position. They are also used to provide spatial localization, as in slice selection. To a lesser extent, such pulses induce a phase shift which depends on the proton's motion, although at worst the effect is to cause a displacement of the apparent position of moving material from its actual position.

Consider, however, the bipolar magnetic field gradient pulse shown in Fig. 2. Such pairs of unipolar gradient pulses arise naturally in conventional MR imaging pulse sequences. An extra lobe of opposite sign is customarily added to the slice selection and/or readout pulse to restore phase coherence. For a stationary proton, the first half of the bipolar pair gives rise to a particular position-dependent phase shift. The second pulse of the pair induces an equal and opposite phase shift, canceling the effect of the first pulse. The stationary proton therefore emerges unaffected. A proton moving along the gradient, however, is at a different position during the first pulse than during

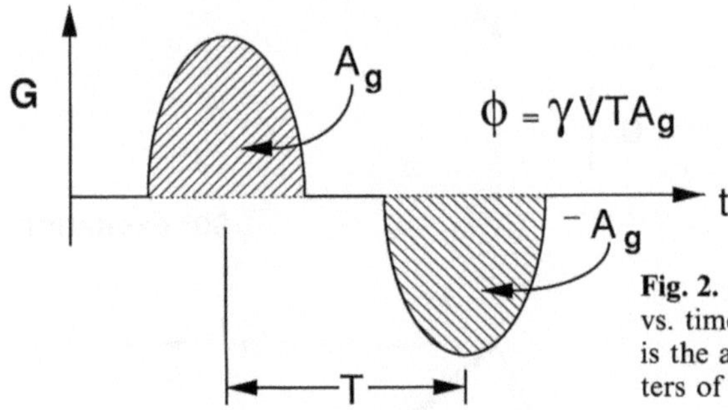

Fig. 2. Plot of magnetic field gradient G vs. time t for a bipolar gradient pulse. A_g is the area under each lobe, and the centers of the lobes are separated by time T

the second, so the cancellation is incomplete. Under the influence of a bipolar gradient pulse the moving proton acquires a net phase shift which, to first order, is proportional to its velocity. It is easily shown [3] that the amount of phase shift induced for a given velocity depends on the strength, duration, and separation of the pulses comprising the bipolar pair. The equation shown in Fig. 2 is approximate: a bipolar gradient pulse produces phase shifts proportional to not only velocity but also acceleration and higher orders of motion. With more lobes in such a composite pulse it is possible to compensate for these extra terms and achieve purer velocity sensitivity, or to enhance them to yield sensitivity to, for example, acceleration [4].

In conventional imaging, the extra phase shift given to moving protons by bipolar portions of the pulse sequence may result in undesirable flow artifacts which can compromise the diagnostic value of the image. One such artifact is a loss of signal in regions containing a wide range of velocities; the resulting wide range in phase shifts can cause cancellation of signals from different parts of a volume element (voxel), reducing the net signal as shown in Fig. 3. The effect is noticeable in regions of complex flow patterns including turbulence, and in laminar flow near the vessel wall where the velocity gradient is steep. Another phase-related artifact is ghosting due to velocity variations during a scan. These artifacts can be mitigated by adding an extra bipolar gradient pulse on the same axis as the first, but reversed in time so that the induced phase shift cancels the first. This technique is referred to as flow compensation, or first-order gradient moment nulling.

Flow-Dephasing Techniques

The signal loss due to flow-induced dephasing referred to in the preceding paragraph forms the basis of several early phase-sensitive techniques for MR flow imaging. In systole/diastole angiography [5] two projection or thick-slice images are acquired, one sampled at a time of rapid and rapidly changing flow and high-velocity gradient near the vessel wall (systole) and one at a relatively quiescent time (diastole). The systole image shows considerable signal loss due

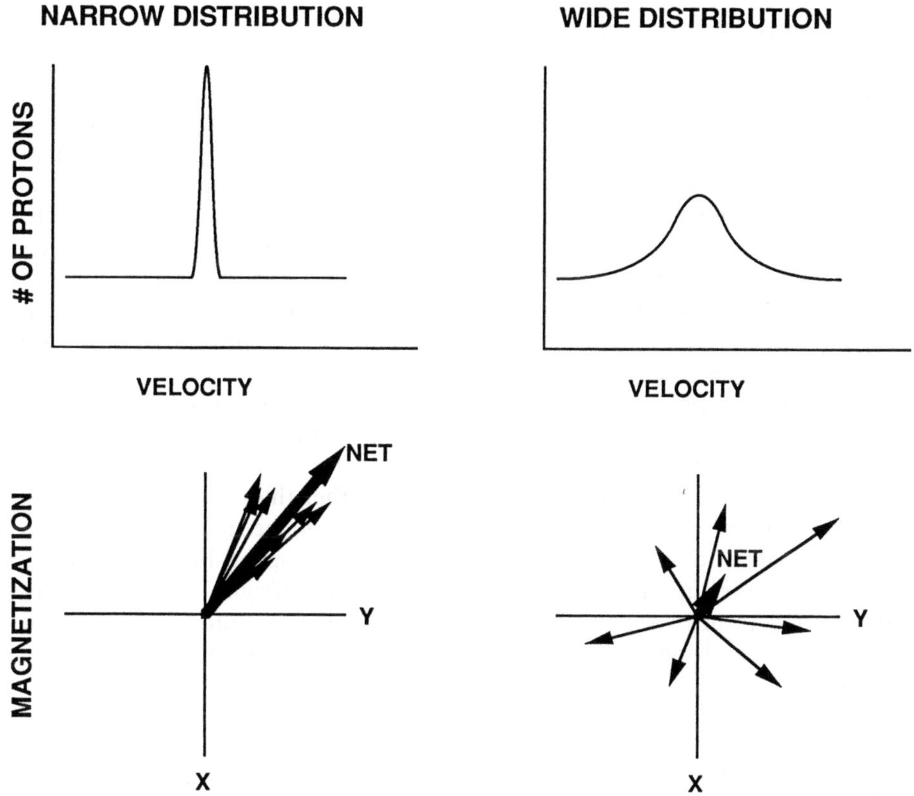

Fig. 3. Effect of velocity distribution within a voxel on the MR signal. *Left:* the distribution of spin velocities is narrow and most of the signal is retained as shown by the *arrow, NET. Right:* the distribution is wide and signal is lost due to cancellation among the velocity components

to dephasing, while the diastole image is less affected. Subtraction of the two separately reconstructed magnitude images yields an angiogram. While useful results have been obtained in the aorta, pelvis and legs, this technique cannot be used for nonpulsatile venous flow.

Another dephasing-based technique involves acquisition of two images, one with flow compensation and one without [6, 7]. In the image without flow compensation, volume elements containing a range of velocities will lose signal due to dephasing, while in the flow-compensated image little or no such signal loss will occur. Again, subtraction of these two magnitude images gives an angiographic image. This technique has been successful to some extent in imaging the carotid bifurcation [7].

In general, flow-dephasing techniques exhibit less-than-optimum flow contrast and suppression of stationary tissue, and they cannot directly give quantitative flow information. At the time of this writing they are not widely used.

Phase-Contrast Flow Techniques

An alternative to flow-dephasing is to directly utilize the phase shifts produced by protons moving along a bipolar gradient. If two data sets are acquired under differing flow-encoding conditions and subtracted, only the flow signal remains. One way to do this is to utilize phase shifts due to bipolar gradients inherent in conventional implementations of slice selection or frequency encoding. Another is to apply a bipolar gradient during a first acquisition and then revese the order of the positive and negative lobes during the second. Data from the two acquisitions are subtracted, before or after Fourier reconstruction, as complex numbers rather than as magnitudes. Quantitative flow data are then directly available in the phase image, and the magnitude image provides an angiographic rendition of the flow [3, 8]. Care must be taken in a phase display to account for and remove phase shifts due to local or global magnetic field inhomogeneities and resonance offsets.

Methods which rely on discrete phase shifts rather than on dephasing have come to be known as phase-contrast methods. In these methods each measurement is sensitive to flow along the direction of the applied bipolar gradient: typically inferior/superior, anterior/posterior, and left/right. Though not always needed to answer the clinical question, total flow information can be obtained by acquiring and combining two or three images sensitive to these flow directions.

Two-dimensional Phase Contrast

Phase-contrast flow imaging has been implemented in many forms, not all of them angiographic. Early versions displayed phase shifts or magnitudes from through-plane or in-plane flow in a thin slice [9–12]. Recent refinements of this technique include phase correction and cine (time-, or cardiac phase-resolved) acquisition and display [13]. Phase-contrast acquisition of angiograms in full projection through the anatomy and displaying magnitude rather than phase images were at first accomplished with cardiac gating [3] in order to obtain consistent flow-induced phase shifts, and later using ungated fast-scanning methods [14]. The basic phase-contrast pulse sequence is shown in Figure 4.

The fast-scan phase-contrast technique reduces exam time and improves image quality. It is related to gradient-refocused imaging methods using low RF excitation flip angles (GRASS, FLASH). Flow detection in phase-contrast imaging is relatively insensitive to the usual MR imaging contrast parameters T_1 and T_2, so maximum signal-to-noise and flow contrast are obtained by minimizing both the repetition time TR (20–30 ms typical) and the echo time TE (9–12 ms typical), and by setting an appropriate flip angle. Since the initial and time-reversed flow encodings are acquired within 20–30 ms of each other, the procedure is relatively insensitive to small aperiodic patient motions; swallowing and other small motions by the subject will usually not degrade the

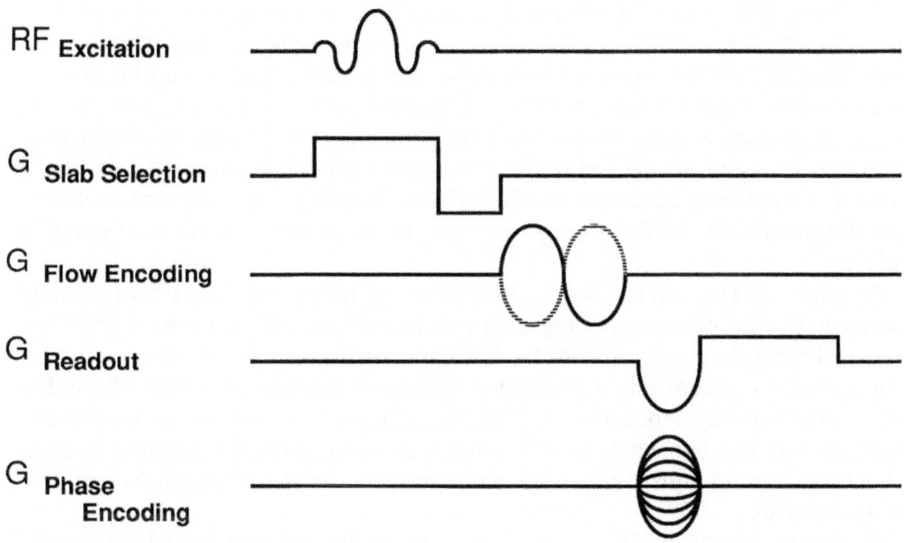

Fig. 4. Basic pulse sequence for phase-contrast MR angiography. The radiofrequency (*RF*) excitation is applied in the presence of a slab selection gradient, and conventional phase-encoding and readout gradients are used as in gradient-echo imaging. A flow-encoding bipolar pulse is applied on any one axis, and the sign of each lobe is inverted on alternate excitations

image noticeably. Figure 5 is an amplitude display of a 2-cm axial slice through the abdomen obtained with the fast-scan phase-contrast technique.

There is, of course, a periodic involuntary motion that can affect the images, namely, pulsatile flow within the vessels being imaged. This pulsatility causes ghost images of the vessels to be distributed along the phase-encoding direc-

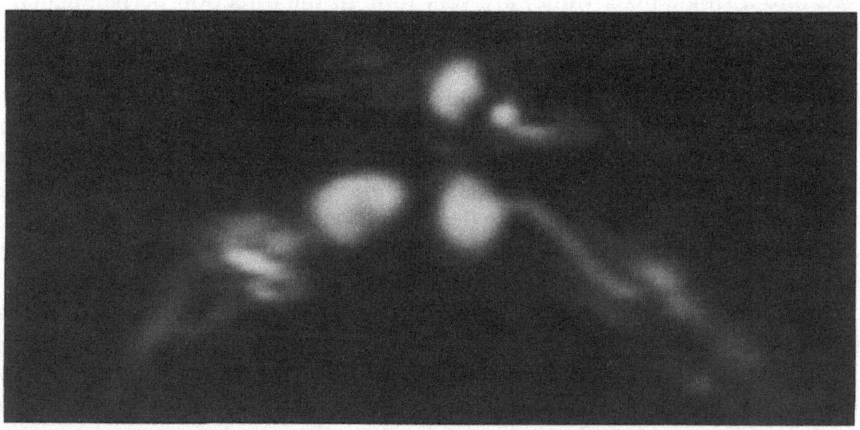

Fig. 5. Phase contrast flow image of a 2-cm axial slice through the abdomen of a normal volunteer. Anterior is *up* in this photo. Images with inferior-superior and left-right flow sensitivity were combined

tion. These ghosts can be effectively eliminated by averaging over a cardiac cycle [14, 15], a procedure sometimes referred to as pseudogating. This is done by setting NEX (the number of excitations per phase-encoding step) times TR to approximately an integer number of cardiac periods, typically one. In this way a high-quality single-flow-direction projective angiogram is obtained in 128 heart beats, or about 2 min; the older gated version required at least twice as many heartbeats. In some cases pulsatility ghosting is not a severe problem, and diagnostically useful angiograms can be acquired without averaging in only 6 s.

Another variant of the basic projective technique again utilizes cardiac gating with one phase-encoding step per heartbeat, but acquires many time frames during the cardiac cycle [16, 17]. Cine, or time-resolved, phase-contrast angiography is essentially a projective analog of the thin-slice cine phase-display techniques described above. It is especially useful in the pelvis and extremities; arterial flow can easily be differentiated on the basis of pulsatility in cases where a suboptimal projective view angle renders the spatial anatomy complex or ambiguous.

Projection phase-contrast techniques are rapid and capable of essentially complete stationary tissue suppression. Good suppression can best be achieved in a system which is relatively free of the effects of eddy currents induced in the magnet structure by the gradients. Eddy currents, which may have components persisting for milliseconds to large fractions of a second after a gradient pulse, produce their own magnetic field gradients within the imaging volume. This has the effect of unbalancing the positive and negative lobes of the bipolar pair, thereby compromising the cancellation of signal from stationary material. One way to reduce the problem is to use actively shielded gradient coils [18].

Projection phase-contrast images are sometimes marred by artifactual signal dropouts, even in patent vessels. This is because, in a direct projection acquisition, the volume elements (within which all portions of the volume are averaged into a single pixel value) are very long in the direction perpendicular to the image plane. Each voxel may thus contain a wide range of velocities, including velocities in opposite directions, which dephase the net magnetization within the voxel, thereby reducing the signal as shown in Fig. 3. This is a problem particularly where vessels overlap in projection and where there are complex flow patterns and vortices, such as in a diseased or even a normal carotid sinus where dephasing effects can mimic or exaggerate true pathology.

The first stratagem for reducing dephasing artifacts is to make the echo time, and therefore the time during which dephasing influences such as complex flow can act, as short as possible. Gradient waveform moment nulling can also be effective, but it has the negative consequence of increasing rather than decreasing TE. Since dephasing is irrecoverable once the data have been acquired, the signal loss associated with it can be reduced by reducing the velocity distribution within a volume element. There are several ways to go about this: all involve adding a third dimension of data acquisition.

The cine projection angiography technique described above does this, in a way, by adding cardiac phase, or time, as the third dimension. Pulsatility in arterial flow implies that the flow patterns which give rise to phase cancellation and signal voids are not necessarily present during the entire cardiac cycle. Thin-slice MR flow imaging experiments in normal volunteers support this [19]. Regions which, because of complex flow, show reduced signal intensity during part or even most of the cycle are bright during others. Examination of the time sequence of images as a cine loop, or combination of these images into a single angiogram (e.g., by selecting for each image pixel location the maximum intensity from among all time frames at that location), will almost always allow differentiation between artifact and pathology. Similarly the nonangiographic technique of time-resolved Fourier flow mapping [19–22] adds both time and velocity resolution, albeit at the sacrifice of one dimension of spatial resolution.

3D Phase-Contrast Angiography

A more direct approach is to add a third dimension of spatial resolution by adding a second spatial phase-encoding axis. By reducing the size of a volume element the range of velocities within each voxel is correspondingly reduced, minimizing signal loss. The resulting technique, known as three-dimensional phase-contrast angiography, has been effective in producing high-quality blood flow images of the head, neck, and abdomen [23–25]. There are other benefits to three-dimensional acquisition. In addition to dephasing by velocity distribution, dephasing signal reduction can occur in regions of local magnetic field inhomogeneities such as are found near tissue – air or tissue – bone interfaces. Again, the small voxel size inherent in 3D acquisition reduces the effect of such dephasing. As in projective fast-scan phase-contrast imaging, pseudogating is effective in reducing pulsatility ghosts, with the number of secondary phase encodings acting in place of NEX.

Unlike the direct projection methods, it is not necessary to know the desired projection direction ahead of time. The 3D phase-contrast angiogram contains information over the entire acquired volume, and using computer postprocessing these data can be presented from any angle. A volume of interest may be defined to limit the data considered in postprocessing: this can greatly speed the processing and analysis of the data and can be used to remove vessel structures that might otherwise obscure the view of the vessels in question.

The 3D phase-contrast exam typically consists of three scans, one for each of three mutually perpendicular flow directions, producing three raw data sets. These scans may be interleaved or sequential and contiguous. After Fourier transformation there are three sets of (typically 32–128) slices. These three sets are combined on a voxel-by-voxel basis to form one set of slices containing total (i.e., independent of flow direction) flow information. This information may be presented in many forms. The most detailed information is obtained by inspecting the individual slices, but this can be tedious and it is difficult to appreciate large-scale structure in such a presentation. Other possibilities include shaded 3D surface presentation and holograms. Projection postprocessing has so far proven to be the presentation of choice because it produces a display familiar to the angiographer. The three-dimensional vascular structure is usually seen to best advantage when projected images at a series of view angles are displayed in rapid succession as if the subject were rotating or rocking back and forth.

There are, in principle, many algorithms for generating projection images from three-dimensional data. We will not consider here the details of casting imaginary rays through a three-dimensional volume. There is, however, a choice of how to generate a projection pixel value from the voxel intensities along each ray, or line-of-sight. One obvious choice is to take the average of all voxel values along each ray; this linear process produces an image with a high signal-to-noise ratio but low contrast. At the other extreme we may take the maximum voxel value along each ray, a process which has come to be called maximum intensity projection (MIP). This method produces images with high contrast but a lower signal-to-noise ratio [26]. Because of its computational simplicity and speed, as well as its high-contrast characteristics, MIP has become the most commonly used projection algorithm for 3D MR angiography. This is especially true for the time-of-flight methods discussed elsewhere in this book, in which MIP is an important factor in obtaining useful flow vs. stationary tissue contrast. Nevertheless, MIP has some important limitations related to its rather incomplete use of the available data. Alternative projection algorithms have been proposed [23, 27] and are being evaluated.

Typical Protocols

Research on and clinical trials of phase-contrast MR angiography are still underway, so we make no attempt here at a comprehensive list of protocols. We will discuss a small sample of protocols which have been used and which show promise of clinical usefulness at the time of this writing.

Saturation of material adjacent to the imaging volume is not a fundamental part of the techniques discussed above, but it has proven useful in their application to particular clinical situations. Saturation is achieved by, prior to each excitation of the imaging volume, applying a high-flip-angle RF pulse to a volume offset from the imaging volume and intentionally dephasing the magnetization produced by this pulse. The desired effect is that protons entering the imaging volume from one or more directions can be selectively suppressed in the angiogram. In favorable geometries, this can be used to remove arterial or venous flow.

Head and Neck

Despite its drawback related to dephasing in long voxels, the projective phase-contrast method still has a potential role in fast evaluation of vessel patency

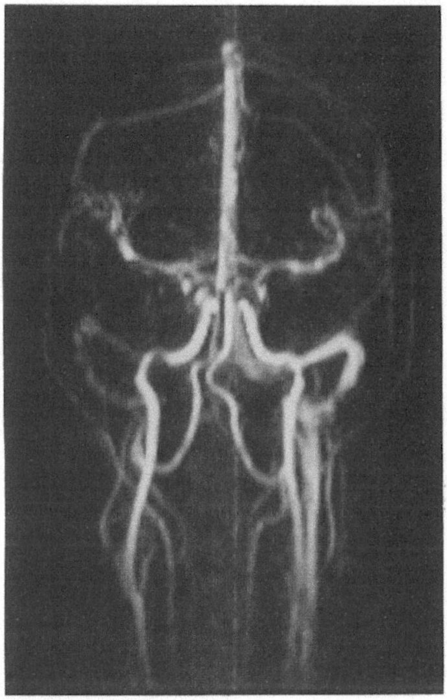

Fig. 6. Projective phase-contrast angiogram of the head of a normal volunteer, showing combined inferior-superior and left-right flow. Total acquisition time was approximately 4 min

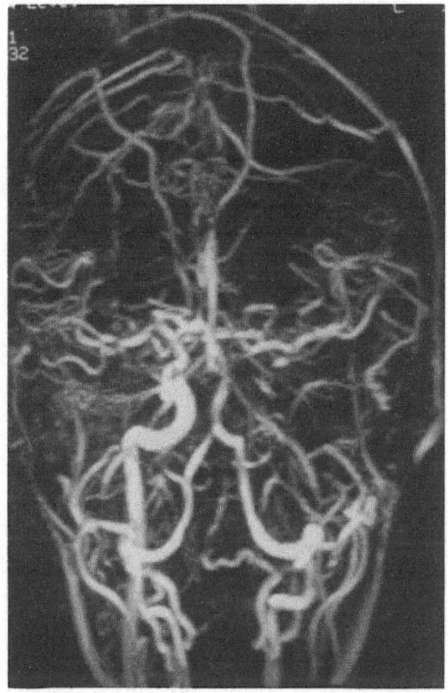

Fig. 7. Three-dimensional phase-contrast angiogram of the head, obtained at Michigan State University. This projection image from a 128-slice data set demonstrates superior sagittal sinus thrombosis and an occluded internal carotid artery

in the head and neck, and as a quick survey tool for medium-to-large aneurisms or arteriovenous malformations. Such a protocol typically entails thick axial slab excitation, anterior-posterior or lateral projection, a field of view of 20–24 cm, flip angle 20°, NEX approximately 20–30 depending on heart rate, and one flow direction for a total exam time of about 2 min. An example combining information from two flow directions is shown in Fig. 6.

A much more detailed data set is obtained by using a whole-head 3D phase-contrast exam. The protocol is similar to the above, except for $256 \times 128 \times 128$ 3D acquisition with a cubic field of view, nonselective (whole-head) excitation, NEX = 2, and three flow directions for a total exam time of 36 min. Postprocessing is performed as required. Figure 7, obtained at Michigan State University, is representative of this protocol.

For the highest contrast, resolution, and freedom from signal loss in localized regions such as the circle of Willis, a restricted-volume 3D phase-contrast protocol with small, cubic voxels may be used. The exam is similar to the previous one, except that the field of view in the axial (Z) direction is reduced

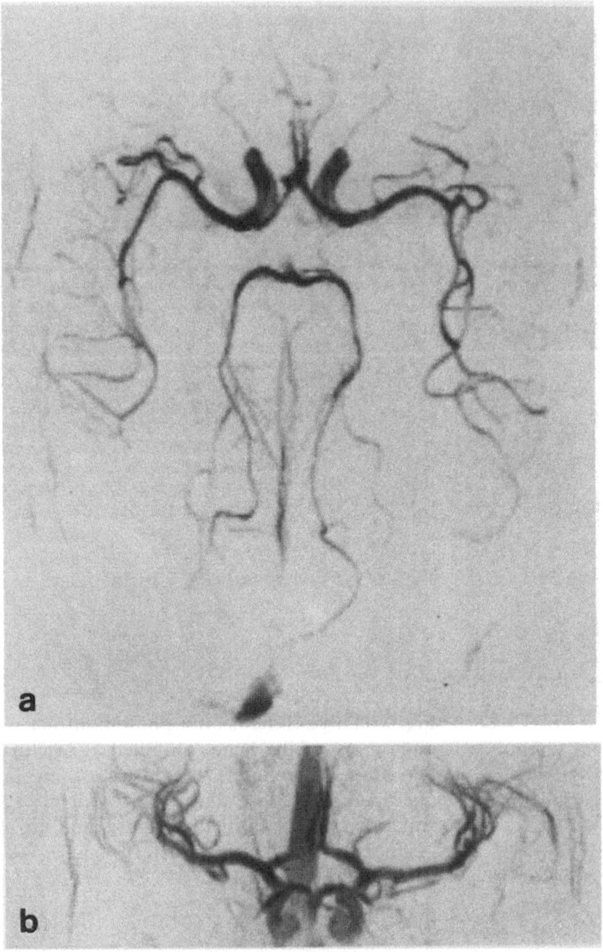

Fig. 8 a, b. Three-dimensional phase-contrast angiogram of the region encompassing the circle of Willis of a normal volunteer. Three sets of data sensitive to orthogonal flow directions were acquired and combined. Voxels are cubic and 0.7 mm on a side. **a** Anterior-posterior projection; **b** inferior-superior projection

to 5 or 6 cm with a corresponding restriction of the excitation slab, the acquisition matrix is $256 \times 256 \times 64$, and the flip angle is increased to about 30°. Spatial resolution is isotropic with cubic voxels 0.7 mm on a side. Figure 8 is an example of this protocol.

Abdomen

For evaluation of the inferior vena cava, portal, renal, splenic and hepatic veins, and possibly the splenic and renal arteries, a 3D phase-contrast protocol

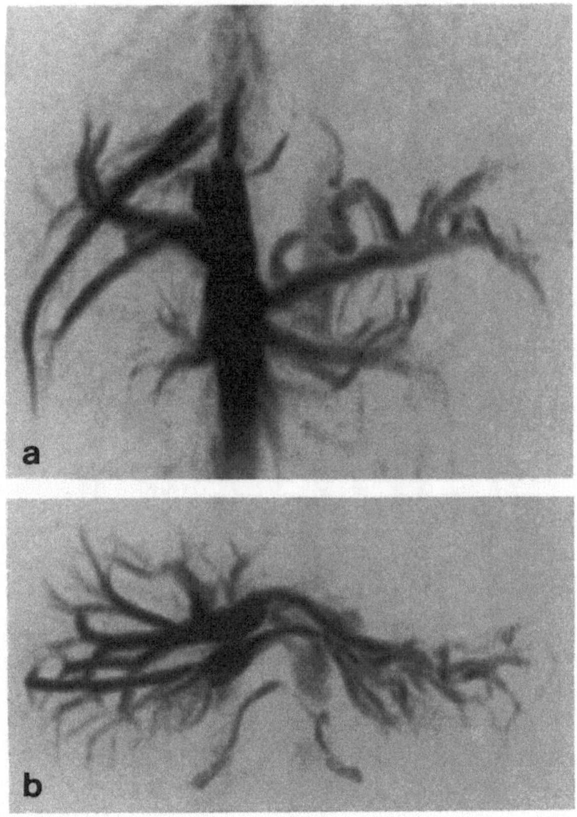

Fig. 9a, b. Three-dimensional phase-contrast angiogram of the abdomen of a normal volunteer. Three sets of data sensitive to orthogonal flow directions were acquired and combined. The field of view is 40 cm. **a** Anterior-posterior projection; **b** inferior-superior projection

is useful [25]. This is similar to the whole-head 3D protocol above, except that the field of view is 32 or 40 cm with suitably limited axial excitation, and for studies of venous flow saturation is used inferior to the imaging volume to suppress signal and pulsatility ghosting from the aorta. An example is shown in Fig. 9.

Extremities

In the legs arterial flow is highly pulsatile, with a significant period of retrograde flow following systole, so a time-resolved protocol is most effective. Thick axial slab excitation is used in a 40-cm field of view, with a flip angle of 15° to 20°, NEX=2, and 20–32 time frames per cardiac cycle. The time frames are viewed in rapid sequence on an image processor screen. Optionally, the time frames may be treated as a stack of slices, and via postprocessing an image containing only steady flow (veins) or only pusatile flow (arteries) may be derived.

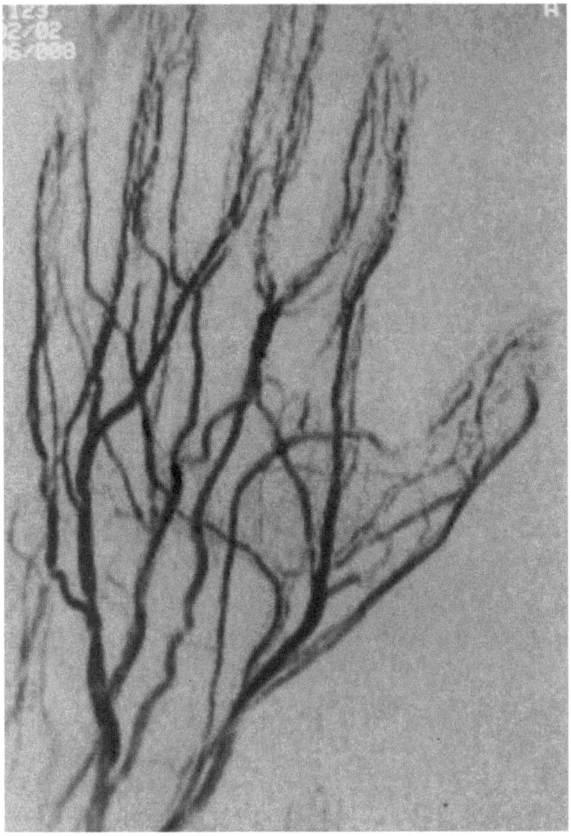

Fig. 10. Projective phase-contrast angiogram of the hand of a normal volunteer. Both directions of flow in the plane of the hand were obtained and combined

Arteriography and venography in the hand is a challenging application. Figure 10 is a two-flow-direction projection phase contrast image of the hand, with a field of view of 16 cm and NEX = 20. Note that many of the vessels seen here are significantly smaller than the image-plane voxel size.

Advantages and Disadvantages

All MR flow-imaging techniques are incomplete or flawed in one respect or another. Since there is such a wide range of phase-sensitive techniques it is difficult to gauge their strengths and weaknesses as a whole. Nevertheless, it can be generally said that an important advantage of phase-contrast MR flow imaging is that when properly executed the *only* source of contrast (i.e., bright pixels) is motion. If a vessel appears to have flow in it, it does. This is not necessarily the case for some time-of-flight techniques, notably those based on inflow of unsaturated protons [27–31]. In these techniques high signal is produced in regions where the magnetization is predominantly longitudinal;

this may be due to rapid relaxation (short T_1) or inflow of fully relaxed spins to the excitation region. Flow-enhancement techniques are in fact short-apparent-T_1 imaging and can give a false-positive result for flow in a vessel, dissection, or aneurism containing short-T_1 material [32].

Phase contrast is capable of imaging even very slow flow, limited only by signal-to-noise and the strength and duration of the flow-encoding bipolar gradient. Additionally, all phase-contrast techniques, whether angiographic or thin-slice oriented, are directly or potentially quantitative [13, 33]. Phase-contrast angiography is, in principle, capable of complete suppression of stationary tissue, and in a sufficiently stable imaging system this is effectively realized.

Finally, the variants of phase-contrast angiography offer the possibility, with nonstandard but straightforward reconstruction, of obtaining both flow and stationary tissue images from one scan, simultaneously and with exact spatial correspondence between the flow and nonflow images [23]. Time-of-flight techniques suppress the stationary tissue by manipulating its longitudinal magnetization in a way that does not permit recovery of a high-quality conventional MR image.

There are, of course, significant drawbacks to phase-contrast methods. One is that, as described earlier, each acquisition is sensitive to flow along one axis; a complete measurement of flow requires three acquisitions. A full 3D phase-contrast exam with $256 \times 128 \times 128$ resolution requires 20–30 min, compared with approximately 12–15 min for a 3D time-of-flight exam at the same resolution. The difference is not as great as one might expect, because while phase contrast benefits from the shortest possible TR, time-of-flight techniques require some time for inflow of material and are usually run with a TR of 50 ms or more. Combining these three images or sets of images into a total flow image adds to processing complexity and time. In some applications the flow direction is known and this is less of a problem.

A more important limitation is the signal loss due to dephasing caused by velocity distributions within a voxel. This is a greater problem in phase-contrast than in inflow methods: all gradient waveforms in a pulse sequence for either technique can be flow compensated *except* the flow-encoding bipolar gradient required for phase contrast. Signal loss due to susceptibility effects, acceleration, or higher orders of motion are equally problematical in time-of-flight and phase-sensitive methods. While the measures described above for reducing dephasing effects (especially 3D acquisition) are reasonably effective, some signal loss often remains and can yield a false-positive result for stenosis; experience and thoroughness in choosing a protocol and examining the images can reduce the chance of error.

Specific phase-contrast variants have strengths and weaknesses that derive by and large from the method of data acquisition. For example, direct projective phase-contrast angiography without cardiac gating is perhaps the fastest MR angiography technique, often capable of demonstrating vessel patency in a single-flow-direction scan as short as 6 s without averaging. Even using averaging over the cardiac cycle to minimize pulsatility ghosts and obtaining all three flow directions, a scan can be completed in 6 min with no need for retrospective projection. In some anatomical regions such as the carotid bifurcation, however, ungated projection phase contrast is highly prone to dephasing signal loss. Direct projective time-of-flight techniques using principles other than inflow enhancement are known [34, 35]; these have a longer minimum scan time than phase contrast and are effective only over a small field of view, but they do a better job of delineating vessel goemetry in regions of complex flow.

Because time-of-flight effects persist for times on the order of T_1, i.e., a few tenths of a second, cardiac-phase-resolved time-of-flight angiography may not be practical and has not yet been demonstrated. The cine variant of phase contrast is effective in reducing pulsatility ghosting and differentiating arterial from venous flow. The time sequence of flow velocities can itself be used as a diagnostic tool, for example in the case of aortic regurgitation. On the negative side, cine acquisitions take longer than ungated scans, and poor results may be obtained when the heart rate is relatively inconstant during the scan.

Three-dimensional phase-contrast angiography appears at this time to be the most diagnostically useful variant, especially in the head and abdomen. Its principal advantage is that it can render high contrast and fine detail, even over a large field of view. If the field of view is restricted, voxels as small as 0.7 mm cubed have been obtained, giving excellent fine-vessel definition and potentially revealing aneurisms as small as 1–2 mm. Arterial and venous flow are imaged equally well, although in favorable geometries one or the other can be suppressed by use of saturation as described above. The principal disadvantage of 3D phase contrast is the long acquisition time: as much as 30 min for three flow directions and 128 slices. In the head and abdomen it is not usually advisable to omit one or two flow directions. Reconstruction is more complex than in other MR angiography methods. Postprocessing is also time consuming but, in principle, no more so than for other volumetric methods; this situation is rapidly being improved by application of commercially available computing resources.

Conclusion

Phase-sensitive magnetic resonance angiography and related flow-imaging techniques offer unique opportuniites for in vivo evaluation of vessel morphology and function. While challenges remain in applying these methods successfully in regular clinical practice, their flexibility and potential for flow quantitation make them competitive with, and in some cases complementary to, other MR flow-imaging techniques.

References

1. Singer JR (1959) Blood flow rates by nuclear magnetic resonance. Science 130:1652–653
2. Hahn EL (1960) Detection of sea-water motion by nuclear precession. J Geophys Res 65:776–777

3. Dumoulin CL, Hart HR (1986) Magnetic resonance angiography. Radiology 161:717–720
4. Xiang OS, Nalcioglu O (1986) A formalism for generating motion-related encoding gradients in NMR imaging. In: Proceedings of the 5th annual meeting of the Society of Magnetic Resonance in Medicine, Montreal, p 100–101
5. Wedeen VJ, Meuli RA, Edelman RR, Frank LR, Brady TJ, Rosen BR (1985) Projective imaging of pulsatile flow with magnetic resonance. Science 230:946–948
6. Axel L, Morton D (1987) A method for imaging blood vessels by phase-compensated/uncompensated difference images. J Comput Assist Tom 12:31–34
7. Masaryk TJ, Ross JS, Modic MT, Lenz G, Haacke EM (1988) Carotid bifurcation: MR imaging, Radiology 166:461–466
8. Wedeen VJ, Rosen BR, Buxton R, Brady TJ (1986) Projective MR angiography and quantitative flow-volume densitometry. Magn Reson Med 3:226–241
9. Bryant DJ, Payne JA, Firmin DN, Longmore DB (1984) Measurement of flow with NMR imaging using a gradient pulse and phase difference technique. J Comput Assist Tomogr 8:588–593
10. Wedeen VJ, Rosen BR, Chesler D, Brady TJ (1985) MR velocity imaging by phase display. J Comput Assist Tomogr 9:530–536
11. O'Donnell M (1985) NMR blood flow imaging using multiecho, phase-contrast sequences. Med Phys 12:59–64
12. Moran PR, Moran RA, Karsteadt N (1985) Verification and evaluation of internal flow and motion: true magnetic resonance imaging by the phase gradient modulation method. Radiology 154:433–441
13. Pelc NJ, Shimakawa A, Glover GH (1989) Phase contrast cine MRI. In: Proceedings of the 8th annual meeting of the society of Magnetic Resonance in Medicine, Amsterdam, p 101
14. Dumoulin CL, Souza SP, Hart HR (1987) Rapid scan magnetic resonance angiography. Magn Reson Med 5:238–245
15. Haacke EM, Lenz GW, Nelson AD (1987) Pseudogating: elimination of periodic motion artifacts in MRI without gating. Magn Reson Med 4:162–174
16. Souza SP, Dumoulin CL (1987) Dynamic magnetic resonance angiography. Dyn Cardiovasc Imaging 1:126–132
17. Dumoulin CL, Souza SP, Walker MF, Yoshitome E (1988) Time-resolved magnetic resonance angiography. Magn Reson Med 6:275–286
18. Roemer PB, Edelstein WA, Hickey JS (1986) Self shielded gradient coils. In: Proceedings of the 5th annual meeting of the society of Magnetic Resonance in Medicine, Montreal, p 1067
19. Souza SP, Steinberg FL, Caro C, Dumoulin CL, Yucel EK (1989) Velocity- and cardiac phase-resolved MR flow imaging. In: Proceedings of the 8th annual meeting of the Society of Magnetic Resonance in Medicine, Amsterdam, p 102
20. Grover T, Singer JR (1971) NMR spin-echo flow measurements. J Appl Phys 42:938–940
21. Feinberg DA, Crooks LE, Sheldon P, Hoenninger J, Watts J, Arakawa M (1985) Magnetic resonance imaging the velocity vector components of fluid flow. Magn Reson Med 2:555–566
22. Hennig J, Muri M, Brunner P, Friedburg H (1988) Quantitative flow measurement with the fast Fourier flow technique. Radiology 166:237–240
23. Dumoulin CL, Souza SP, Walker MF, Wagle W (1989) Three-dimensional phase-contrast angiography. Magn Reson Med 9:139–149
24. Pernicone JR, Siebert JE, Potchen EJ, Pera A, Dumoulin CL, Souza SP (1990) Three dimensional phase-contrast MR angiography in the head and neck. Am J Neuroradiology, volume 11, page 457 through?
25. Vock P, Terrier F, Wegmuller H, Strauch E, Souza SP, Dumoulin CL (1989) MR angiography of abdominal vessels. In: Proceedings of the 8th annual meeting of the Society of Magnetic Resonance in Medicine, Amsterdam, p 1012

26. Cline HE, Dumoulin CL, Lorensen WE, Souza SP, Adams WJ (1990) Connectivity algorithms for MR angiography. Proceedings of the 9th annual meeting of the Society of Magnetic Resonance in Medicine, New York, p 61
27. Amartur SC, Masaryk TJ, Modic MT, Ross JS, Ruggieri PM, Haacke EM, Laub GA (1989) 3DFT time-of-flight magnetic resonance angiography. Dynamic Cardiovasc Imaging 2:170–177
28. Ruggieri PM, Laub GA, Masaryk TJ, Modic MT (1989) Intracranial circulation: pulse sequence considerations in three-dimensional (volume) MR angiography. Radiology 171:785–791
29. Dumoulin CL, Cline HE, Souza SP, Wagle WA, Walker MF (1989) Three-dimensional time-of-flight magnetic resonance angiography using spin saturation. Magn Reson Med 11:35–46
30. Groen JP, de Graaf RG, van Dijk P (1988) MR angiography based on inflow. In: Proceedings of the 7th annual meeting of the Society of Magnetic Resonance in Medicine, San Francisco, p 906
31. Keller PJ, Drayer BP, Fram EK, Williams KD, Dumoulin CL, Souza SP (1989) MR angiography with two-dimensional acquisition and three-dimensional display. Radiology 173:527–532
32. Grist TM, Boyko OB (1989) Spritzer CE, MacFall JR, Dumoulin CL, Souza SP, Keller PJ (1989) Integration of MRI with MR angiography: comparison of projection phase-contrast techniques and flow related enhancement sequences for neurologic imaging. In: Proceedings of the 75th annual meeting of the Radiological Society of North America, Chicago, 1989
33. Walker MR, Souza SP, Dumoulin CL (1988) Quantitative flow measurement in phase-contrast magnetic resonance angiography. J Comput Assist Tomogr 12:304–313
34. Wehrli FW, Shimakawa A, MacFall JR, Axel L, Perman W (1985) MR imaging of venous and arterial flow by a selective saturation-recovery spin-echo (SSRSE) method. J Comput Assist Tomogr 9:537–545
35. Nishimura DG, Macovski A, Pauly JM, Conolly SM (1987) MR angiography by selective inversion recovery. Magn Reson Med 4:193–202

Chapter 8

Three-dimensional Inflow MR Angiography

G. LAUB

Introduction

The understanding of flow phenomena has always been an important area of research in magnetic resonance imaging. On the one hand, flow phenomena are responsible for a number of artifacts which can drastically impair the diagnostic value of the images; on the other hand, there has always been the hope that flow phenomena can be used to develop noninvasive techniques for vascular diagnosis without using contrast agents. Thus, MRI has developed in two directions: an effort to reduce the sensitivity to flow effects in conventional MRI and efforts to display moving fluids, particularly blood. The first goal has been achieved by developing gating methods, motion-compensation techniques [1], and spatial presaturation [2]. Techniques to achieve the second goal in a variety of anatomical regions are still being actively researched using a number of different applications of the same basic principles. The purpose of this chapter is to present the principles of three-dimensional inflow MR angiography, a method which, according to current experience, has the potential to become clinically useful technique [3, 4].

Basic Physics

As discussed in chapters 6 and 7 the most techniques of magnetic resonance angiography are based on: inflow enhancement or phase-shift effects. Inflow enhancement is due to the motion of spins, with a different history, into the slice or volume to be investigated [5, 6]. Phase-shift effects are related to the motion of spins along the direction of magnetic field gradient that are used for the spatial encoding of the spins [7, 8].

Both effects have been described extensively in chapters 6 and 7 and therefore will be summarized only briefly within this context. Figure 1 demonstrates the principle of the inflow phenomenon. The blood flow is assumed to be perpendicular to the imaging plane (or volume in the case of a 3D study). For repetition times shorter than the longitudinal relaxation time T_1, the signal of stationary tissue will be reduced due to partial saturation effects. Blood flow will move spins from outside the slice which have not been subjected to the spatially selective radiofrequency pulses into the imaging volume. These unsat-

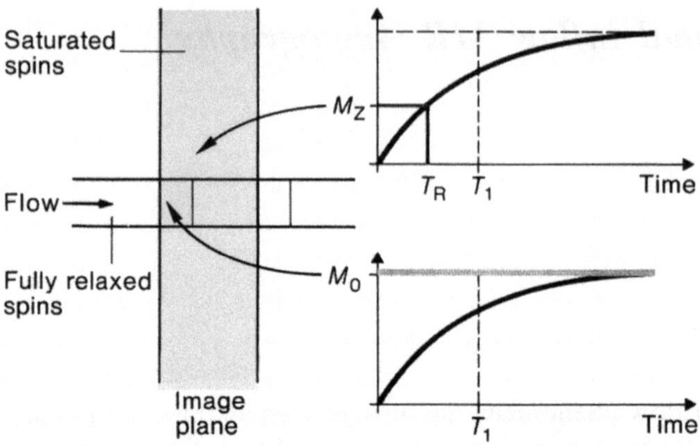

Fig. 1. Basic configuration for demonstration of inflow effects. Longitudinal magnetization (M_z) of spins in the image plane is reduced due to partial saturation for $TR < T_1$. Inflowing spins have equilibrium magnetization (M_0) and will therefore produce more signal intensity

urated or fully relaxed spins have full equilibrium magnetization and therefore, upon entering the slice, will produce a much stronger signal than stationary spins. This effect has also been referred to as "entry slice phenomenon".

The second flow effect – flow-induced phase shifts – is a consequence of the phase memory of the spin system. When traveling along the direction of magnetic field gradients, the excited spins will experience additional phase shifts which depend on the motion of the spins. In general, these flow-induced phase shifts cause many of the well-known flow artifacts, in particular for nonconstant flow situations, such as in the arteries. The physical origin of the flow-induced phase shifts lies in the fact that moving spins, as long as magnetic field gradients are on, will experience different local magnetic field strenghts on their way through the magnet. As a result, a net phase shift will be accumulated depending on the spin's velocity and gradient timing.

Some methods for magnetic resonance angiography make use of these flow-induced phase shifts to allow the display of blood vessels. Moreover, it is also possible to evaluate the flow velocity in blood vessels, using the known time dependence of the magnetic field gradients [9]. A problem common to all of these methods is their inherent sensitivity to pulsatile flow in the arteries. In order to obtain consistent data it is generally necessary to use cardiac gating for data acquisition. This need for cardiac synchronization severely increases the data acquisition time, particularly if three-dimensional data are to be obtained.

The problem of pulsatile flow can be reduced to an acceptable degree by using flow-compensation techniques in the applied sequences. The purpose is to compensate for the flow-induced phase shifts by using additional gradient pulses in such a way that all spins moving at a constant velocity will be

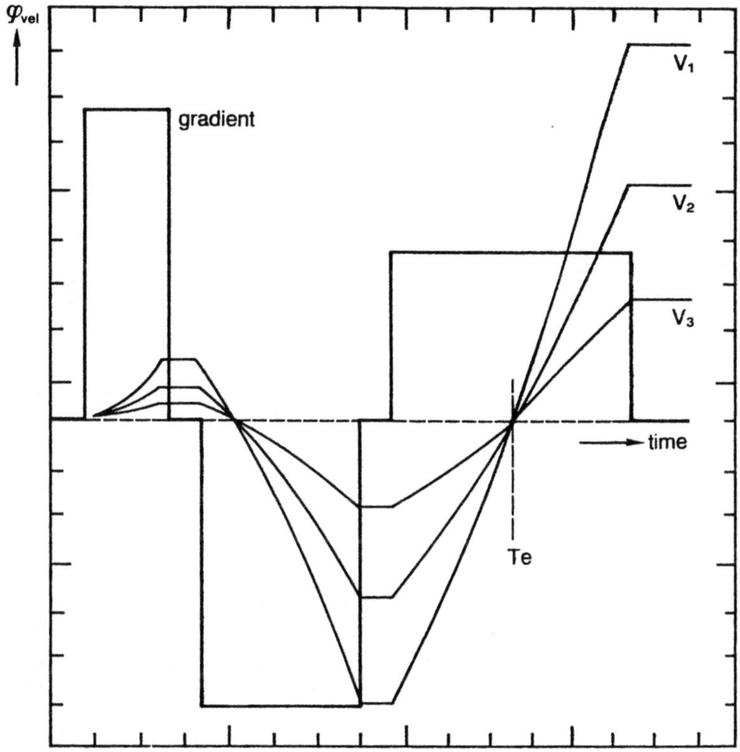

Fig. 2. Evolution of the phase of transverse magnetization for spins moving at three different velocities. The gradient waveform is flow compensated, resulting in spin refocusing at the time of the echo, independent of the actual velocity

rephased at the echo. Figure 2 shows a typical solution for a first-order flow-compensated gradient waveform. The evolution of the phase of the spin's transverse magnetization is shown for three different velocities. As can be seen from Fig. 2, the flow-induced phase shifts will be zero at the time of the echo independent of the flow velocity. The gradient waveform shown in Fig. 2 represents the readout gradient of a gradient-echo sequence (e.g., FLASH or FISP sequence). A sequence like this will be called flow compensated along the readout direction. Likewise, flow compensation can be obtained along the slice-select or phase-encoding direction. With this type of pulse sequence it is possible to eliminate most of the flow artifacts related to the phase-shift phenomenon without cardiac sychronization. The inflow phenomena described before can be employed for an enhanced display of blood vessels only if the flow-induced phase shifts are completely compensated for, as demonstrated in Fig. 3. The large blood vessels in the neck are shown in an axial slice, with and without flow compensation. In both cases signal enhancement due to inflow of unsaturated spins is visible. Without flow compensation, however, the pul-

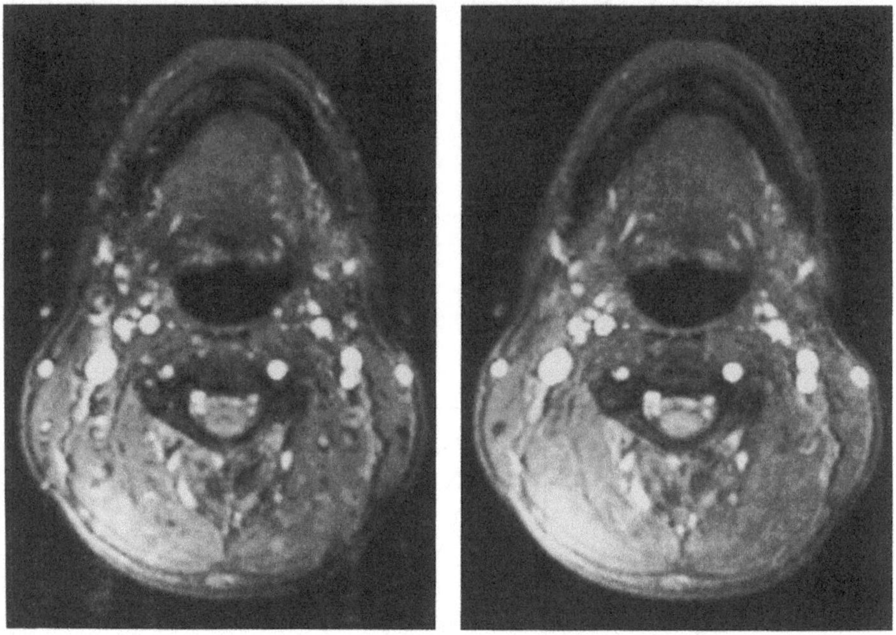

Fig. 3. Axial slice of the neck vessels. Without flow compensation (*left*) severe flow artifacts are observed. With flow compensation (*right*) the vessel lumen can be identified correctly

satile nature of the blood flow causes severe artifacts, in this case smearing and ghosting of the blood signal along the phase-encoding direction. Only if flow compensation is used will a correct interpretation of the vessel lumen be possible.

In principle, with this technique any vessel segment can be imaged by cutting through the vessel perpendicularly to the flow direction. With repetitive increments of the slice position an image of the complete vessel tree can be reconstructed. Therefore, this method is particularly suitable for relatively straight vessel segments, as indicated in Fig. 4. Still, good resolution will require relatively narrow slices with the corresponding loss in signal-to-noise ratio. In practice, 2-mm cuts with a pixel size of 1×1 mm are possible using this two-dimensional technique.

Improvement of the spatial resolution without a reduction of the signal-to-noise ratio can be achieved with three-dimensional imaging techniques. As shown in Fig. 4, the whole volume is excited simultaneously and will then be subdivided into thin partitions or slices by using an additional phase-encoding scheme in the slice-select direction. Unlike in 2D imaging, where the slice resolution is defined by the excitation profile of the radio-frequency pulse, the slice resolution is defined by spatially encoding magnetic field gradients and can be less than 1 mm. In 3D or volume imaging complete compensation of the flow-induced phase shifts is also necessary to avoid flow artifacts in the form

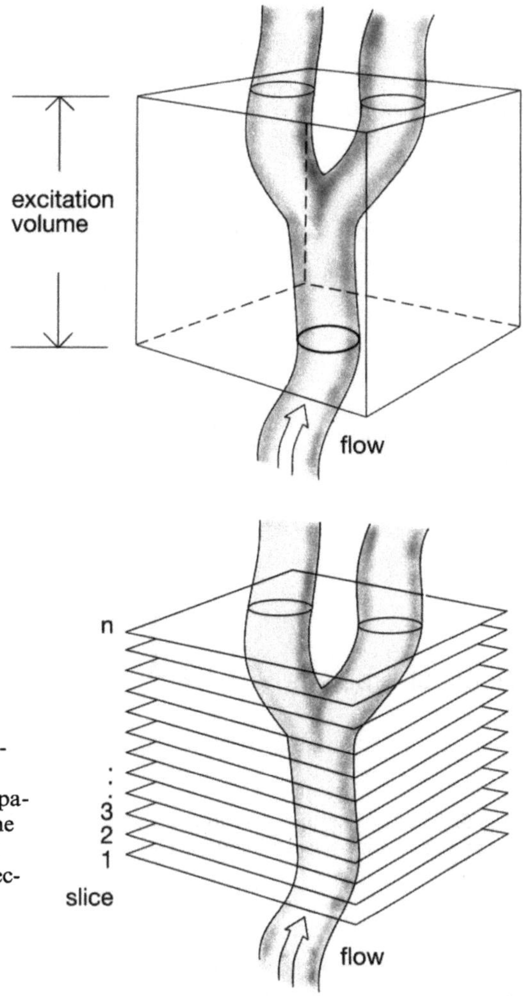

Fig. 4. Two-dimensional (*left*) vs three-dimensional (*right*) acquisition techniques. In 2D each slice is acquired separately, whereas in 3D the whole volume is excited at once and then subdivided into single partitions (or slices) by a second phase-encoding gradient

of signal loss and ghosting. For this purpose flow compensation is applied in conjunction with short echo times. The reason for using the shortest possible echo times is related to deviations from constant spin velocities, which again causes additional phase shifts. These phase shifts caused by higher-order motion of the spins (such as acceleration or turbulence) are most effectively reduced by shortening the echo times. A typical pulse sequence is shown in Fig. 5. The FISP sequence shown is flow compensated in the readout and slice-select directions. With gradients up to 10 mT/m, echo times of about 7 ms have been realized. Similar to two-dimensional sequences but with a better spatial resolution, the signal enhancement due to inflow of unsaturated spins can be observed. As an example, Fig. 6 shows a representation of the carotid bifurcation with a slice thickness of 1 mm and a pixel size of 0.8×0.8 mm. The orientation of the images shown is axial. The total volume covered simulta-

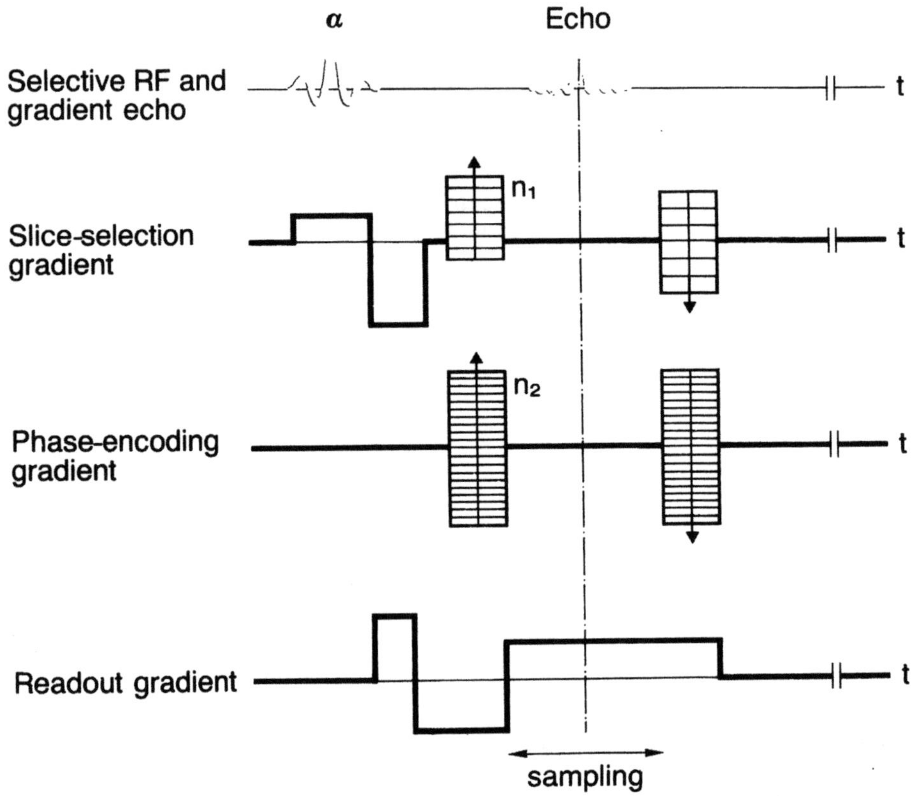

Fig. 5. Timing diagram of a three-dimensional FISP sequence with flow compensation in the readunt and slice-select directions

neously with the three-dimensional technique is 64 × 200 × 200 mm, using a matrix size of 256 × 256 × 64 slices. The acquisition time TA for this type of sequence is calculated according to

$$TA = TR \times N1 \times N2 \times m$$

where TR denotes the pulse repetition time of the sequence, and N1 and N2 are the number of phase-encoding steps in two orthogonal directions; m represents the number of excitations per phase-encoding step, which typically can be set to one for high-field imaging systems, such as 1.0 T or 1.5 T. For lower field strengths one has to compensate for the intrinsically lower signal-to-noise ratio related to the reduced magnetic field strength by using more than one excitation. In all of the examples shown here we have used one excitation and repetition times between 20 and 40 ms. The typical matrix size was 64 × 256 × 256 or 64 × 256 × 192, resulting in acquisition times between 4 and 11 min.

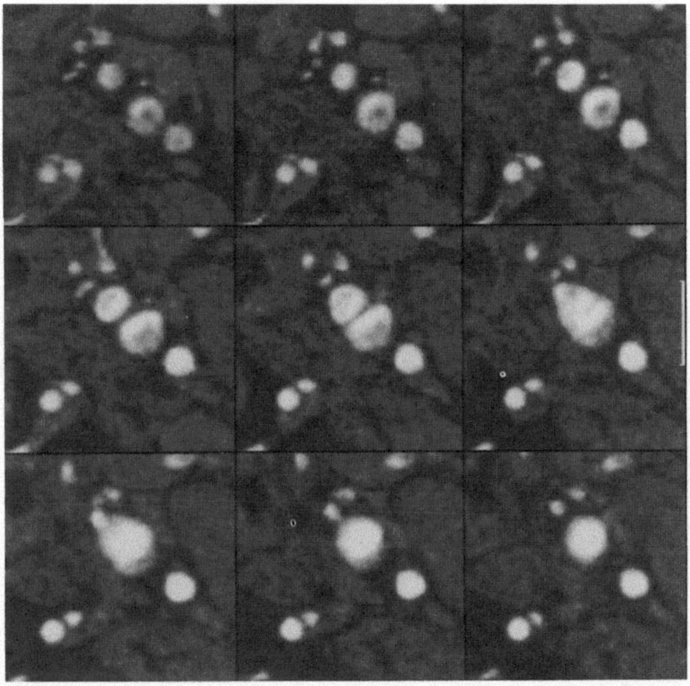

Fig. 6. Three-dimensional acquisition of the carotid bifurcation. Nine images from 64 slices are shown. Each slice has a thickness of 1 mm and represents 256 × 256 pixels with spatial resolution of 0.8 × 0.8 mm

Postprocessing of Three-dimensional Data Sets

The individual slices of a three-dimensional data set together represent the complete spatial information on the volume covered. For the observer, this form of representation requires experience in order to obtain the correct three-dimensional spatial impression. Obviously, postprocessing methods should be used to extract two-dimensional projections of certain structures (e.g., the vessel tree) from the three-dimensional volume information. With these methods spatial impression can be obtained in two ways – by showing a sequence of projective images with different projection angles or by coding of the depth information onto the surface of the displayed objects [10, 11].

Since the surfaces of most vessels are relatively small, the first method – multiple projections with different angles – has proven more useful in practice. The starting point for this method must be a data set in which the structures to be extracted are associated with a characteristic range of signal-intensity levels. In this case a projective image can be calculated by penetrating the data volume with a set of parallel projection rays and selecting along each of these rays only the data point that represents the intensity maximum (maximum

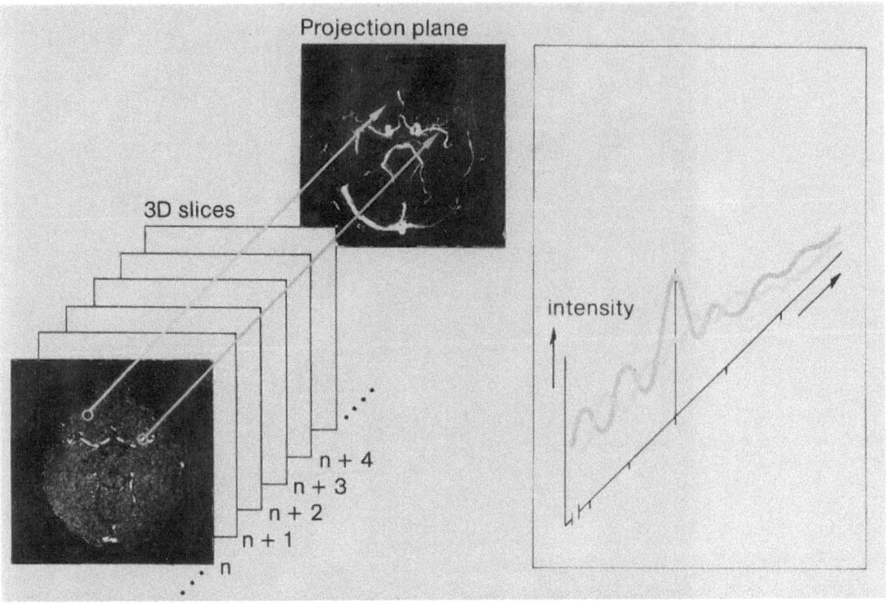

Fig. 7. The maximum intensity projection (MIP) method

intensity projection, MIP). The inflow enhancement and the pulse-sequence parameters chosen appropriately (flip angle, pulse repetition time, and flow-compensation on) ensure that the maximum intensity is always associated with a blood vessel, as long as the projection ray intersects at least one. All of the other projection rays will just pick up a background pixel out of the three-dimensional data set. The principle of MIP is shown in Fig. 7 for two projection rays; one is hitting a vessel while the second gives only background signal intensity. As a result, Fig. 8 demonstrates a complete projection image calculated from one 3D data set.

By varying the projection angle, multiple projective images can be obtained which allow the observer to obtain the correct spatial impression of the three-dimensional information. One way to display this type of information is the stereo image technique. Using slightly different projection angles for each eye, the human cognitive system can generate the correct spatial impression. But, since three-dimensional magnetic resonance angiography allows the calculation of many projective images at arbitrary angles, a slightly different technique has been found more useful. By displaying a number of projections with projection increments of only about 2° in a rapid fashion, the impression of a continuously rotated object will be generated which allows a correct three-dimensional visualization of such complex structures as a vessel tree.

Figure 9 shows four images from such a set of projections. The carotid and vertebral arteries of a healthy volunteer are demonstrated. The measurement time for the original 3D data set (64 × 256 × 256 voxels) was about 10 min.

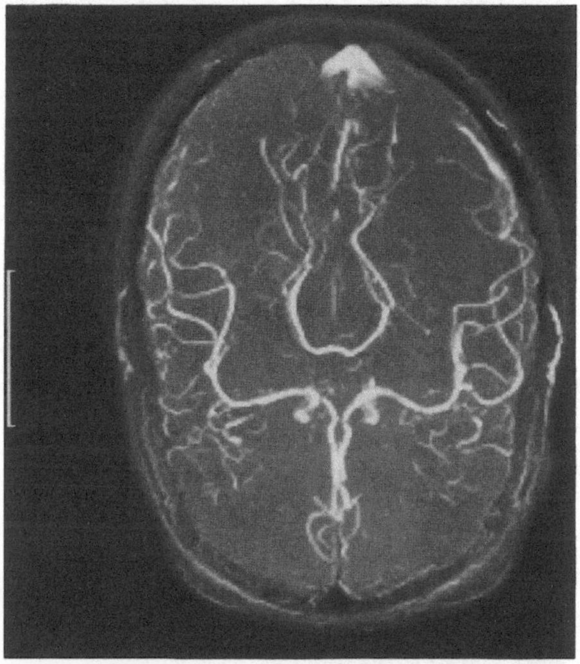

Fig. 8. Retrospectively calculated projection image of the intracranial vasculature

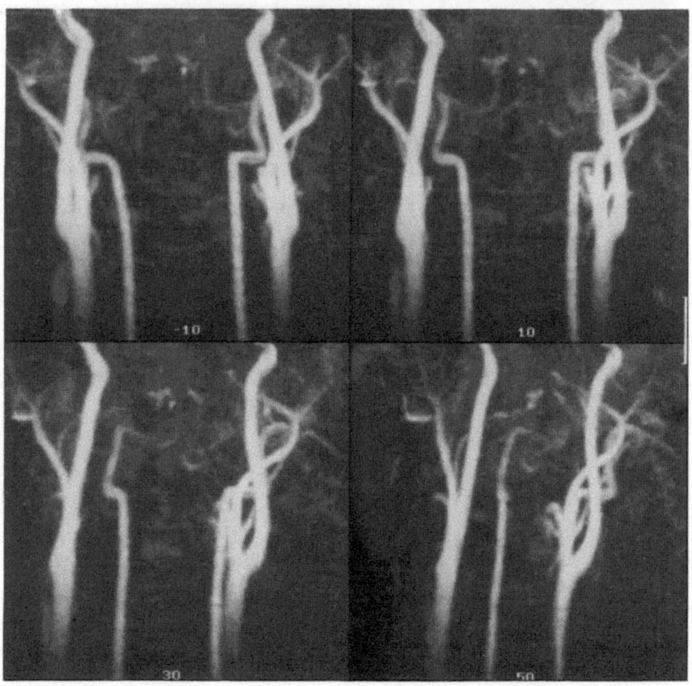

Fig. 9. Four projections at different viewing angles from a three-dimensional data set of a normal volunteer

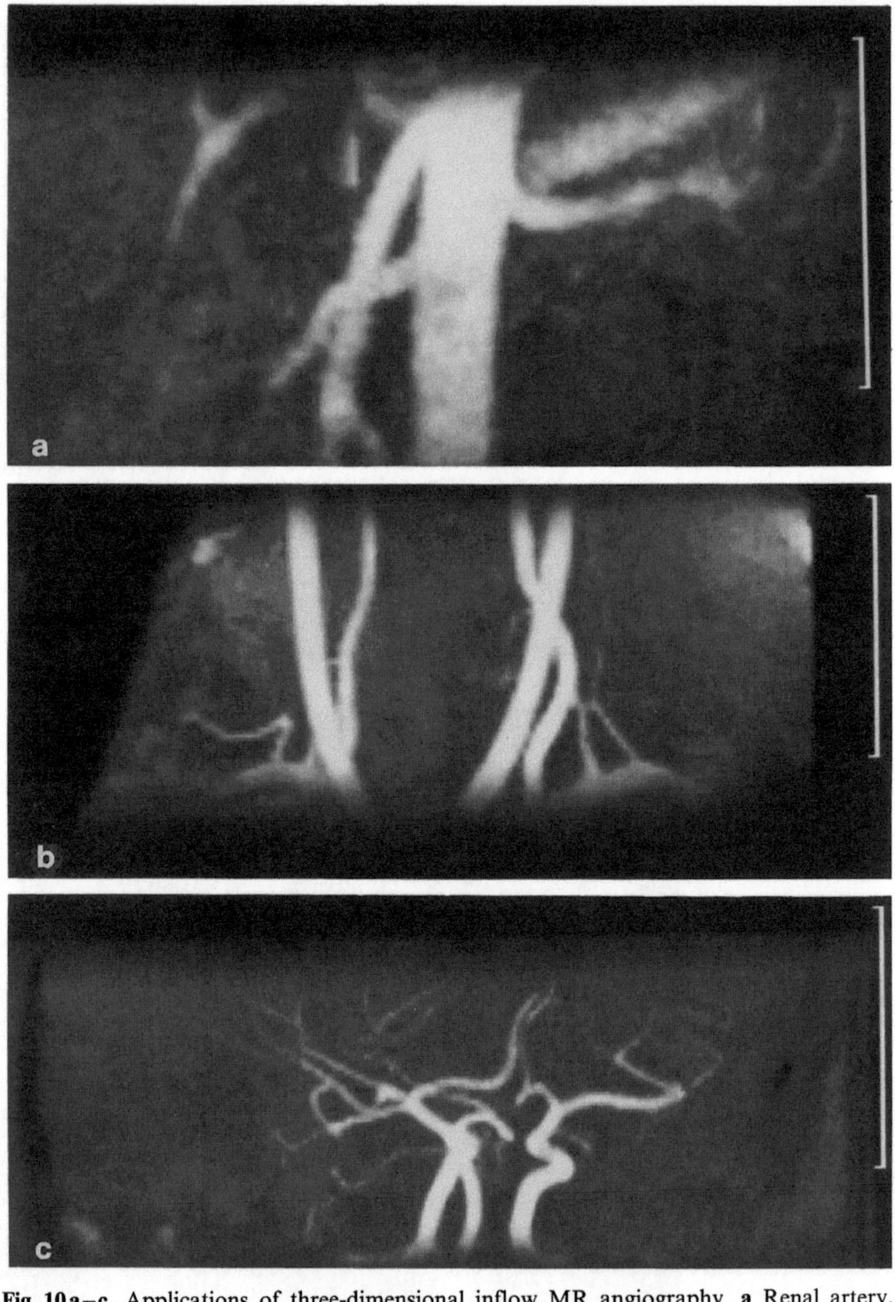

Fig. 10 a–c. Applications of three-dimensional inflow MR angiography. **a** Renal artery, **b** carotid and vertebral arteries, **c** intracranial vasculature. All images were acquired using 3D inflow techniques with MIP processing. (Courtesy of University Hospitals Cleveland, Dr. J. Lewin)

After a reconstruction time of approximately 1 min for all of the 64 slices, projective images can be calculated at a rate of one every 5 s.

Discussion

The results of preliminary clinical studies indicate that this MR angiography technique can provide accurate, reproducible flow images in different anatomical regions of the body. Examples of this are shown in Fig. 10. This technique is noninvasive and can be performed in conjunction with two-dimensional spin-echo or gradient-echo imaging, with only a 5- to 10-min extension of the examination time.

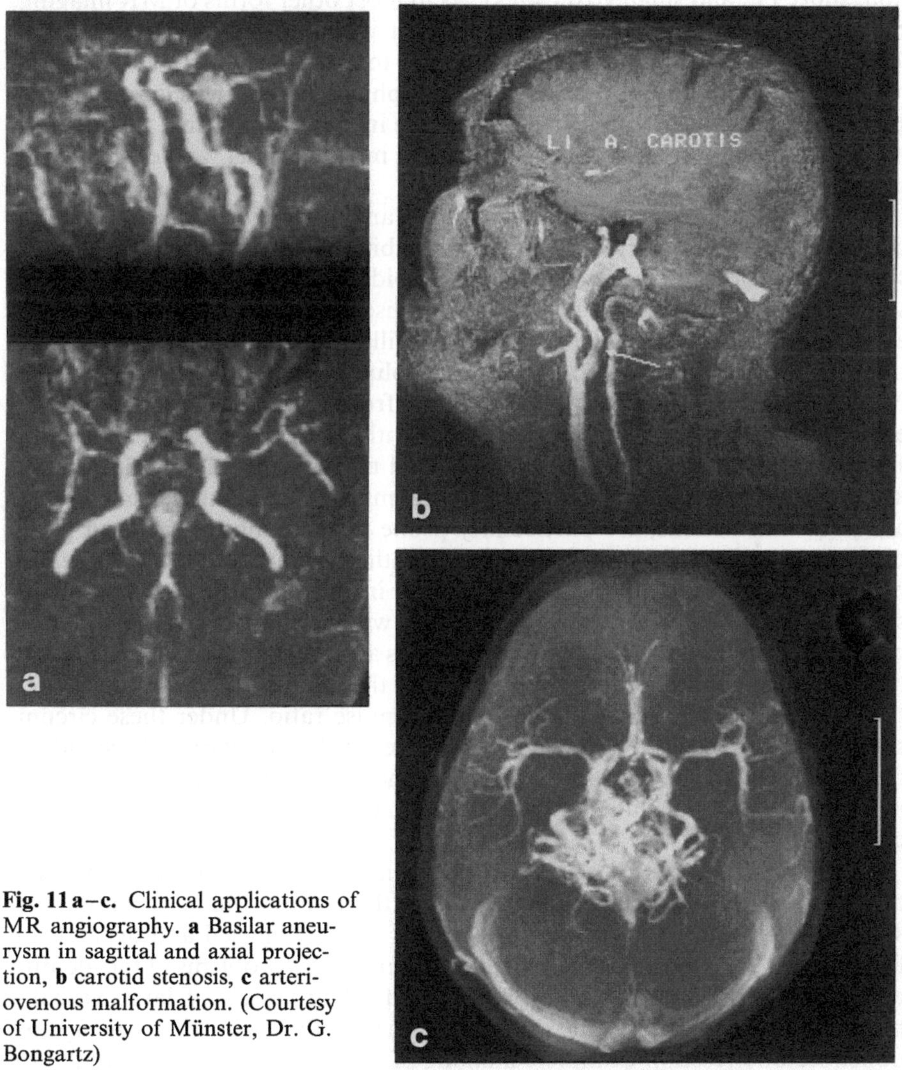

Fig. 11 a–c. Clinical applications of MR angiography. **a** Basilar aneurysm in sagittal and axial projection, **b** carotid stenosis, **c** arteriovenous malformation. (Courtesy of University of Münster, Dr. G. Bongartz)

The unique advantages of this technique include the capacity to provide multiple projections of anatomically complex vascular abnormalities with a single data acquisition. It appears that this capability may increase both the sensitivity and specificity of MR angiography in some patients with cerebrovascular disease when MR angiography is performed in conjunction with routine spin-echo imaging of the brain. This is demonstrated by Fig. 11 showing three different clinical applications of this technique.

The rationale for the MR angiography methods presented here is that flow-related enhancement and vessel contrast can be maximized with appropriate volume orientation, TR, and flip angle, while signal loss secondary to phase dispersion can be minimized through gradient manipulations (flow compensation, short TE, and small voxel sizes). As in most other forms of MR imaging, changes involving these operator-dependent variables have consequences in terms of examination time, contrast, signal-to-noise ratio, and, thus, spatial resolution. While the inflow MR angiography technique was originally designed to be as widely applicable as possible in a clinical setting, the inherent problems and limitations of the method are more obvious with certain types of diseases.

Although magnetic susceptibility artifacts are not a problem with this method (because of the short echo time in combination with small voxel sizes), spatial resolution and intravascular signal void have been most troublesome in some cases of suspected vascular occlusive disease and represented sources of error. Intravascular signal loss appeared as filling defects, stenoses, or discontinuities within arteries in both healthy volunteers and patients with cerebrovascular disease. Such signal loss results from higher-order motion (acceleration or turbulence) and is most frequently seen in the anterior carotid siphon, the internal carotid bifurcation, and the bend of the middle cerebral artery. High flow states, such as those present in the feeding vessels of large arteriovenous fistulae, are particularly prone to these motion-induced phase dispersions and signal losses. To eliminate this problem, the echo time will probably have to be further reduced, with the incorporation of a flow-compensation scheme. However, such techniques will be limited by the available gradient strengths and gradient rise/fail times and will have consequences with respect to bandwidth and sampling times of the applied sequences; therefore, they will also affect the available signal-to-noise ratio. Under these circumstances, spatial resolution may be inadequate for the evaluation of second-order vessels. With the premium placed on short echo times, adequate spatial resolution may necessitate new reconstruction techniques or improvements in coil design.

Unlike subtraction MR angiography techniques, where the signal intensity depends only on the actual spin's velocity, the inflow volume technique described here is more limited in the region of interest imaged and depends on flow-related enhancement for contrast. This produces problems in the identification of distant vascular territories (e.g., feeding vessels of large vascular malformations or tumors) and in lesions with slow flow (e.g., giant aneurysms, dolichoectatic vessels). Complete imaging of the adult intracranial circulation requires approximately 128 partitions covering 120–150 mm. Even with the shortest TRs possible (20 ms), overall imaging time reaches the limits of patient tolerance to stay still. Additionally, with thicker imaging volumes and shorter TRs, saturation of the inflowing spins (and thus signal loss) becomes a potential problem in the most superior portions of the imaging volume. Even with the relatively modest imaging volumes that are used for normal studies, slow-flow lesions (e.g., giant intracranial aneurysms, venous angiomas, and dolichoectatic vessels) were poorly or never visualized secondary to saturation of moving spins residing too long in the volume of excitation. These problems are inherent in the nature of this angiography technique and will likely prove more difficult to solve than the limitations of spatial resolution or persistent motion-induced signal losses.

Finally, while it is possible to detemine the direction and velocity of blood flow with phase images derived from these acquisitions, inflow MR angiography is primarily an anatomically imaging examination. Unlike conventional angiography, it is not possible to derive pathophysiological information such as arteriovenous circulation time, patterns of vascular filling, collateral supply, neovascularity, or early-draining veins. Although often not essential in diagnosis, such important clinical information is lacking particularly in cases of large-vessel occlusion, arteriovenous fistulae, and neoplasm, in which the treatment of patients is affected. In some cases, the intelligent implementation of saturation pulses with such sequences may permit examination of selected vascular territories [12].

Conclusion

While the acceptance of the three-dimensional inflow MR angiography technique will depend on its sensitivity and specificity in disease diagnosis, as determined by means of prospective clinical trials relative to other screening modalities, the preliminary clinical results suggest that this MR angiographic method can serve as a screening tool for the identification of normal vasculature as well as stenosis and/or occlusions produced by atherosclerotic disease. It can be performed in conjunction with MR imaging of the brain, with only a small increase in examination time, and provides both vascular and parenchymal evaluation of cerebrovascular disease in a single setting.

References

1. Pattany PM, Phillips JJ, Chiu LC et al. (1987) Motion artifact suppression technique (MAST) for MR imaging. J Comput Assist Tomogr 11:369–377
2. Edelman RR, Atkinson DJ, Silver MS et al. (1988) FRODO pulse sequences: a new means of eliminating motion, flow, and wraparound artifacts. Radiology 166:231–236
3. Masaryk TJ, Modic MT, Ross JS et al. (1989) Intracranial circulation: preliminary clinical results with three-dimensional (volume) MR angiography. Radiology 171 (3):793–799

4. Masaryk TJ, Modic MT, Ruggieri PM et al. (1989) Three-dimensional (volume) gradient-echo imaging of the carotid bifurcation: preliminary clinical experience. Radiology 171 (3):801–806
5. Nishimura DG, Macovski A, Pauly JM (1986) Magnetic resonance angiography. IEEE Trans Med Imaging 5:140–151
6. Wehrli FW, Shimakawa A, Gullberg GT et al. (1986) Time-of-flight MR flow imaging: selective saturation recovery with gradient refocussing. Radiology 160:781–785
7. Moran PR, Moran RA, Karstaedt N (1985) Verification and evaluation of internal flow and motion. True magnetic resonance imaging by the phase gradient modulation method. Radiology 154:433–441
8. Dumoulin CL, Hart HR (1986) Magnetic resonance angiography. Radiology 161:717–720
9. Nayler GL, Firmin DN, Longmore DB (1986) Blood flow imaging by cine magnetic resonance. J Comput Assist Tomogr 10:715–722
10. Laub GA, Kaiser WA (1988) MR Angiography with gradient motion refocussing. J Comput Assist Tomogr 12:377–382
11. Koenig HA, Laub GA (1988) The processing and display of three-dimensional data in magnetic resonance imaging. Electromedica 56:42–49
12. Edelman RR, Wentz KU, Matile HP et al. (1989) Intracerebral arterio-veneous malformations: evaluation with selective MR angiography and venography. Radiology 173:831–837

Part IV

Chapter 9

Color Doppler Flow Imaging of the Carotid and Vertebral Arteries

W. Steinke and M. Hennerici

Introduction

Diagnostic ultrasound techniques are the primary noninvasive tests for assessment of cerebrovascular disease in patients with transient ischemic neurological deficits or stroke. Asymptomatic subjects with atherosclerosis of peripheral or coronary arteries, particularly if they undergo cardiovascular or extensive abdominal surgery, are also routinely studied to detect coincident disease in the carotid and vertebral arteries [68]. In addition, long-term prospective studies of the natural history of extracranial cerebrovascular disease have become feasible with the introduction of Doppler and duplex sonography [9, 24, 26, 63, 76].

Continuous wave (CW) and pulsed wave (PW) Doppler sonography incorporated in duplex systems provide information about normal and pathological intravascular hemodynamics. These are accurate methods for the detection and classification of carotid stenoses greater than 50% [2, 22, 36, 37, 61–63, 78, 83]. On the other hand, high-resolution B-mode imaging visualizes the echomorphology of nonstenotic plaques (luminal narrowing < 40%) in detail [16, 25, 80, 83].

Color Doppler flow imaging (CDFI) preserves the advantages of conventional Doppler and duplex sonography but has additional capacities. Color-coded intravascular Doppler signals facilitate the identification of the extracranial cerebral vessels, thus significantly reducing the examination time [19, 53] and greatly increasing the rate of technically satisfactory studies done by less experienced sonographers. Since the color-flow patterns indicate the location of maximal intrastenotic flow velocity, interobserver reproducibility is high for both the placement of the Doppler sample volume and the classification of carotid stenosis [53]. However, the use of CDFI should not be restricted to the detection of stenotic lesions in the carotid or vertebral arteries. This ultrasound technique displays gray-scale echotomograms of the vascular morphology and color-coded blood flow patterns simultaneously in real time. Thus, still poorly understood morphologic-hemodynamic interactions and their significance for the dynamics of atherosclerosis and the occurrence of ischemic cerebral events can be studied.

Carotid Artery

Examination Technique

Since the cervical extracranial arteries lie close to the skin surface, transducers operating at an ultrasound frequency of 7.5–10.0 MHz with a relatively short insonation depth can be used, providing a high-resolution gray-scale image and satisfactorily high frame rates. The examination should be performed in a systematic fashion with the patient in a relaxed supine position, the neck slightly hyperextended, and the head rotated to the contralateral side. The transducer head is positioned anterolaterally in the cervical region along the longitudinal axis of the carotid artery. In many patients, the vessel can be better visualized if the probe is placed lateral to the sternocleidomastoid muscle and the insonation beam is directed medially from this position. This approach regularly displays the jugular vein superior to the carotid artery with a particularly good visualization of the anterior vessel wall.

Sequential longitudinal and transverse B-mode echotomograms begin at the most proximal accessible segment of the common carotid (CCA) and continue through the bifurcation up the internal carotid (ICA) as far as possible under the mandible. If both branches of the bifurcation cannot be imaged simultaneously, the CCA is displayed together with either the ICA or the external carotid artery (ECA). Particular attention should be paid to the echomorphologic characterization of pathological vascular wall structures during this phase of the examination. Color-coded Doppler signals are then superimposed for the evaluation of distinct blood flow patterns in the carotid system including:

a) extent and distribution of separated and reversed flow at the level of the normal bifurcation,
b) location and degree of turbulence associated with nonstenotic lesions is documented, and
c) characterization of the flow patterns of stenoses.

Stenotic flow patterns are characterized with respect to the spatial extent and duration of abnormal color Doppler signals, secondary to pre- and post-stenotic hemodynamics. In addition, the surface morphology of the stenotic lesions outlined by color-coded blood flow, as well as the percentage of diameter and area reduction should be determined from transverse and longitudinal sections by dividing the residual arterial lumen contrasted by color signals into the normal prestenotic width of the vessel at the same site.

The Doppler sample volume is then placed in the CCA, ICA, and ECA to confirm vessel identification according to typical Doppler frequency profiles (Fig. 1). Systolic and diastolic flow velocities are compared with the contralateral carotid system. Doppler frequency spectra are assessed from intrastenotic areas of maximal Doppler shift frequency indicated by color fading and aliasing, or by a "green tag" function that is available in some systems. The severity of stenoses is at present most reliably assessed by measuring the peak systolic

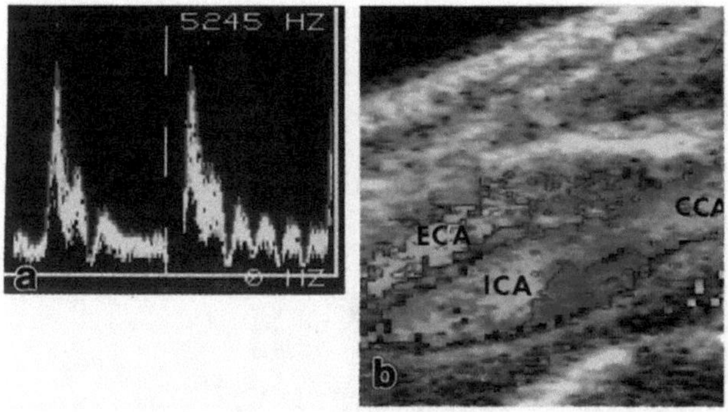

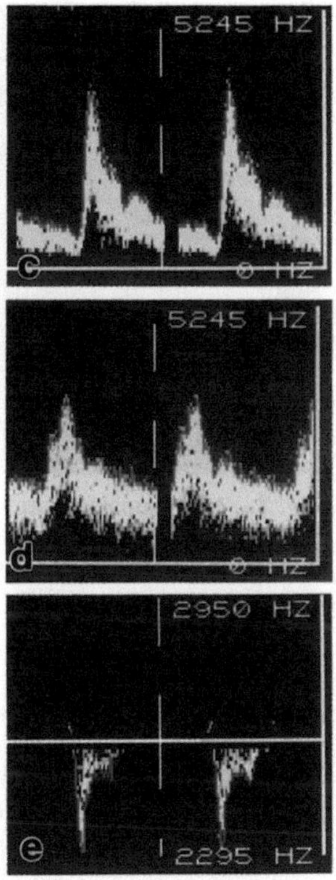

Fig. 1. Normal carotid bifurcation shows blue-coded flow separation at the internal carotid bulb (**b**) and characteristic PW-Doppler spectra (**a, c, d, e**). High systolic and low diastolic flow in the ECA (**a**) represents a high-resistance Doppler waveform, transmitted oscillations from tapping the superficial temporal artery further confirm the vessel identification. In contrast, flow in the ICA is high in diastole (**d**) reflecting supply to a low-resistance territory. PW-Doppler signals from the blue-coded area indicate reversed flow (**e**). The CCA Doppler waveform reflects the two vascular beds it supplies, but it is dominated by the low-resistance flow to the brain (**c**)

velocity and the dispersion of the Doppler frequency spectra [36, 62, 77]. Diastolic and mean flow velocities, as well as various velocity ratios and pulsatility indices, can also be determined from the Doppler frequency spectra; however, their diagnostic significance is less well established.

Normal Findings

Flow Patterns

Atherosclerosis in the carotid system is predominantly located at the bifurcation, in particular at the origin of the ICA. On the assumption that distinct flow patterns are associated with atherogenesis, experimental studies have investigated the significance of characteristic hemodynamics in the carotid bifurcation [41, 48, 66]. Such studies demonstrated reversed flow, boundary layer separation, and areas of low shear stress at the outer wall of the carotid sinus. Although studies using single or multigate pulsed Doppler sonography have confirmed these findings [28, 39, 49, 51, 57], conventional ultrasound techniques could not adequately assess such complex flow patterns in the normal carotid bifurcation.

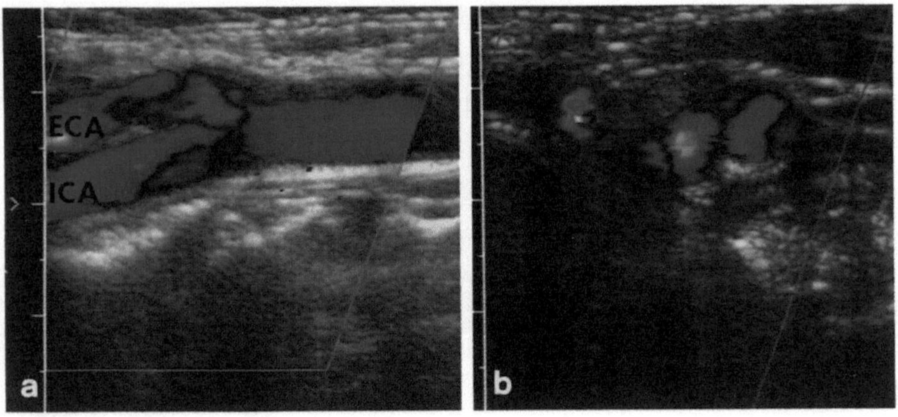

Fig. 2a, b. Normal carotid bifurcation showing that blue-coded flow separation is not restricted to the outer wall of the ICA, but also occurs in the ECA and adjacent to the flow divider. Longitudinal (**a**) and corresponding cross-section distal to the flow divider (**b**)

Since CDFI, on the other hand, demonstrates the spatial and temporal distribution of blood flow direction and velocity, flow separation and changes during the cardiac cycle can be visualized. Blue-coded Doppler signals indicate reversed and separated flow, which is most frequently located at the common and internal carotid bulb opposite the flow divider [52, 81] (Fig. 1b, e). The maximal spatial extent of the separation zone is normally seen immediately after the systolic peak. The extent and distribution of this flow separation is more variable than previously assumed [44, 71]. It is frequently found in the ECA as well (Fig. 2), and in some cases a particular horseshoe pattern is seen with the blue-coded Doppler signals extending from the ICA into the ECA around the flow divider. If the flow velocity significantly decreases during diastole, or if stagnation occurs, color signals cannot be generated in this area due to extremely low or absent Doppler shift resulting in a black area without color-coded blood flow (Fig. 3). Zones of secondary flow tend to be larger if the diameter of the carotid bulb is more dilated compared with the adjacent proximal vessel segment (Fig. 4a). The branching angle and other variables of the bifurcation geometry contribute to the variability of separated flow patterns, but the significance of these parameters has not yet been defined.

Although flow separation appears to be a significant factor for the initiation of atherosclerosis in the human carotid bifurcation, it represents a normal physiological phenomenon and was found in 93.6% and 99% of cases in two large CDFI series. The absence of separated flow may indicate early atherosclerotic disease with a smooth plaque filling the carotid bulb and causing secondary flow to disappear [33, 44] (Fig. 4b). However, boundary layer separation may also be absent if there is no common or internal carotid bulb. On the other hand, display of blue-coded flow in the carotid bifurcation may

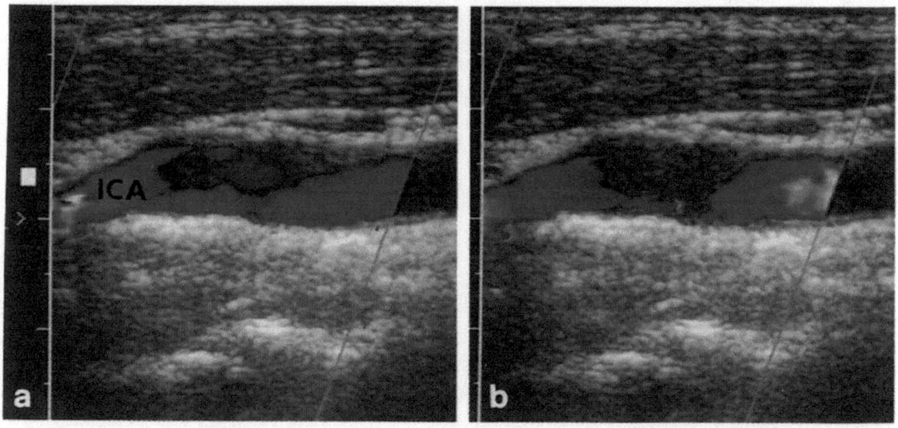

Fig. 3. Normal internal carotid bulb with area of blue-coded flow separation (**a**). Due to extremely slow flow in the separation zone during diastole, Doppler shift is too low to create color signals, resulting in a large color-free zone (**b**)

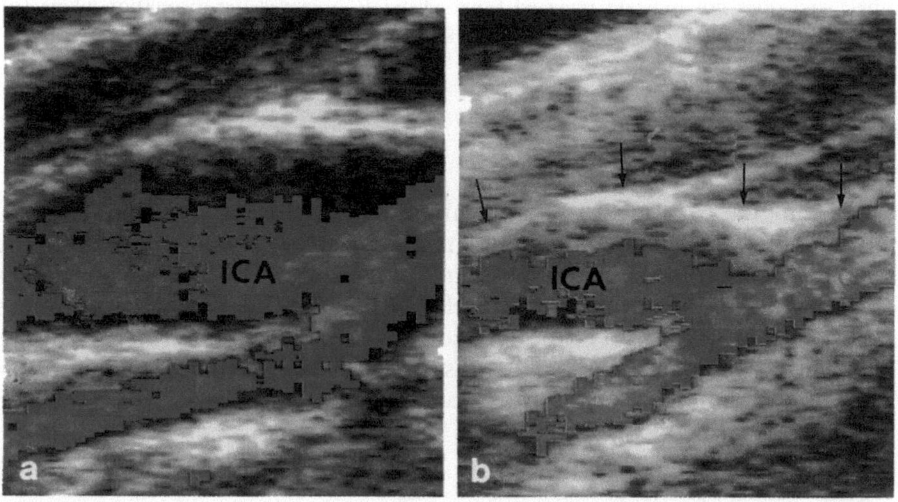

Fig. 4a, b. Carotid bifurcations. Extended zone of flow reversal (*blue*) in a dilated internal carotid bulb (**a**). Absent flow separation due to a smooth plaque (*arrows*) completely filling the carotid bulb (**b**)

be caused by plaques, depending on the location and configuration of the lesion. A correct interpretation of blue-coded flow signals in the carotid bifurcation therefore requires a consideration of the vascular morphology displayed in the gray-scale echotomogram.

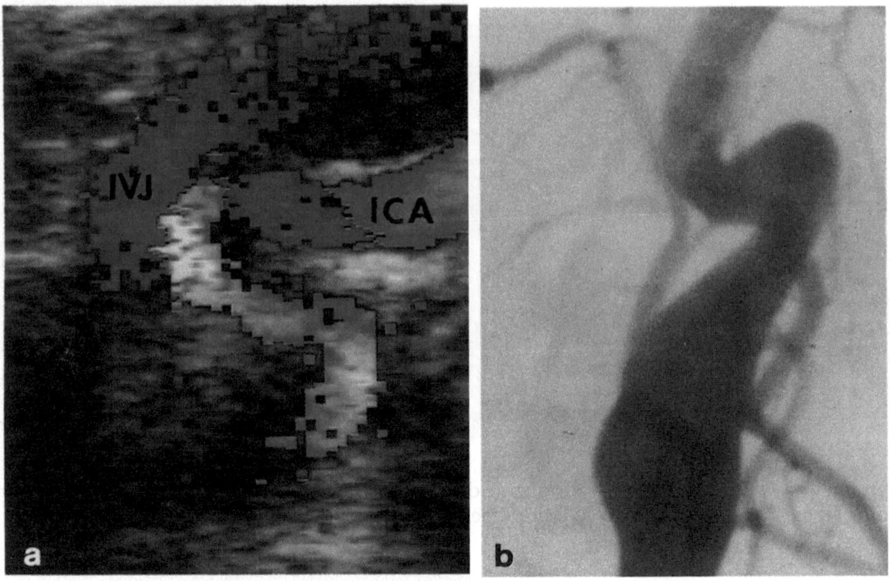

Fig. 5. Coiling of the ICA. Blue-coded flow in the proximal segment is due to the insonation angle and does not indicate reversed flow. Distal ICA leaves the insonation plane (**a**); **b** corresponding angiogram. *IVJ*, Internal jugular vein

Coiling

Coiling of the carotid artery is a variant of the normal anatomy (Fig. 5). Interpretation of increased flow velocity and reversed flow direction due to the changing Doppler angle along the course of the vessel is difficult in some cases since stenotic lesions, which may develop within the coiling, contribute to the increase in Doppler shift. Color-coded flow signals facilitate imaging of the vessel in a two-dimensional plane, thus providing hemodynamic and morphologic information for the assessment of relevant atherosclerotic disease in the tortuous vessel segment.

Atherosclerotic Disease

Atherosclerotic Plaques

The echogenicity of carotid plaques approximates distinct histomorphologic features and characterizes different stages of atherosclerosis [16, 25, 80]. *Homogeneous* echoes are typical of uncomplicated lesions consisting of dense fibrotic tissue. These plaques represent early atherosclerotic disease and are only infrequently seen in significant stenoses. *Heterogeneous* lesions are indicative of advanced disease with accumulation of cholesterol crystals, calcification, necrosis, and intraplaque hemorrhage [6, 13, 16, 80]. Although well-defined echolucent areas in the core of the lesion have been attributed to plaque hemorrhage, such areas may in some cases represent lipid aggregates [80]. Both homogeneous and heterogeneous high-level echoes with associated *acoustic shadowing* are caused by plaque calcification. Echo shadows are a major obstacle for adequate visualization of vascular structures and color-coded Doppler signals, particularly, if the calcific lesion is located at the anterior vessel wall.

Heterogeneous plaques are frequently associated with an irregular or ulcerative surface structure, while most of the homogeneous lesions are smooth (Fig. 6a). Although B-mode imaging had a high sensitivity for the detection of ulcerations in carotid artery specimen [16, 25], it demonstrated a lower in vivo accuracy in a comparison with findings at carotid endarterectomy [11]. The quantitative assessment of the plaque extent by B-mode echotomography also had important limitations, including a low interobserver reproducibility [40, 58, 65].

Since color-coded Doppler signals delineate the intravascular lumen, CDFI markedly improves the detection and characterization of carotid plaques [14, 45, 72]. The differentation of smooth and ulcerated surfaces is facilitated, the plaque configuration and the residual vessel lumen can be assessed more accurately when contrasted with color-coded blood flow (Fig. 6). Large echolucent components of an atherosclerotic lesion consisting mainly of adjacent thrombotic material or large intraplaque hemorrhage can be detected indirectly by sparing the color-coded blood flow (Fig. 6b). The hemodynamic disturbance associated with non-stenotic plaques is considerably variable; however, blue-coded turbulent flow is typically found at irregular surfaces or within ulcer craters (Fig. 6c, d) and is less often seen in smooth flat plaques (Fig. 6a). Depending on the lesion configuration and the vessel geometry, turbulence may even be absent in extended nonstenotic atherosclerotic lesions (Fig. 6b).

An association between heterogeneous plaque echogenicity and the occurrence of ischemic cerebral events has repeatedly been reported in studies using conventional duplex sonography assuming that heterogeneous echoes mainly represent intraplaque hemorrhage [6, 75]. On the other hand, heterogeneous lesions are more unstable and frequently tend to progress while homogeneous plaques are constant over a long period or even decrease in size [24]. Evaluation of the morphologic-hemodynamic interaction of non-stenotic plaques by CDFI revealed that progression of atherosclerosis was associated with turbulent flow at the lesion site [23]. A correlation of nonstenotic ulcerative plaques and ischemic cerebral symptoms has been reported in a series of carotid endarterectomies [35]; however, no systematic studies using conventional ultrasound techniques investigated the significance of small ulcerative lesions as for cerebral embolism. Preliminary data from one CDFI study suggest that turbulence at nonstenotic plaques with irregular surface were more frequently absent in asymptomatic patients [72]. Further investigation using the CDFI technique is needed to analyze the clinical significance of hemodynamic patterns associated with small carotid plaques in prospective trials.

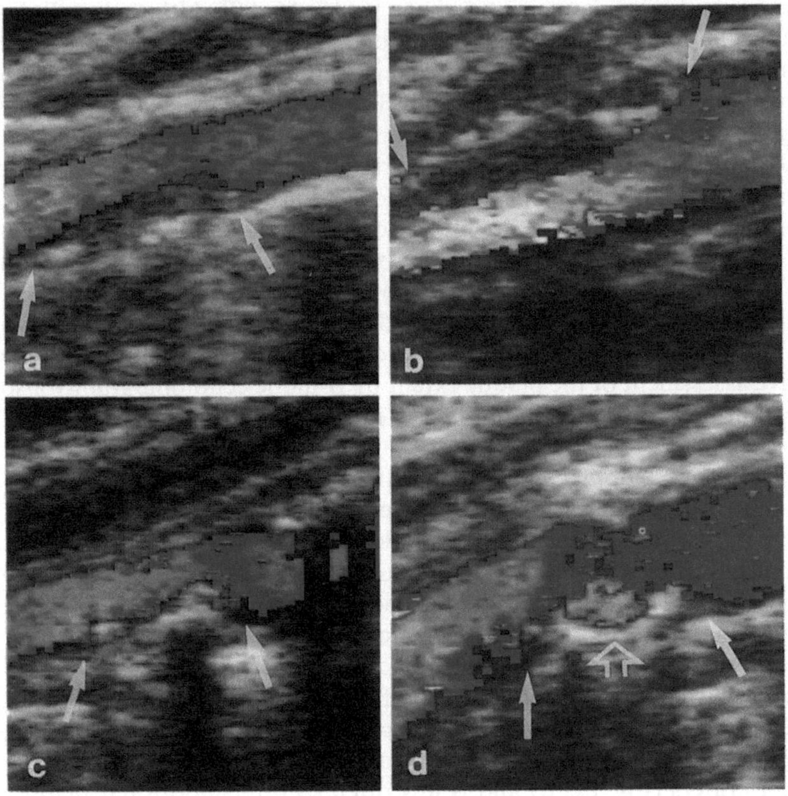

Fig. 6a–d. Nonstenotic carotid artery plaques (*arrows* indicate plaque extent). **a** Homogeneous smooth plaque with medium-level echogenicity. Minimal turbulence at the proximal portion. **b** Large smooth lesion at the origin of the ICA without associated flow disturbance. Extended echolucent distal portion of the plaque indicative of adjacent thrombotic material or intraplaque hemorrhage. **c** Partially calcified heterogeneous plaque with minimal turbulence (*blue*) at the irregular surface. **d** Heterogeneous plaque with a central ulceration filled with turbulent flow (*open arrow*)

Internal Carotid Artery Stenosis

The combination of B-mode echotomography and Doppler sonography in duplex instruments considerably improves the accuracy of the noninvasive diagnosis and classification of carotid stenosis. B-mode imaging in duplex systems commonly serves as a road map to place the PW-Doppler sample volume. The degree of stenosis is then classified according to distinct parameters of the Doppler frequency spectrum. However, instead of Doppler shift frequencies equivalent flow velocity values after correction of the Doppler insonation angle according to the flow direction in the vessel segment are calculated.

Intravascular color signals considerably facilitate the examination procedure and contribute important additional information for a correct grading of carotid stenoses. To achieve optimal diagnostic results the classification of mild (40–60%), moderate (61%–80%) and severe (81%–90%) ICA stenoses by CDFI (Figs. 7–10) is based on three sources of information provided by the system: (a) the Doppler frequency spectrum, (b) typical features of the color flow patterns, and (c) measurement of the residual vessel lumen. Table 1 shows the CDFI criteria for different categories of ICA stenosis in detail.

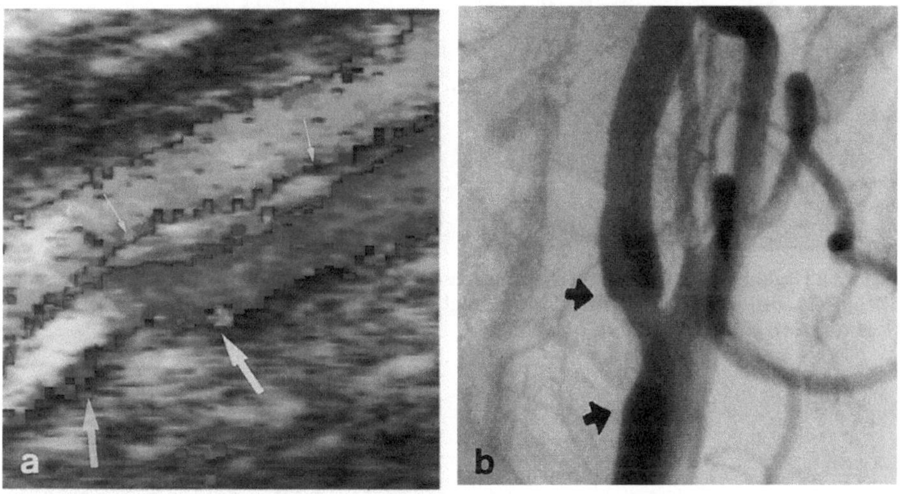

Fig. 7. **a** Low-grade (40%–60%) stenosis at the origin of the ICA due to a smooth plaque, demonstrating distal color fading and minimal proximal turbulence. **b** Smooth flat plaque at the opposite vessel wall is not obvious in the corresponding angiogram

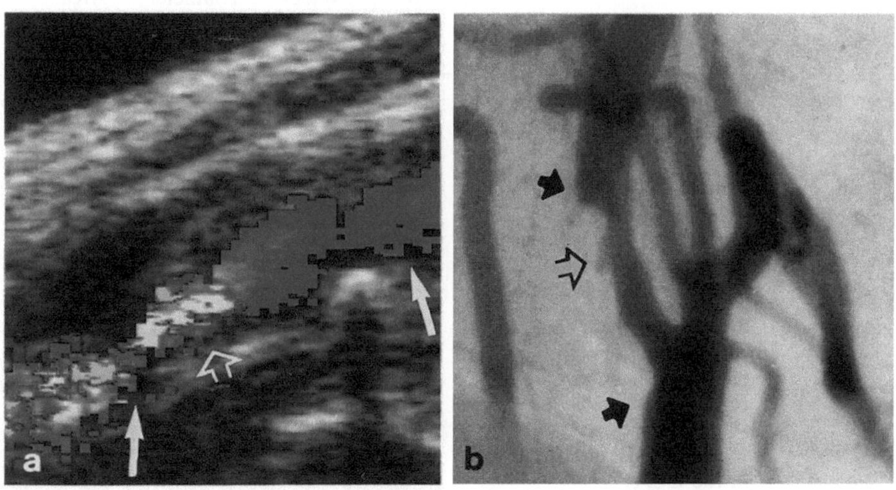

Fig. 8. **a** Moderate (61%–80%) ICA stenosis due to a long-segment heterogeneous plaque with a flat ulcerative niche in its distal portion (*open arrow*). Marked poststenotic turbulence. **b** Corresponding angiogram

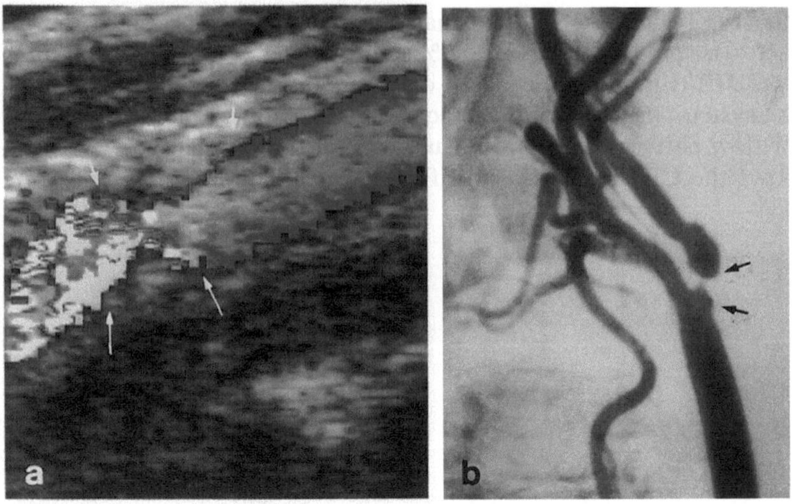

Fig. 9. **a** High-grade (81%–90%) stenosis caused by a homogeneous smooth circular plaque at the ICA origin. Circumscribed area of high stenotic flow, poststenotic turbulence, and flow reversal. **b** Corresponding angiogram

Table 1. Classification of internal carotid artery stenosis by color Doppler flow imaging

Degree of color flow pattern stenosis		Doppler spectrum		Luminal narrowing	
		Frequency[a] (kHz)	Velocity (cm/s)	Diameter	Area[b]
Low grade	– Color fading only in diastole – Long segment of color fading – Minor turbulence	>4.0	>120	40%–60%	64%–83%
Medium grade	– Color fading more circumscribed – Turbulence and reversed poststenotic flow – Increased flow in diastole	>4.0	>120	61%–80%	84%–95%
High grade	– Short segment of marked color fading or aliasing – Severe poststenotic flow reversal and mixed turbulence ("mosaic pattern") – Decreased prestenotic flow in the CCA	>8.0	>240	81%–90%	96%–99%

[a] Assuming a concentric stenosis
[b] Emission frequency 5 MHz

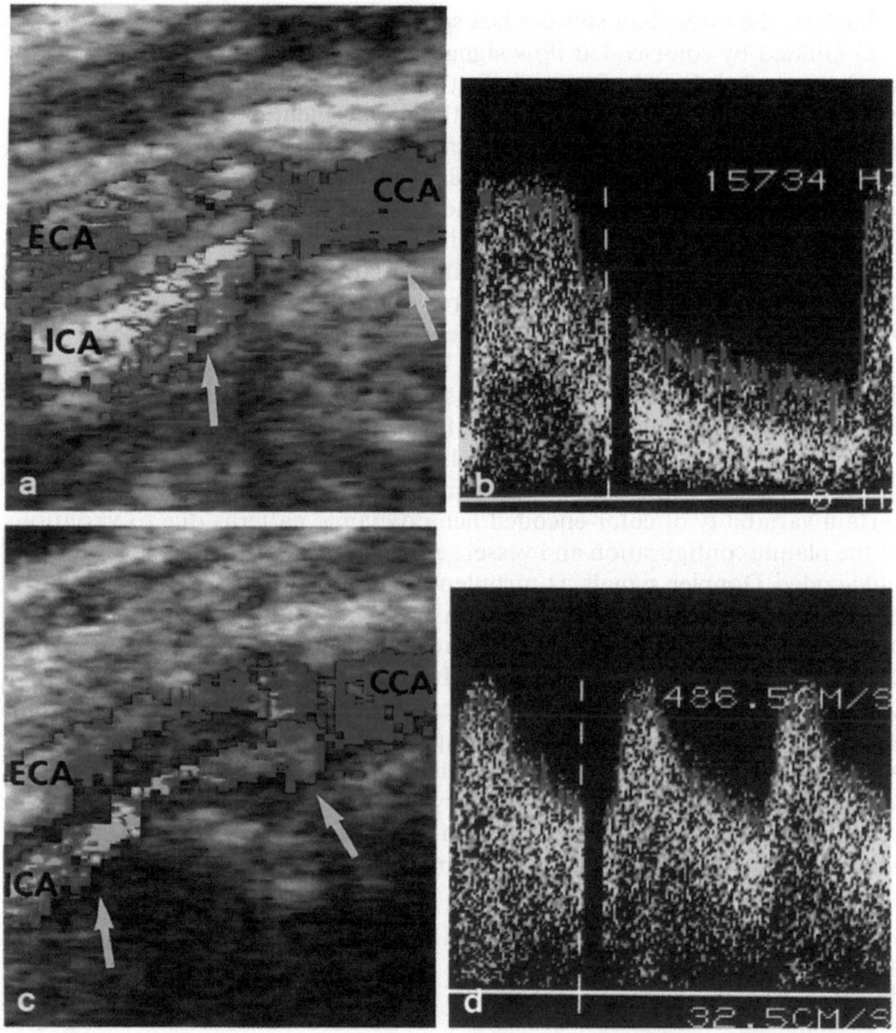

Fig. 10a–d. High-grade (81%–90%) ICA stenosis due to a relatively smooth plaque with medium-level echogenicity. **a** Color flow patterns demonstrate an area of jetstream (*white*) and poststenotic flow reversal adjacent to the vessel wall. **b** Corresponding PW-Doppler spectrum from the site of maximal Doppler shift shows an extremely high peak systolic frequency approaching 16 kHz. **c** Subtotal ICA stenosis with a minimal residual lumen and extended poststenotic blue-coded signals indicating aliasing, turbulence, and reversed flow. **d** Corresponding PW-Doppler spectrum demonstrates an angle-corrected systolic peak velocity of 486 cm/s

Each of the three data sources has specific advantages and limitations.

a) Guided by color-coded flow signals the PW-Doppler frequency spectra can be recorded directly from relevant pre-, intra-, and poststenotic regions without time-consuming search for optimal placement of the Doppler sample volume [53, 72]. The Doppler spectrum is particularly important since it can be recorded in most cases, even if plaque calcification obscures adequate visualization of color flow patterns and of the residual vessel lumen. Among various parameters determined from the Doppler spectrum the peak systolic frequency (velocity) is the most frequently used criterion for the classification of the severity of disease because of its reliability and the ease of measurement. The separation of ECA from ICA stenoses may be difficult in some cases because of similar Doppler profiles in high-grade obstructions; however, the ECA can be identified by tapping the superficial temporal artery repetitively or by displaying branches originating from the ECA (Fig. 11).

b) Typical color flow patterns are characteristic for different categories of stenoses [19, 60, 72] although this semiquantitative analysis is limited by a certain variability of color-encoded hemodynamic patterns due to variations of the plaque configuration and vessel geometry. In addition, interpretation of blue-coded Doppler signals as turbulent or reversed flow or as aliasing phenomen may be difficult, but in some instruments identification of areas with aliasing of the Doppler signal is facilitated separating them from nonaliasing color signals by an intermediate color-free zone. The peak systolic frequency cannot be determined directly from the color-encoded Doppler information, since each color pixel represents the approximate mean Doppler frequency shift, which is lower than the peak. This is especially so, if turbulence results in broadening of the Doppler frequency spectrum [46, 83].

c) The residual vessel lumen and plaque configuration can reliably be assessed in sequential longitudinal and transverse sections since the intravascular

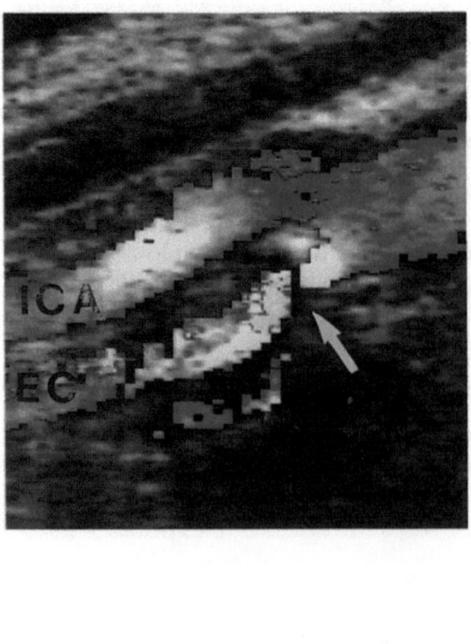

Fig. 11. Circumscribed high-grade ECA stenosis with branching superior thyroid artery

surface is contrasted by color flow signals [14, 19]. The percentage area reduction in cross-sections is higher than the relative diameter reduction assuming a concentric stenosis (Table 1). However, since the plaque configuration is more variable, it is mandatory to assess multiple sequences of longitudinal and transverse sections at the lesion site to determine the correct degree of luminal narrowing [10, 83].

The overall accuracy of CDFI for the classification of carotid stenosis varied between 71% and 96% compared with angiography [14, 53, 72]. However, the capacity to separate subtotal stenoses from total occlusions is still controversial with regard to the small numbers of patients included in the available studies [14, 32, 72]. A discussion of the clinical significance of carotid stenosis is beyond the scope of this chapter. It should be mentioned, however, that long-term prospective studies of the natural history of asymptomatic carotid disease demonstrated the associated stroke risk to be lower than previously assumed from symptomatic patients [9, 26, 63]. On the other hand, the subgroup of patients with rapid progression of a previously asymptomatic carotid stenosis had a higher incidence of ischemic cerebral events [26, 76]. In a recent series of high-grade carotid obstructions studied by CDFI, ulcerated lesions were found more frequently in symptomatic patients while the color-coded Doppler patterns were not significantly different in symptomatic and asymptomatic stenoses [30].

Carotid Occlusion

B-mode sonography without additional Doppler information has been demonstrated to be unreliable for the diagnosis of carotid occlusion [40]. In contrast, CW-Doppler and duplex sonography provided a significantly higher diagnostic accuracy, although the separation of subtotal stenosis from occlusions remained difficult. The introduction of CDFI raised hopes that this new noninvasive technique would overcome the diagnostic problem of ICA "pseudoocclusions." However, as mentioned above, it cannot yet be definitely concluded from the available data whether CDFI is superior to conventional ultrasound techniques in the diagnosis of this condition.

The presence of an *ICA occlusion* can be inferred if the color-flow signals in the CCA demonstrate a low systolic blood flow and a marked decrease or absence of color-coded flow during systole. The Doppler frequency spectrum confirms the significant decrease or absence of diastolic Doppler shift, in particular in comparison with an unaffected contralateral CCA. In ICA occlusions the Doppler profile of a patent ECA is frequently "internalized," showing low-resistance characteristics which indicate collateral flow to the brain via the ophthalmic artery. Color flow signals in the occluded ICA are absent, but there may be blue-coded flow reversal in the residual stump ("stump flow": Fig. 12a). In old occlusions the vessel lumen can merely be visualized distal to the orifice in the gray-scale image, but in some cases of acute thrombotic ICA occlusions echopoor intravascular material fills the lumen in the absence of color-flow signals (Fig. 12b, c).

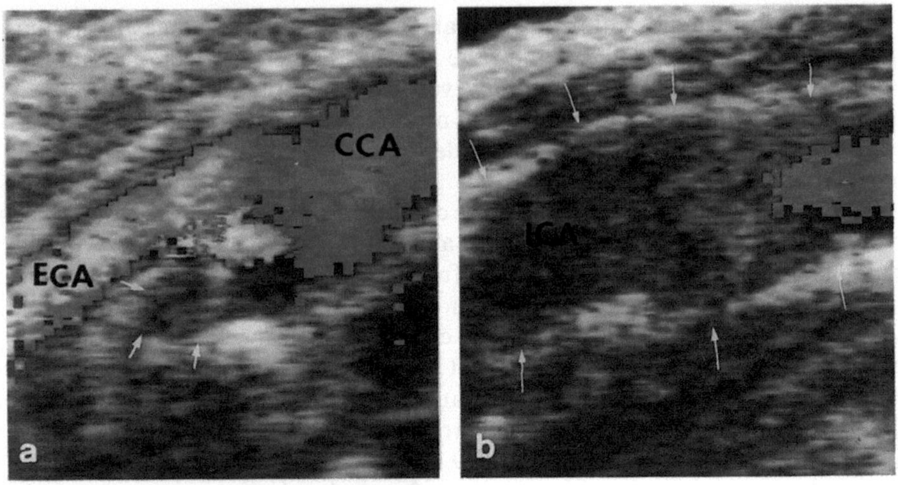

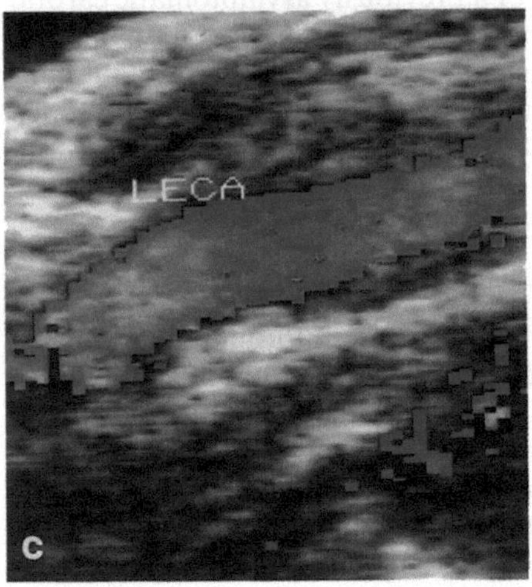

Fig. 12. **a** Old ICA occlusion with blue-coded flow reversal at the residual stump (*arrows*). **b** Acute thrombotic ICA occlusion demonstrating echolucent homogeneous intravascular material (*arrows* indicate the vascular wall). Red-coded flow signals in the distal CCA can be seen (*right*, **b**) and normal flow in the ECA (**c**)

Since the patients' management may be different in subtotal stenoses and occlusions, further development of the CDFI technique is necessary to improve diagnostic accuracy. At the present time it is not yet clear whether special programs designed to display very slow blood flow, will overcome this diagnostic difficulty. In contrast to what was previously assumed, ICA occlusion is not a stable and relatively safe condition. A prospective study of the spontaneous history of asymptomatic extracranial cerebrovascular disease demonstrated a 4.4% annual incidence of stroke in the period after occlusion of the ICA [55]. From the type of infarct seen on CT scans in another study it was concluded that embolism to the carotid territory is probably the predominant cause of cerebral ischemia in ICA occlusions [59].

CCA occlusion is a relatively infrequent finding, but the correct diagnosis is of particular importance because of the high incidence of related strokes [56]

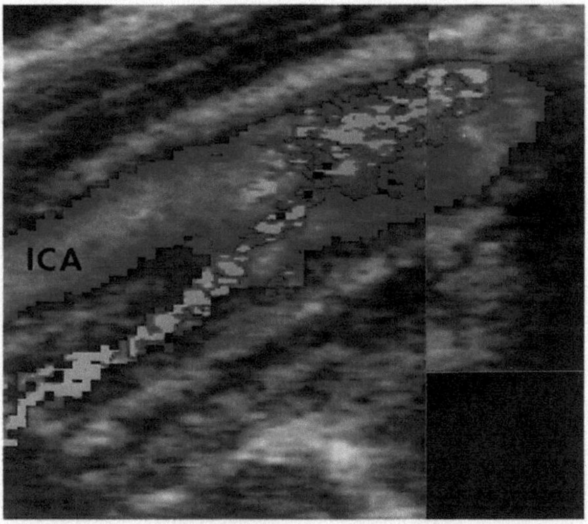

Fig. 13. Common carotid occlusion. Retrograde blue high-velocity flow in the ECA filling the carotid bifurcation and providing orthograde red flow in the ICA

and the possibility of surgical treatment. Standard Doppler and duplex sonography are usually accurate in the detection of a CCA occlusion; however, it is more difficult using these techniques to assess the collateral flow and the patency of the ICA, which is a precondition for surgery. CDFI has been demonstrated to be very helpful in demonstrating the collateral supply to the ICA coming mainly from the ipsilateral vertebral artery via the proatlantal artery and ECA branches [56, 72]. The typical display shows blue-coded flow signals in the ECA indicating retrograde flow direction to the bifurcation and orthograde filling of the ICA in the absence of flow signals in the CCA (Fig. 13).

Carotid Endarterectomy

CW-Doppler sonography and duplex scanning are widely used to assess the patency of the carotid system after endarterectomy in the early postoperative phase and to detect recurrent stenosis in follow-up examinations [3, 21, 64]. In contrast, surgically induced alterations of the vascular morphology have rarely been studied in detail by high-resolution B-mode sonography [15, 43].

CDFI examinations can be performed a few days postoperatively usually by insonating from the posterior lateral position. Although there may still be considerable swelling of the neck, it can be regularly assessed whether the endarterectomized vessel is patent or not. In addition, alterations of the vascular geometry, wall abnormalities, and associated hemodynamic disturbances can be displayed [74]. Since the lumen is frequently wider after endarterectomy extended flow separation occurs, predominantly located adjacent to the arterial wall in systole and extending to the center of the vessel during diastole (Fig. 14a, b). Early after the operation, but also at follow-up examinations

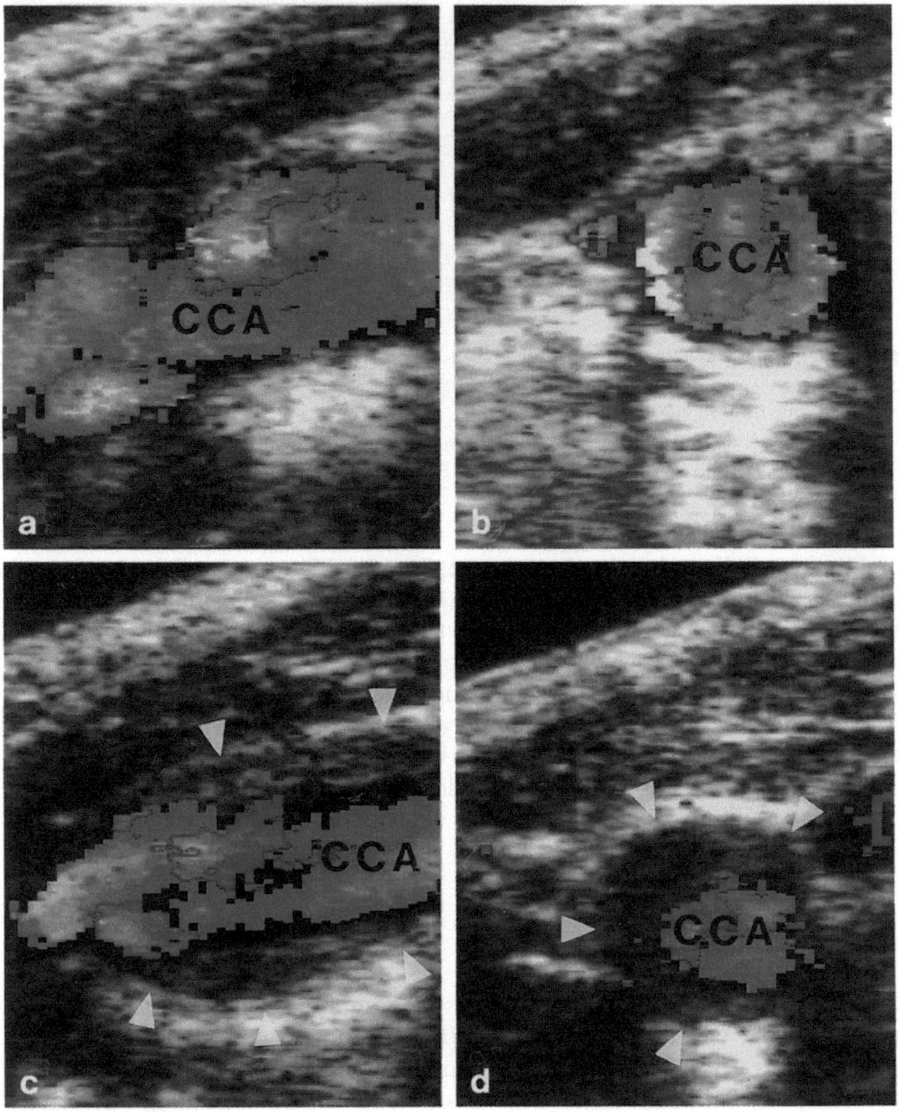

Fig. 14. a–d. Color flow imaging a few days after carotid endarterectomy. **a** Large areas of blue-coded flow separation at the vessel wall in the endarterectomized segment. **b** Corresponding cross-section. **c** Echolucent smooth matrix probably representing thrombotic material (*white arrowheads*) does not compromise the vessel lumen. **d** Corresponding cross-section demonstrates the half-moon-like shape of the thrombus (*white arrowheads*)

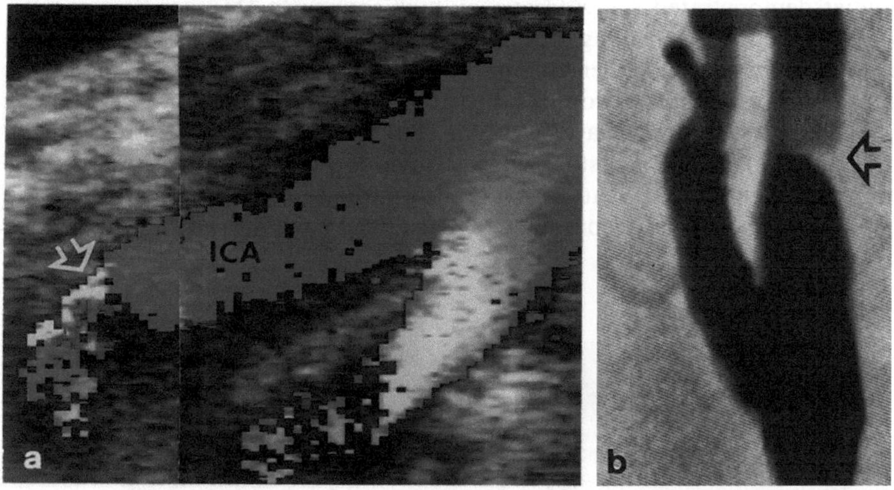

Fig. 15a, b. Recurrent internal carotid artery stenosis 2 years after endarterectomy. **a** Circumscribed moderate stenosis at the presumed distal end of the endarterectomy (*open arrow*). **b** Corresponding angiogram

months or years later, it is not uncommon to find large echopoor intravascular matrix with a smooth surface in these areas of very slow separated blood flow, probably representing thrombotic material (Fig. 14c, d). Surprisingly, in most cases the thrombus does not produce a relevant luminal narrowing, and the majority of patients remain asymptomatic [74].

It is still controversial whether neointimal hyperplasia at the distal and proximal end of the endarterectomy or progressive atherosclerotic disease is the predominant cause for recurrent stenosis after carotid endarterectomy (Fig. 15) [34, 50]. Since the interaction between postoperative vascular morphology and altered blood flow patterns probably represent a major pathogenetic factor for both neointimal hyperplasia and atherosclerosis, CDFI provides the capacity to investigate the significance of morphologic and hemodynamic features for the development of a restenosis.

Dissection

Although the incidence of carotid dissection is not exactly known, the studies published in recent years indicate a more frequent occurrence than previously thought [8, 47, 70]. According to the different pathogenesis and sonographic features two major categories of extracranial carotid artery dissection can be distinguished.

ICA dissections typically occur after minor trauma or spontaneously and are associated with a characteristic clinical syndrome including focal cerebral deficits, headache, neck pain, and ipsilateral Horner's syndrome. CW-Doppler studies demonstrated a low-frequency high-resistance Doppler flow pattern which could be traced along the course of the ICA in the neck [29]. Gray-scale echotomograms were either unremarkable or demonstrated a tapering lumen with occasional visualization of a floating intimal flap (Fig. 16a, b). Marked blue-coded flow reversal at the origin of the ICA in systole and absent or minimal blood flow in diastole without obvious abnormalities of the vascular morphology at the bifurcation is a typical finding in CDFI (Fig. 16c, d), the color-flow pattern corresponding to the already mentioned high-resistance

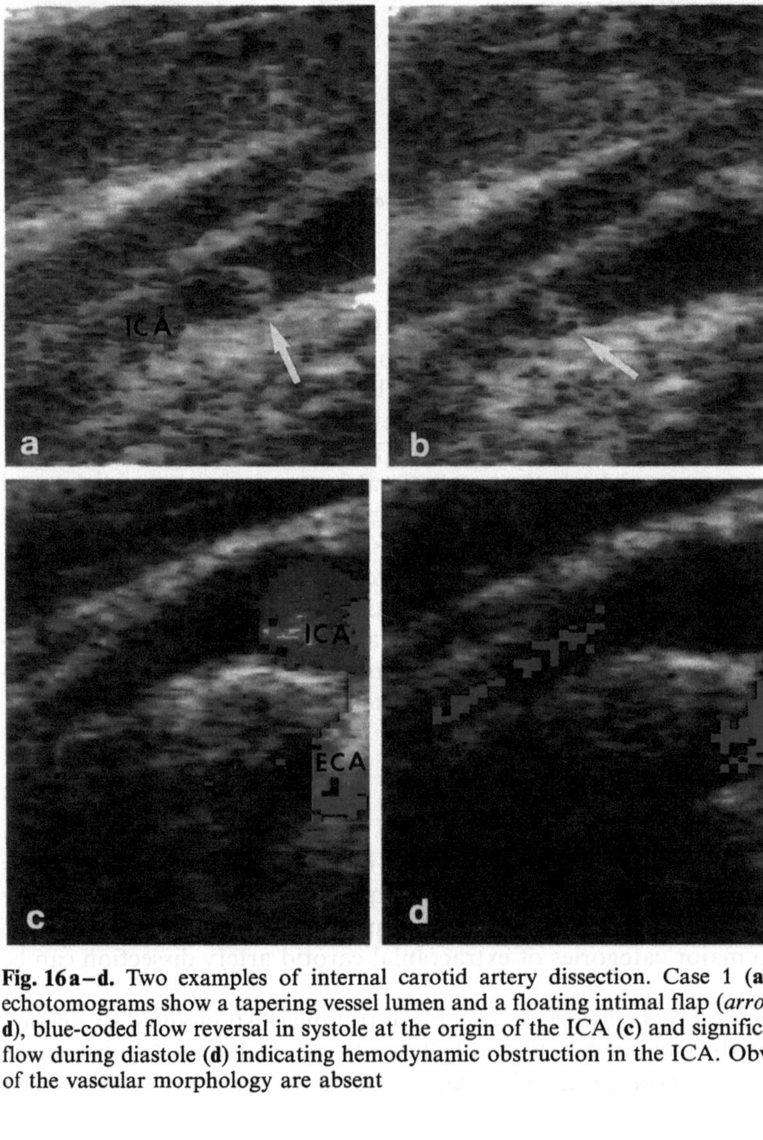

Fig. 16a–d. Two examples of internal carotid artery dissection. Case 1 (**a, b**), B-mode echotomograms show a tapering vessel lumen and a floating intimal flap (*arrow*). Case 2 (**c, d**), blue-coded flow reversal in systole at the origin of the ICA (**c**) and significantly reduced flow during diastole (**d**) indicating hemodynamic obstruction in the ICA. Obvious changes of the vascular morphology are absent

Doppler profile (Fig. 17a). Follow-up examinations demonstrate recanalization of the ICA and final normalization of the hemodynamics within days to weeks in the majority of patients (Fig. 17) [29, 70].

CCA dissections are commonly associated with aortic arch dissection [31]. Ischemic cerebral symptoms occur less frequently, and Horner's syndrome, headache, or neck pain have not been described in the available clinical case reports. The B-mode echotomogram typically demonstrates an intravascular double lumen separated by the dissecting intima (Fig. 18a) [4]. Different Doppler frequency spectra in both lumina were reported from studies by duplex sonography [38, 82]. CDFI easily displays the flow patterns in the true and false lumen and the site of communication between them [7, 73] (Fig. 18).

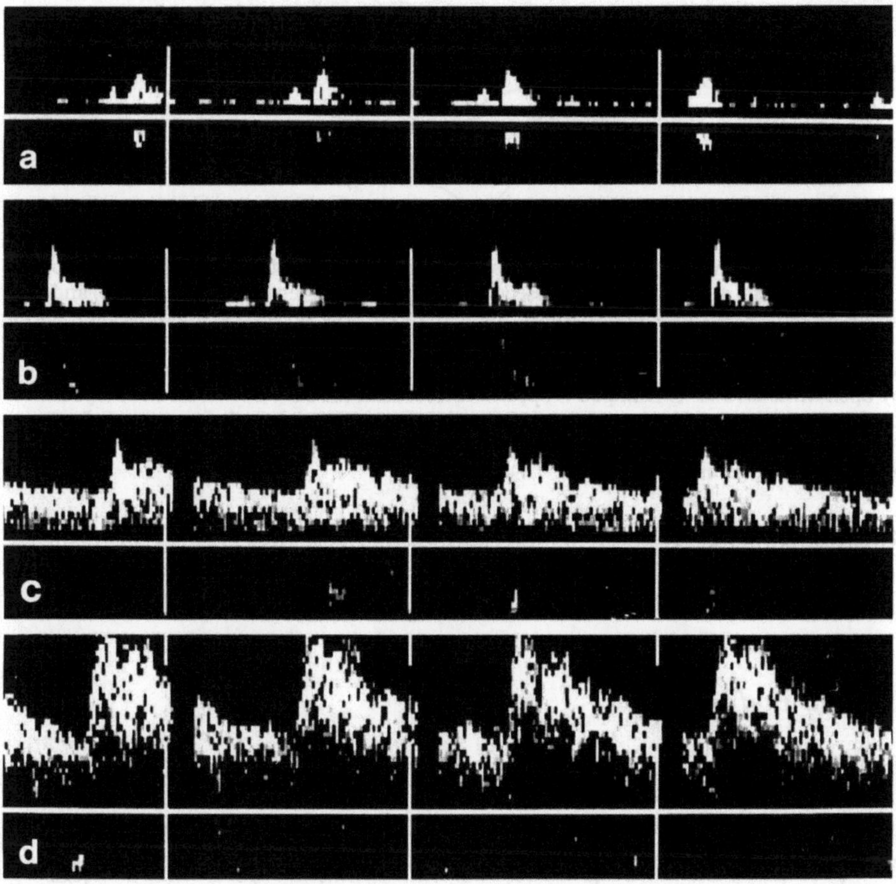

Fig. 17a–d. Sequential PW-Doppler spectra in internal carotid dissection (case 2 of Fig. 16) assessed over a period of 4 weeks. **a** Initial characteristic bidirectional high-resistance low-flow pattern. **b** At day 10 absent early systolic flow reversal. **c** After 3 weeks restitution of diastolic flow but still decreased systolic flow. **d** At week 4 normal PW-Doppler spectrum, indicating complete recanalization of the ICA

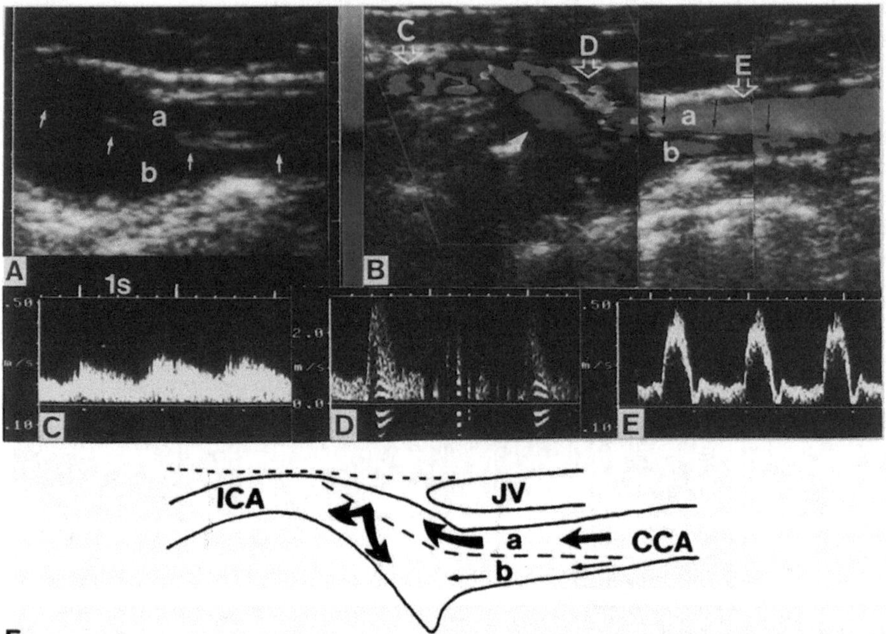

Fig. 18 A–F. Common carotid artery dissection in a patient with aortic arch dissection. **A** Gray-scale image demonstrates the dissecting intima (*arrows*) separating the false (*a*) and true (*b*) vessel lumen. **B** Color-flow patterns show marked flow reversal at the ICA bulb and a "mosaic pattern" indicating high-grade stenosis and turbulence at the distal reentry of the false lumen. **C, D, E** PW-Doppler frequency spectra assessed at the locations indicated in **B** by open *white arrows*. **C** Damped Doppler waveform in the ICA due to the hemodynamical significance of the CCA dissection. **D** Stenosis with typical musical murmurs. **E** Reduced flow in the false lumen of the CCA. **F** Schematic drawing corresponding to **B** illustrates the hemodynamics (from Steinke et al., 1990, ref. 73)

Regardless of the complex hemodynamics of CAA dissections CDFI allows reliable assessment of a hemodynamically significant decrease in blood flow in the ICA (Fig. 18c).

Carotid Body Tumors

Patients with tumors of the carotid body usually present with an asymptomatic pulsating neck mass. Duplex sonography has been used to diagnose carotid paraganglioma, however, the assessment of the characteristic hypervascularity and the separation from avascular or hypovascularized cervical masses such as lymph node metastases, salivary gland tumors, or branchial cysts was difficult [17, 42]. The CDFI diagnosis of carotid body tumors is based on both the typically widened configuration of the bifurcation and the visualization of hypervascular tissue between the ICA and ECA [67, 69] (Fig. 19). Palpable pulsations due to aneurysmal dilatation of the carotid artery (Fig. 20) or

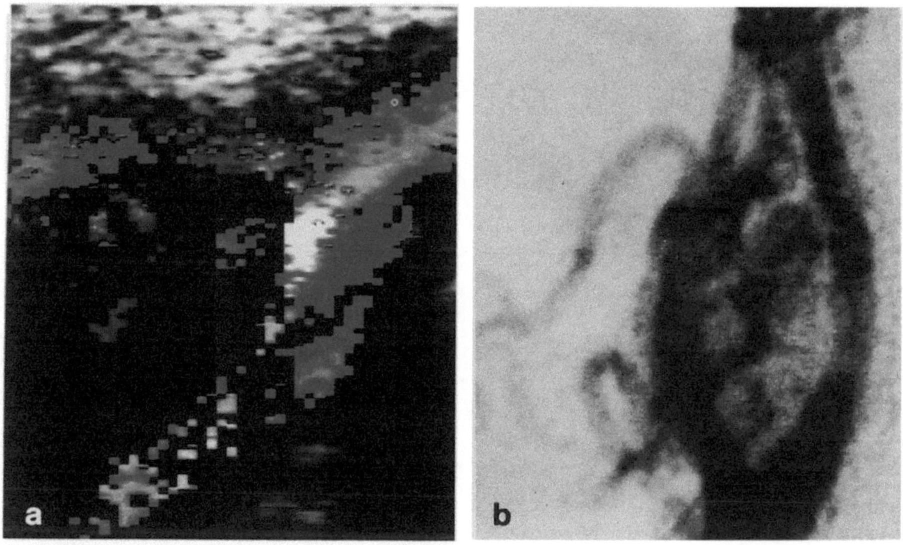

Fig. 19 a, b. Carotid body tumor. **a** Characteristic widening of the bifurcation angle and hypervascularity between the internal and external carotid arteries. **b** Corresponding angiogram (from Steinke et al., 1989, ref. 69)

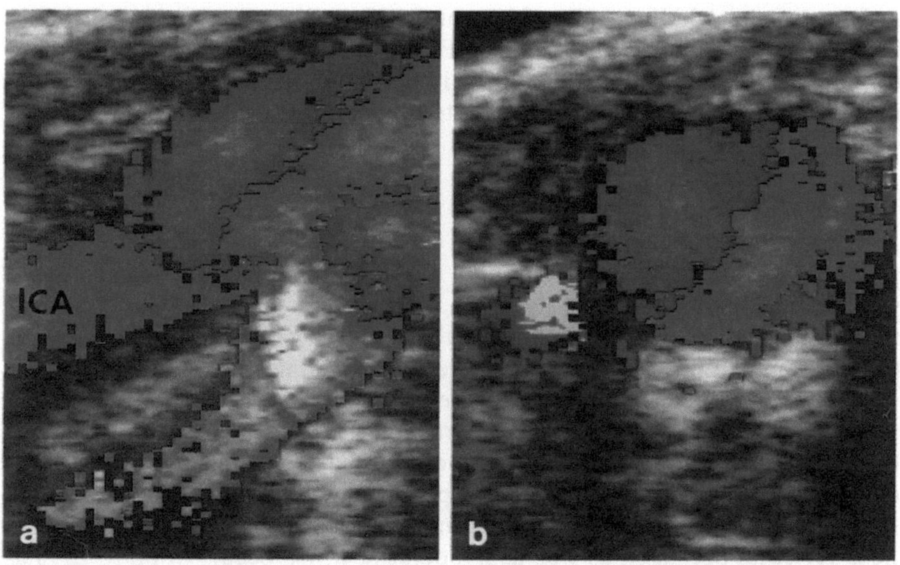

Fig. 20 a, b. Aneurysmal dilatation of the carotid bifurcation. **a** Large area of separated and reversed flow at the common and internal carotid bulb. **b** Corresponding cross-section

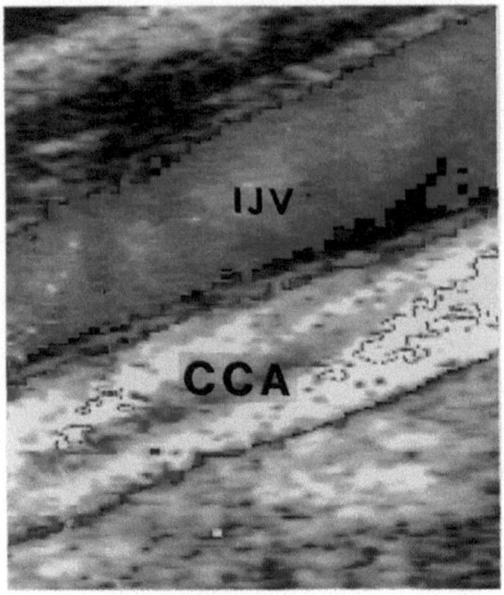

Fig. 21. Marked color fading in the common carotid artery indicating significantly increased flow ("super-flow") in a patient with an arteriovenous fistula below the base of the skull supplied by branches of the external carotid artery. IJV, Internal jugular vein

arteriovenous superflow caused by a fistula between the carotid artery and the internal jugular vein (Fig. 21) can easily be distinguished from carotid body tumors.

Since the risk of surgical treatment of carotid paraganglioma is size dependent, early diagnosis is important [20]. On the other hand, due to the high perioperative morbidity surgery in asyptomatic patients is recommended only if growth of the tumor is observed. CDFI provides the appropriate ultrasound technique to diagnose and follow carotid body tumors, and therefore angiography should be restricted to immediate preoperative evaluation.

Vertebral Artery

CW- and PW-Doppler criteria for the evaluation of vertebral artery disease were defined many years ago [1, 5, 22]; however, the diagnosis and classification of stenosis and occlusion of the vertebral artery has been found to be more difficult compared with the carotid system. The technically limited assessment of adequate Doppler signals from the ostium of the vertebral artery, the predominant site of extracranial vertebral atherosclerosis, and from the course of the vessel through the transverse foramina has contributed to these difficulties [1, 12]. From the still limited experience with CDFI it can be assumed that the capacity to visualize the proximal pretransverse portions and to diagnose vertebral occlusion as well as hypoplasia, which occurs in up to 25% as a normal variant, is improved. Additionally the atlas loop can frequently be displayed [79] (Fig. 22).

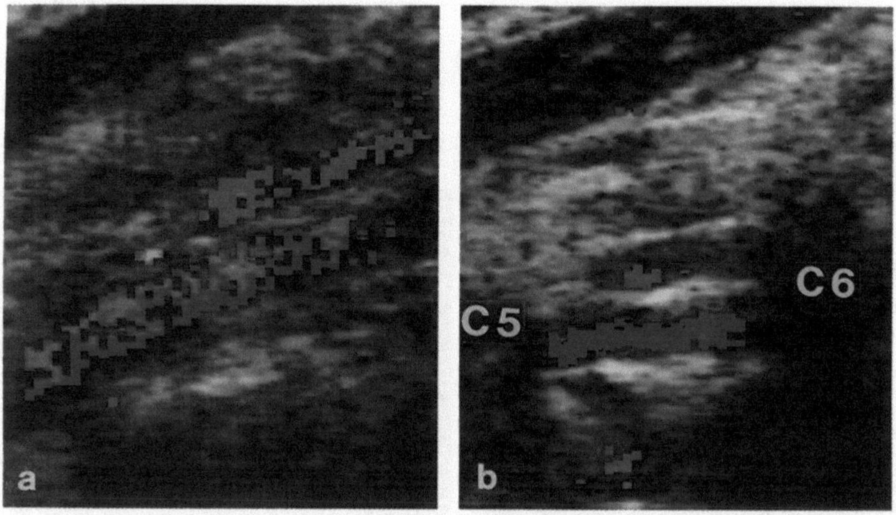

Fig. 22 a, b. Normal vertebral artery (*red*) and vein (*blue*). Proximal pretransverse course (**a**) and intertransverse segment (**b**) between the fifth (*C5*) and sixth (*C6*) cervical vertebra

Examination Technique

When the examination of the carotid artery is completed, the probe is shifted laterally from the CCA while maintaining a longitudinal position until the cervical vertebrae and the intertransverse segment of the vertebral artery are visualized. Additional adjustment of the insonation angle is often necessary for adequate display of vessel lumen and intravascular color-coded flow signals. The Doppler sample volume is positioned in the intertransverse vessel lumen to record Doppler frequency spectra. The resistive index is calculated as 1 – (end-diastolic Doppler frequency/peak systolic frequency). Then the vertebral artery is followed in proximal direction and the pretransverse section below the sixth cervical vertebra is traced to its ostium in the subclavian artery, where another PW-Doppler spectrum is recorded. The atlas loop of the vertebral artery can be displayed with the probe positioned below the mastoid process and the insonation beam directed toward the contralateral orbit.

Normal Findings

The normal Doppler frequency spectrum of the vertebral artery commonly recorded from an intertransverse and pretransverse segment demonstrates a relatively high diastolic Doppler shift, indicating a low-resistance blood flow pattern (Fig. 23). The reported normal resistive indices determined from the PW-Doppler spectrum of a CDFI system ranged from 0.62 to 0.75 [79]. Since

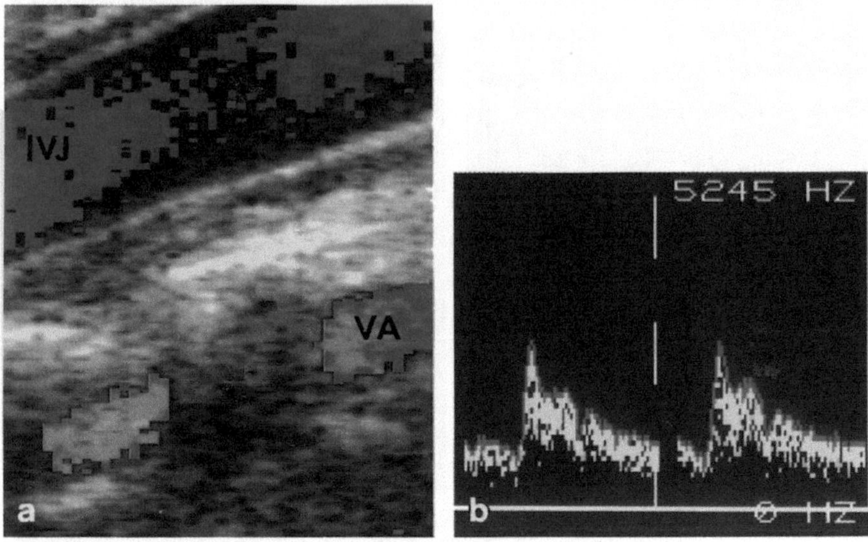

Fig. 23. **a** Normal vertebral artery proximal and distal to the sixth cervical vertebra. In this projection the internal jugular vein (*IJV*) is displayed superior to the vertebral artery. **b** Typical low-resistance Doppler waveform of a normal vertebral artery assessed in the inter-transverse segment

the peak systolic frequency ranges widely in normal subjects, we agree with Ackerstaff [1], that a systolic Doppler shift frequency below 4 kHz in the absence of turbulence is considered normal regardless of broadening of the Doppler frequency spectrum. If hemodynamically significant carotid disease is present, the peak systolic frequency may even exceed 4 kHz in a normal vertebral artery, indicating collateral supply from the posterior to the anterior circulation.

Difference in vessel diameter between the two vertebral arteries determined from the B-mode echotomogram and the width of the intravascular color-coded Doppler signals are not uncommon and comparison between both sides may provide differentiation between hypoplasia and diseased arteries (e.g., atherosclerosis or dissection). However, the lumen may not be visualized in hypoplastic vertebrals, and only a thin string of red-coded Doppler signals indicates patency of the vessel. The systolic and diastolic frequency in hypoplastic vertebrals is decreased, occasionally demonstrating an early systolic drop in the Doppler frequency spectrum ("intermediate flow pattern").

Pathological Conditions

Color flow patterns of vertebral stenoses are similar to those observed in the carotid artery. Typically the segment of color fading becomes more circum-

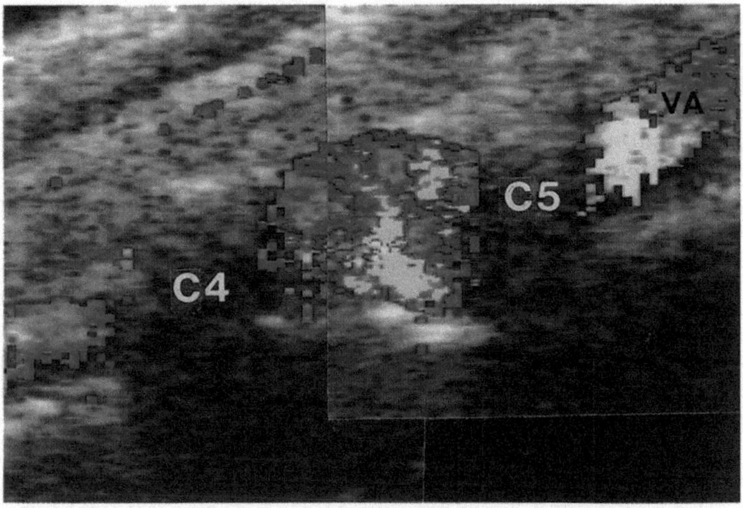

Fig. 24. Color fading proximal to the fifth cervical vertebra (*C5*) and marked dilatation of the adjacent intertransverse segment with severe turbulence and flow reversal indicate vertebral artery stenosis and a small poststenotic aneurysm. The site of maximal luminal narrowing is located in the transverse foramen and cannot be displayed

scribed and the poststenotic turbulence and flow reversal more severe with increasing degree of obstructions (Fig. 24). The peak systolic frequency assessed from the site of maximal intrastenotic fading of the color signals is greater than 4 kHz. An abnormally high-resistance flow pattern with the resistive index exceeding 0.80 may be indicative of a hemodynamically significant obstruction of the vertebral artery distal to the atlas loop such as dissection or high-grade atherosclerotic stenosis and occlusion. In acute complete vertebral occlusions at the origin from the subclavian artery the vessel lumen may still be visible; however, color-flow signals are absent. Retrograde blue-coded flow does not indicate vertebral artery disease but severe obstructions in the proximal subclavian or innominate artery (see next section).

Subclavian and Innominate Arteries

Although the innominate artery and the subclavian arteries can be visualized by CDFI, adequate examination may be difficult if the stenotic lesion is located very far proximal. The normal Doppler waveforms from these arteries demonstrate a brief early diastolic period of reversed flow after a steep systolic peak; later diastolic frequency shift is usually low or absent, indicating high peripheral resistance. Significant proximal obstructions produce a marked decrease in the systolic Doppler frequency, and the early diastolic flow reversal may diseappear [1, 22, 27, 54].

The diagnosis of a hemodynamically relevant stenosis or occlusion of the innominate or subclavian artery may be supported by abnormal flow profiles in the vertebral and carotid arteries.

1. Detection of blue-coded retrograde flow in the vertebral artery is diagnostic for a significant obstruction of the proximal *subclavian artery*. Deflation of a blood pressure cuff, which had been inflated above systolic pressure on the ipsilateral arm, causes an immediately increasing retrograde blood flow velocity in the vertebral artery, indicative of subclavian steal. Since both the vertebral artery and vein run side by side within the transverse foramina, blue-coded Doppler signals in the vein must be distinguished from reversed arterial blood flow. Since only few patients with this condition develop related focal cerebral symptoms, surgical treatment of this rather benign flow abnormality is not generally recommended [27]. An "intermediate" flow pattern similar to that seen in hypoplastic vertebral arteries may also be found in normal vertebral arteries distal to a hemodynamically relevant obstruction. This is characterized by a waveform with initially reversed flow

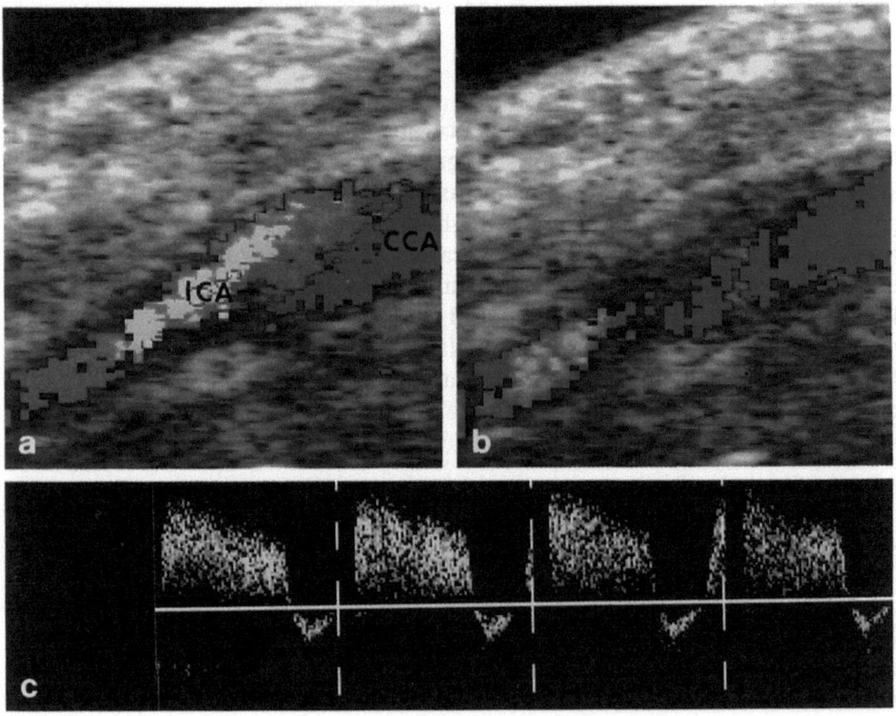

Fig. 25 a–c. Characteristic flow pattern in the carotid artery ipsilateral to an occlusion of the innominate artery. **a** Blue-coded reversed flow in the internal carotid artery in early systole. **b** Slow orthograde flow in diastole. **c** Corresponding Doppler Spectrum shows early systolic flow reversal and moderate flow during diastole

in systole. Inflation of the blood pressure cuff usually makes the systolic drop disappear, while it is more prominent immediately after deflation.

2. In high-grade stenosis or occlusion of the innominate artery the right vertebral similarly demonstrates reversed flow and subclavian steal; however, the Doppler flow profile in the ipsilateral carotid artery is also abnormal. In the CCA and frequently in the ICA a brief period of early systolic blue-coded flow reversal and a decreased diastolic flow is found, resulting in a typical Doppler frequency profile (Fig. 25). During inflation of a blood pressure cough on the right arm the early systolic drop in the Doppler waveform may disappear but become even more marked after deflation. Since hemodynamically significant lesions of the innominate artery may lead to a relevant reduction in cerebral blood supply, the incidence of cerebrovascular events is higher than in patients with carotid or vertebral artery disease [54].

References

1. Ackerstaff RGA, Hoeneveld H, Slowikowski JM, Moll FL, Eikelboom BC, Ludwig JW (1984) Ultrasonic duplex scanning in atherosclerotic disease of the innominate subclavian and vertebral arteries. A comparative study with angiography. Ultrasound Med Biol 10:409–418
2. Arbeille P, Lapierre F, Patat F, Benhamou AC, Alison D, Dusorbier CH, Pourcelot L (1984) Évaluation du degré des sténoses carotidiennes par l'analyse spectrale du signal Doppler. Arch Mal Coeur 77:1097–1107
3. Bandyk DF, Moldenhauer P, Lipchik E, Schreiber E, Pohl L, Cato R, Towne JB (1988) Accuracy of duplex scanning in the detection of stenosis after carotid endarterectomy. J Vasc Surg 8:696–702
4. Bashour TT, Crew JP, Dean M, Hanna ES (1985) Ultrasonic imaging of common carotid artery dissection. J Clin Ultrasound 13:210–211
5. Bendick PJ, Jackson VP (1986) Evaluation of the vertebral arteries with duplex sonography. J Vasc Surg 3:523–530
6. Bluth EI, Kay D, Merritt CRB, Sullivan M, Farr G, Mills NL, Foreman M, Sloan K, Schlater M, Stewart J (1986) Sonographic characterization of carotid plaque: detection of hemorrhage. AJR 146:1061–1065
7. Bluth EI, Shyn PB, Sullivan MA, Merritt CRP (1989) Doppler color flow imaging of carotid artery dissection. J Ultrasound Med 8:149–153
8. Bogousslavsky J, Despland PA, Regli F (1987) Spontaneous carotid dissection with acute stroke. Arch Neurol 44:137–140
9. Chambers BR, Norris JW (1986) Outcome of patients with asymptomatic neck bruits. N Engl J Med 315:860–865
10. Comerota AJ, Cranley JJ, Cook SE (1981) Real-time B-mode carotid imaging in diagnosis of cerebrovascular disease. Surgery 89:718–729
11. Comerota AJ, Katz ML, White JV, Grosh JD (1990) The preoperative diagnosis of the ulcerated carotid atheroma. J Vasc Surg 11:505–510
12. Davis PC, Nilsen B, Braun IF, Hoffman JC (1986) A prospective comparison of duplex sonography vs angiography of the vertebral arteries. AJNR 7:1059–1064
13. O'Donnell TF, Erdoes L, Mackey WC, McCullough J, Shepard A, Heggerick P, Isner J, Callow AD (1985) Correlation of B-mode ultrasound imaging and arteriography with pathologic findings at carotid endarterectomy. Arch Surg 120:443–449
14. Erickson SJ, Mewissen MW, Foley WD, Lawson TL, Middleton WD, Quiroz FA, Macrander SJ, Lipchik EO (1989) Stenosis of the internal carotid artery: assessment using color Doppler imaging compared with angiography. AJR 152:1299–1305

15. Glover JL, Bendick PJ, Dilley RS, Jackson VP, Reilly MK, Dalsing MC, Robinson RJ (1985) Restenosis following carotid endarterectomy. Arch Surg 120:678–684
16. Goes E, Janssens W, Maillet B, Freson M, Steyaert L, Osteaux M (1990) Tissue characterization of atheromatous plaques: correlation between ultrasound image and histological findings. J Clin Ultrasound 18:611–617
17. Gritzmann N, Herold C, Haller J, Karnel F, Schwaighofer B (1987) Duplex sonography of tumors of the carotid body. Cardiovasc Intervent Radiol 10:280–284
19. Hallam MJ, Reid JM, Cooperberg PL (1989) Color-flow Doppler and conventional duplex scanning of the carotid bifurcation: prospective, double-blind, correlative study. AJR 152:1101–1105
20. Hallett JW, Nora JD, Hollier LH, Cherry KJ, Pairolero PC (1988) Trends in neurovascular complications of surgical management for carotid body and cervical paragangliomas: a fifty-year experience with 153 tumors. J Vasc Surg 7:284–291
21. Healy DA, Zierler RE, Nicholls SC, Clowes AW, Primozich JF, Bergelin RO, Strandness DE (1989) Long-term follow-up and clinical outcome of carotid restenosis. J Vasc Surg 10:662–669
22. Hennerici M, Neuerburg-Heusler D (1988) Gefäßdiagnostik mit Ultraschall. Thieme, Stuttgart
23. Hennerici M, Steinke W (1989) Untersuchungen zur Entwicklung extrakranieller Karotisplaques mit der farbkodierten Duplexsonographie. In: Kessler C (ed) Plättchenfunktion und Gefäßwand. TM Verlag, Hameln, p 207
24. Hennerici M, Steinke W (1991) Carotid plaque developments – aspects of hemodynamic and vessel wall interaction. Cerebrovasc Dis 1:142–148
25. Hennerici M, Reifschneider G, Trockel U, Aulich A (1984) Detection of early atherosclerotic lesions by scanning of the carotid artery. J Clin Ultrasound 12:455–464
26. Hennerici M, Hülsbömer HB, Hefter H, Lammerts D, Rautenberg W (1987) Natural history of asymptomatic extracranial arterial disease. Results of a long-term prospective study. Brain 110:777–791
27. Hennerici M, Klemm C, Rautenberg W (1988) The subclavian steel phenomenon: a common vascular disorder with rare neurologic deficits. Neurology 38:669–673
28. Hennerici M, Bürrig KF, Daffertshofer M (1989) Flow pattern and structural changes at carotid bifurcation in hypertensive cynomolgus monkeys. Hypertension 13: 315–321
29. Hennerici M, Steinke W, Rautenberg W (1989) High-resistance Doppler flow pattern in extracranial carotid dissection. Arch Neurol 46:670–672
30. Hennerici M, Steinke W, Rautenberg W, Mohr JP (1991) High-grade carotid stenosis in Doppler color flow imaging. Neurology 41 [Suppl 1]:26P
31. Hirst AE, Johns FJ, Kime SW (1958) Dissecting aneurysm of the aorta: a review of 505 cases. Medicine 37:217–279
32. Hübsch P, Schwaighofer B, Karnel F, Braunsteiner A, Frühwald F, Pichler W, Tratting S (1988) Farbkodierte Doppler-Sonographie der Karotiden. Fortsch Röntgenstr 149:189–192
33. Houi K, Mochio S, Isogai Y, Miyamoto Y, Suzuki N (1990) Comparison of color flow and 3D image by computer graphics for the evaluation of carotid disease. Angiology 41:305-312
34. Imparato AM, Bracco A, Kim GE, Zeff R (1972) Intimal and neointimal fibrous proliferation causing failure of arterial reconstructions. Surgery 72:1007–1017
35. Imparato AM, Riles TS, Mintzer R, Baumann FG (1983) The importance of hemorrhage in the relationship between gross morphologic characteristics and cerebral symptoms in 376 carotid artery plaques. Ann Surg 197:195–203
36. Jacobs NM, Grant EG, Schellinger D, Byrd MC, Richardson JD, Cohan SL (1985) Duplex carotid sonography: criteria for stenosis, accuracy, and pitfalls. Radiology 154:385–391
37. Johnston KW, Baker WH, Burnham SJ, Hayes AC, Kupper CA, Poole MA (1986) Quantitative analysis of continuous-wave Doppler spectral broadening for the diagnosis of carotid disease: Results of a multicenter study. J Vasc Surg 4:493–504
38. Kotval PS, Babu SC, Fakhry J, Cozzi A, Barakat K (1988) Role of the intimal flap in arterial dissection: sonographic demonstration. AJR 150:1181–1182

39. Ku DN, Giddens DP, Phillips DJ, Strandness DE (1985) Hemodynamics of the normal carotid bifurcation: in vitro and in vivo studies. Ultrasound Med Biol 11:13–26
40. O'Leary DH, Bryan FA, Goodison MW, Rifkin MD, Gramiak R, Ball M, Bond MG, Dunn RA, Goldberg BB, Toole JF, Wheeler HG, Gustafson NF, Ekholm S, Raines JK (1987) Measurement variability of carotid atherosclerosis: real-time (B-mode) ultrasonography and angiography. Stroke 18:1011–1017
41. LoGerfo FW, Nowak MD, Quist WC (1985) Structural details of boundary layer separation in a model human carotid bifurcation under steady and pulsatile flow conditions. J Vasc Surg 2:263–269
42. Mäkäräinen H, Päivänsalo M, Hyrynkangas K, Leinonen A, Siniluoto T (1986) Sonographic patterns of carotid body tumors. J Clin Ultrasound 14:373–375
43. Marosi L, Ehringer H, Piza F, Wagner O (1984) Die frühpostoperative Morphologie der Arteria carotis nach Endarteriektomie: Systematische prospektive Untersuchungen mit einem hochauflösenden Ultraschall-Duplex-Echtzeit-Darstellungssystem. Ultraschall 5:202–214
44. Middleton WD, Foley WD, Lawson TL (1988) Flow reversal in the normal carotid bifurcation: color Doppler flow imaging analysis. Radiology 167:207–209
45. Middleton WD, Foley WD, Lawson TL (1988) Color-flow Doppler imaging of carotid artery abnormalities. AJR 150:419–425
46. Mitchell DG (1990) Color Doppler imaging: principles, limitations, and artifacts. Radiology 177:1–10
47. Mokri B, Sundt TM, Houser OW, Piepgras DG (1986) Spontaneous dissection of the cervical internal carotid artery. Ann Neurol 19:126–138
48. Motomiya M, Karino T (1984) Flow patterns in the human carotid artery bifurcation. Stroke 15:50–56
49. Nicholls SC, Phillips DJ, Primozich JF, Lawrence RL, Kohler TR, Rudd TG, Strandness DE (1989) Diagnostic significance of flow separation in the carotid bulb. Stroke 20:175–182
50. Palmaz JC, Hunter G, Carson SN, French SW (1983) Postoperative carotid restenosis due to neointimal fibromuscular hyperplasia. Radiology 148:699–702
51. Philips DJ, Greene FM, Langlois Y, Roederer GO, Strandness DE (1983) Flow velocity patterns in the carotid bifurcations of young, presumed normal subjects. Ultrasound Med Biol 9:39–49
52. Polak JF, O'Leary DH, Quist WC, Creager MA, LoGerfo FW (1990) Pulsed and color Doppler analysis of normal carotid bifurcation flow dynamics using an in-vitro model. Angiology 41:241–247
53. Polak JF, Dobkin GR, O'Leary DH, Wang AM, Cutler SS (1989) Internal carotid artery stenosis: accuracy and reproducibility of color-Doppler-assisted duplex imaging. Radiology 173:793–798
54. Rautenberg W, Hennerici M (1988) Pulsed Doppler assessment of innominate artery obstructive diseases. Stroke 19:1514–1520
55. Rautenberg W, Mess W, Hennerici M (1990) Prognosis of asymptomatic carotid occlusion. J Neurol Sci 98:213–220
56. Rautenberg W, Steinke W, Schwartz A, Hennerici M (1990) Common carotid artery occlusion – clinical and diagnostic aspects. J Neurol 237:157
57. Reneman RS, Van Merode T, Hick P, Hoeks APG (1985) Flow velocity patterns in and distensibility of the carotid artery bulb in subjects of various ages. Circulation 71:500–509
58. Ricotta JJ, Bryan FA, Bond MG, Kurtz A, O'Leary DH, Raines JK, Berson AS, Clouse ME, Calderon-Ortiz M, Toole JF, DeWeese JA, Smullens SN, Gustafson NF (1987) Multicenter validation study of real-time (B-mode) ultrasound, arteriography, and pathologic examination. J Vasc Surg 6:512–520
59. Ringelstein EB, Zeumer H, Angelou D (1983) The pathogenesis of strokes from internal carotid artery occlusion. Diagnostic and therapeutical implications. Stroke 14:867–875
60. Rittgers SE, Shu MCS (1990) Doppler color-flow images from a stenosed arterial model: interpretation of flow patterns. J Vasc Surg 12:511–522

61. Rittgers SE, Thornhill BM, Barnes RW (1983) Quantitative analysis of carotid artery spectral waveforms: diagnostic value of parameters. Ultrasound Med Biol 9:255–264
62. Robinson ML, Sacks D, Perlmutter GS, Marinelli DL (1988) Diagnostic criteria for carotid duplex sonography. AJR 151:1045–1049
63. Roederer GO, Langlois YE, Jager KA, Primozich JF, Beach KW, Phillips DJ, Strandness DE (1984) The natural history of carotid arterial disease in asymptomatic patients with cervical bruits. Stroke 15:605–613
64. Sanders EACM, Hoeneveld H, Eikelboom BC, Ludwig JW, Vermeulen FEE, Ackerstaff RGA (1987) Residual lesions and early recurrent stenosis after carotid endarterectomy. J Vasc Surg 5:731–737
65. Schenk EA, Bond MG, Aretz TH, Angelo JN, Choi HY, Rynalski T, Gustafson NF, Berson AS, Ricotta JJ, Goodison MW, Bryan FA, Goldberg BB, Toole JF, O'Leary DH (1988) Multicenter validation study of real-time ultrasonography, arteriography, and pathology: pathologic evaluation of carotid endarterectomy specimens. Stroke 19:289–296
66. Schmid-Schönbein H, Wurzinger LJ (1988) Vortex transport phenomena of the carotid trifurcation: interaction between fluid-dynamic transport phenomena and hemostatic reactions. In: Hennerici M, Sitzer G, Weger HD (eds) Carotid artery plaques. Karger, Basel, p 64
67. Shulak JM, O'Donovan PB, Paushter DM, Lanzieri CF (1989) Color flow Doppler of carotid body paraganglioma. J Ultrasound Med 8:519–521
68. Steinke W, Al-Deeb S, Hennerici M (1987) Prävalenz extrakranieller Gefäßprozesse und Risikoprofil bei Patienten mit peripherer und koronarer Arteriopathie. VASA 16:283–290
69. Steinke W, Hennerici M, Aulich A (1989) Doppler color flow imaging of carotid body tumors. Stroke 20:1574–1577
70. Steinke W, Aulich A, Hennerici M (1989) Diagnose und Verlauf von Carotisdisektionen. Dtsch Med Wschr 114:1869–1875
71. Steinke W, Kloetzsch C, Hennerici M (1990) Variability of flow patterns in the normal carotid bifurcation. Atherosclerosis 84:121–128
72. Steinke W, Kloetzsch C, Hennerici M (1990) Carotid artery disease assessed by color Doppler flow imaging. AJNR 11:259–266
73. Steinke W, Schwartz A, Hennerici M (1990) Doppler color flow imaging of common carotid artery dissection. Neuroradiology 32:502–505
74. Steinke W, Kloetzsch C, Hennerici M (1991) Doppler color flow imaging after carotid endarterectomy. Eur J Vasc Surg (in press)
75. Sterpetti AV, Schultz RD, Feldhaus RJ, Davenport KL, Richardson M, Farina C, Hunter WJ (1988) Ultrasonographic features of carotid plaque and the risk of subsequent neurologic deficits. Surgery 104:652–660
76. Taylor LM, Loboa L, Porter JM (1988) The clinical course of carotid bifurcation stenosis as determined by duplex scanning. J Vasc Surg 8:255–261
77. Taylor DC, Strandness DE (1987) Carotid artery duplex scanning. J Clin Ultrasound 15:635–644
78. Trockel U, Hennerici M, Aulich A, Sandmann W (1984) The superiority of combined continuous wave Doppler examinations over periorbital Doppler for the detection of extracranial carotid disease. J Neurol Neurosurg Psychiatry 47:43–50
79. Trattnig S, Hübsch P, Schuster H, Polzleitner D (1990) Color-coded Doppler imaging of normal vertebral arteries. Stroke 21:1222–1225
80. Wolverson MK, Bashiti HM, Peterson GJ (1983) Ultrasonic tissue characterization of atheromatous plaques using a high resolution real time scanner. Ultrasound Med Biol 9:599–609
81. Zierler RE, Phillips DJ, Beach KW, Primozich JF, Strandness DE (1987) Noninvasive assessment of normal carotid bifurcation hemodynamics with color-flow ultrasound imaging. Ultrasound Med Biol 13:471–476
82. Zirkle PK, Wheeler JR, Gregory RT, Snyder SO, Gayle RG, Sorrell K (1984) Carotid involvement in aortic dissection diagnosed by duplex scanning. J Vasc Surg 1:700–703
83. Zwiebel WJ, Knighton R (1990) Duplex examination of the carotid arteries. Semin Ultrasound CT MRI 11:97–135

Chapter 10

Color Doppler Imaging of Abdominal Vessels

K. HAAG and P. LANZER

Introduction

Color Doppler flow imaging (CDFI) is a new ultrasonographic modality for assessing intraabdominal vascular pathology. By generating two-dimensional flow maps superimposed on B-mode images, CDFI facilitates the identification of intraabdominal vascular pathology and therefore increases the realibility of duplex examinations while decreasing the examination time. The quality of CDFI data depends greatly on the operator's skill and experience. The sonographer performing abdominal CDFI should be thoroughly familiar with the normal B-mode sonography, standard duplex techniques, and the anatomy of the abdominal organs and their vasculature. Correct data interpretation also requires an understanding of vascular pathology and hemodynamic principles. An operational understanding of color Doppler imaging, instrumentation, and technology is also needed.

Establishing a routine abdominal CDFI protocol will aid in producing consistent and reproducible results. To avoid respiratory effects on venous hemodynamics it is preferable to acquire all quantitative data during breath holding. Fasting also facilitates the examination. Gentle pressure on the transducer avoids interference with venous hemodynamics. The examinations begin with conventional B-mode sonography to define the topography and morphology of the abdominal organs. The operator notes the position, size, and texture of the parenchymal organs and documents any structural abnormality [1]. CDFI is then used to identify the vessels of interest and to determine their flow characteristics. Whenever possible, the vessels are examined in both longitudinal and transverse directions along the standard imaging planes [2]. To minimize errors the vessels should be examined along their complete course. Basic imaging planes and transducer orientations for an abdominal CDFI examination have recently been outlined. Quantitative flow measurements are accomplished at present using the standard Doppler methods [3]. To perform reliable Doppler and gray-scale imaging-related measurements the examiner should be familiar with the potential sources of errors [4, 5] and with the essentials of splanchnic hemodynamics [6]. Quantitative assessments of flow directly from the CDFI data should be possible in the future [7].

Portal System

Potential indications for a CDFI examination of the portal system include:

- Portal hypertension, ascites of unknown etiology, esophageal varices
- Liver cirrhosis
- Venoocclusive disease and Budd-Chiari syndrome
- Splenomegaly
- Portosystemic shunts
- Space-occupying lesions in the liver
- Abdominal trauma
- Gastrointestinal bleeding without endoscopically confirmed cause
- Abdominal angina

Examination Technique

A complete examination of the portal circulation should include evaluation of the celiac trunk and the superior and inferior mesenteric arteries. The topographic anatomy (see Chap. 1; [1, 8]) and hemodynamics [9] of the portal venous system should be reviewed and well understood.

The examination begins by locating the portal vein. The ultrasound beam is directed obliquely in the upper right quadrant between the umbilicus and the costophrenic angle. In this orientation the portal vein and its right main branch can be visualized in its long axis from the venous confluence to its division into the right and left branches (Fig. 1). A large anatomic angle

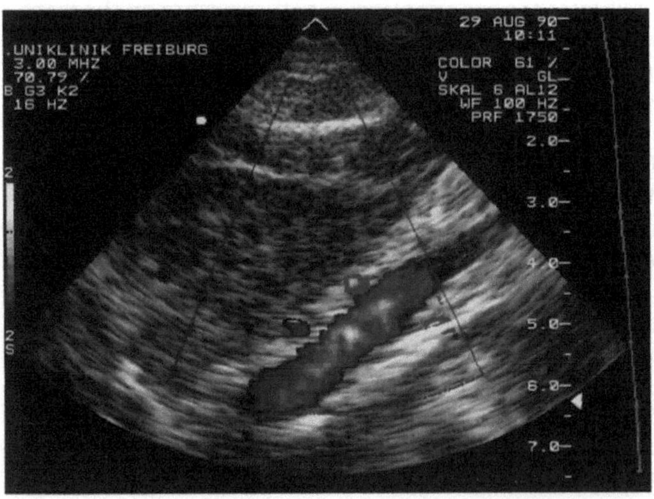

Fig. 1. Normal portal vein. ***Blue*** indicates hepatopetal flow from the venous confluence to the hepatic portal. *White* within the vessel represents a higher flow velocity in the central portions of the vessel. The *red spot* ventral to the portal vein corresponds to the proper hepatic artery. (All CFDM images were produced by Ultramark 9, ATL, Solingen, FRG)

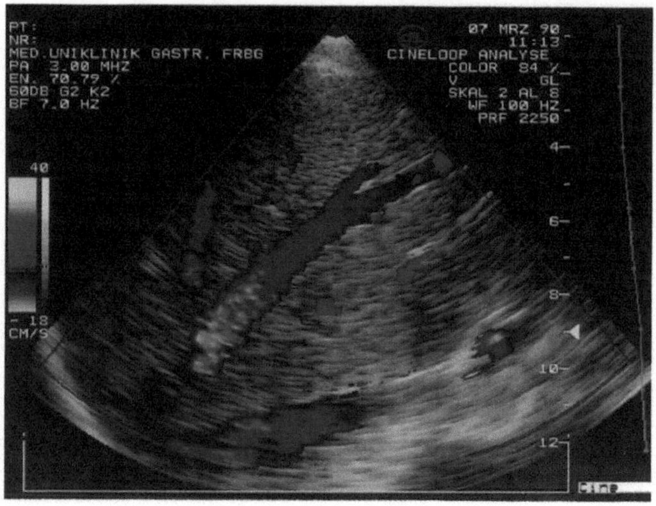

Fig. 2. Right lateral view of the hepatic veins of the right liver lobe

between the course of the vein and the ultrasound beam makes a more caudal approach necessary. In patients with ascites, bowel gas, or abdominal dressings the vein can be imaged laterally from the right side using an oblique transducer orientation. The left main portal branch is imaged from ventral epigastric with a transverse transducer orientation. Higher order portal vein branches are identified within the liver parenchyma.

The hepatic veins are imaged from the upper right quadrant with a transverse transducer orientation at the midclavicular line. Imaging of the left veins is often facilitated from a more left lateral orientation while the patient holds his breath in deep inspiration. The veins of the right liver lobe are frequently better visualized from a more right lateral oblique position (Fig. 2).

The splenic vein is typically examined from the epigastrium longitudinally in a coronal plane or transversally in a left parasagittal plane. This approach allows visualization of the vessel from the end at the portal vein to the tail of the pancreas (Fig. 3). The hilar segment can be visualized from the left lateral using the spleen as an echogenic window. Blood flow should be also assessed in several parenchymal branches if splenic infarction is suspected.

The superior mesenteric vein can be imaged following its oblique course by orienting the transducer from left caudal to right lateral in the upper right quadrant. Frequently large angles of insonation require moving the transducer more cranially or caudally from its original position to obtain Doppler flow measurements. The inferior mesenteric vein displays a more variable anatomic course and may frequently be identified at the confluence with the splenic vein or the superior mesenteric vein. The left and right gastric veins can be identified close to the porta hepatis where they empty into the portal vein, with the left vein occasionally joining the splenic vein. The recognition of a retrograde

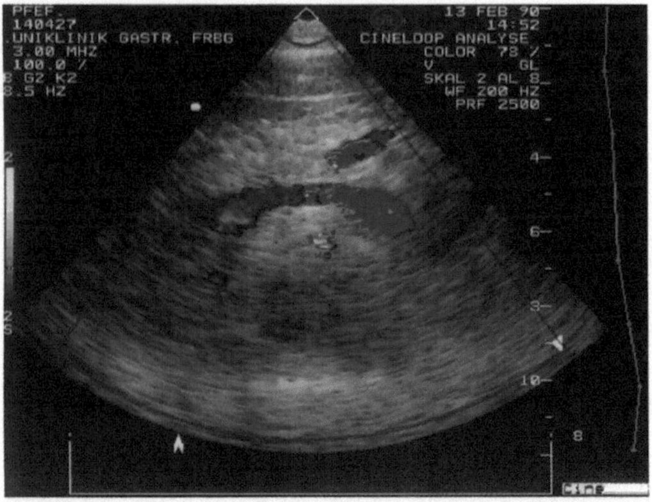

Fig. 3. Splenic vein. *Red* and *blue* correspond to blood flow towards and away from the transducer, respectively

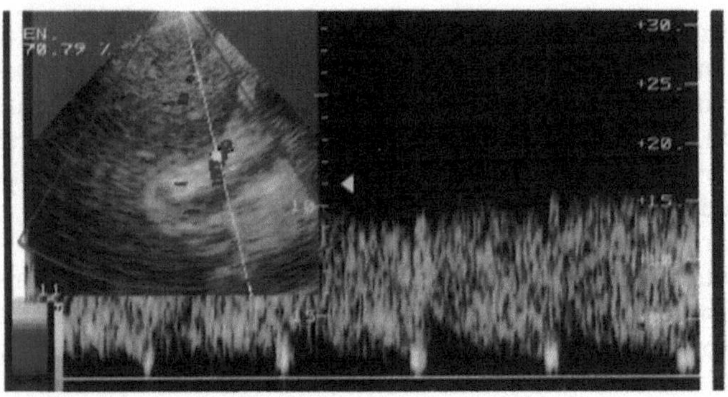

Fig. 4. Right gastric vein in a patient with liver cirrhosis and a portal hypertension. *Red* represents the hepatofugal retrograde blood flow

gastric venous flow is important in the diagnosis of portal hypertension (Fig. 4). Infrequently, downstream filling of a thrombosed portal vein via the right gastric vein can be observed.

The umbilical vein joins the left branch of the portal vein, where it is easily identified (Fig. 5).

Varicose veins of the gastric cardia and the esophagus are recognized as tortuous vascular structures at the level of the epigastrium and the upper left quadrant. When there is splenic obstruction, the short gastric veins can be visualized as short stumps originating from the splenic vein close to the hilus and moving upward toward the greater curvature of the stomach. Other small

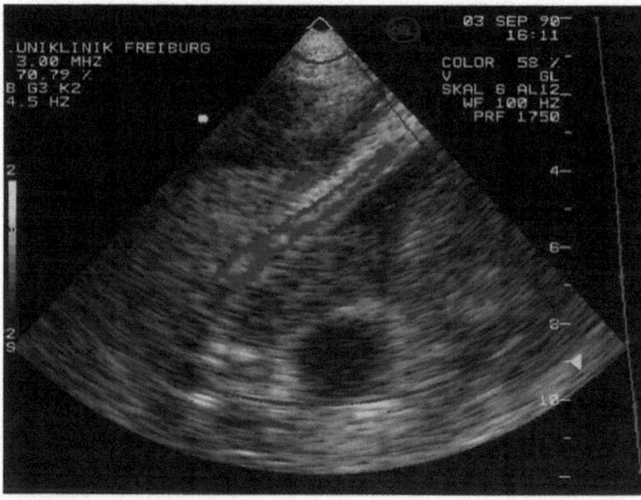

Fig. 5. Collateral umbilical circulation in a patient with liver cirrhosis, portal hypertension and ascites

vessels providing collateral circulation in the presence of a portal hypertension move from the hilar segment of the splenic vein downward and into the retroperitoneum. Other portocaval collateral pathways occasionally documented by CDFI are the veins of the gallbladder walls emptying into the veins of the liver capsula. Small veins originating in the territory of the inferior mesenteric vein and joining the hemorrhoidal plexus of the inferior and middle rectal vein may also occasionally be seen with CDFI.

In addition to its use in examining the native portal vasculature, CDFI is becoming an important method for assessing the functional status of surgical

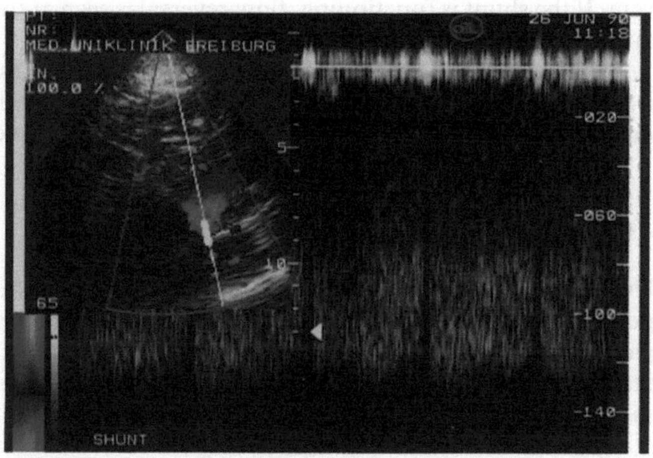

Fig. 6. Portocaval shunt. The patency of the shunt and a high flow velocity are demonstrated (V_{max} = 130 cm/s)

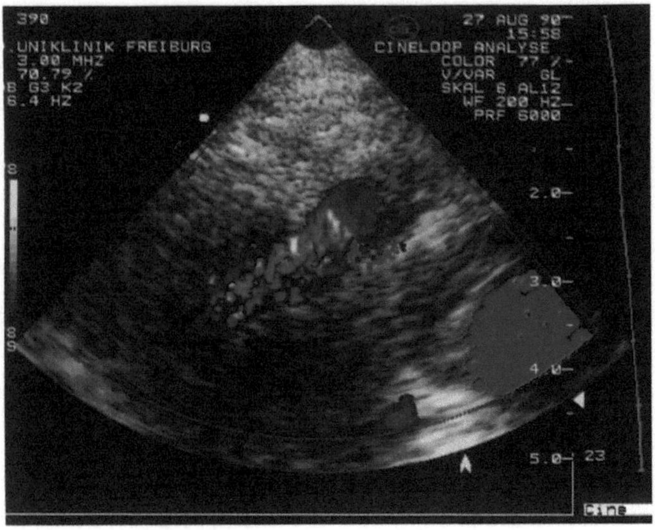

Fig. 7. Mesocaval shunt. A mesocaval shunt in a transverse section caudally to the renal arteries is demonstrated. The patency of the shunt is documented by the turbulent flow (*mosaic color*) from the superior mesenteric vein to the cava inferior. The aorta is located at the right bottom (*red*)

portosystemic shunts [10]. Proximal and distal splenorenal shunts are best seen in a cross-sectional view with the transducer in the left lateral or anterior position. Portocaval shunts are visualized using techniques described for portal vein imaging in short- and long-axis views along the hepatoduodenal ligament (Fig. 6). Mesocaval shunts are visualized along the superior mesenteric vein. If the shunt is functioning, flow reversal is seen at the level of its proximal anastomosis (Fig. 7).

To allow correct interpretations of the CDFI findings in the portal system it is important to perform all examinations consistently in fasting patients during breath holding in midexpiration or during a quiet respiration.

Pathologic Findings

Normal Doppler waveforms and the flow directions within the veins of the portal system must be recognized to allow a correct assessment of portal pathology [11]. Recognition of abnormal flow patterns using CDFI greatly facilitates and shortens the examination by quickly determining the optimal sampling site for Doppler-related measurements.

Liver Cirrhosis and Portal Hypertension. The CDFI findings should be interpreted within the context of the entire diagnostic picture. Individual findings

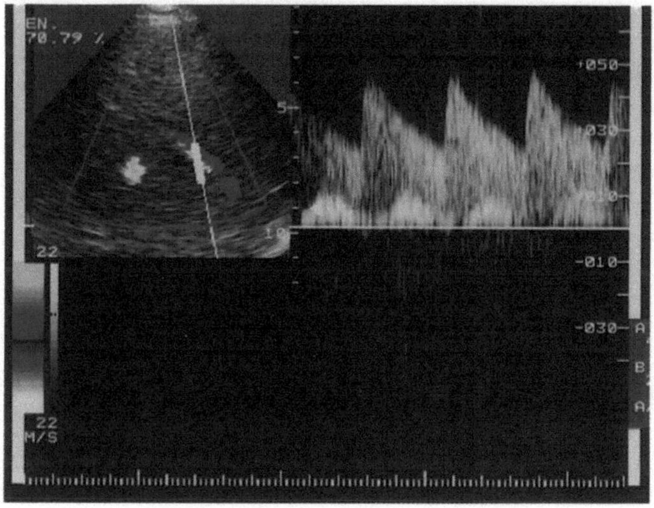

Fig. 8. Predominance of arterial blood flow over the portal venous blood flow in a patient with liver cirrhosis

are often nonspecific and misleading. The effect of liver cirrhosis on portal circulation is variable, reduced maximum flow velocity in the portal vein, increased portal vein diameter, and decreased flow rates [12–14] as well as normal caliber of the portal vein [15] have been reported. In our own experience the flow rates (normal values in our laboratory, 600–1000 ml/min) in patients with cirrhosis can be normal, decreased or even elevated. We observe high flow rates most frequently in patients with early stages of cirrhosis, splenomegaly and a patent umbilical vein. In patients with advanced cirrhosis the flow rates tend to decline, reaching subnormal levels.

In the majority of our patients with an advanced liver cirrhosis the arterial Doppler flow signal "dominates" over the portal venous Doppler flow signal whereas in healthy subjects the portal flow signal "dominates" (Fig. 8). The dominance is assessed using a large-range gate spanning both the vein and the accompanying artery. The dominant flow is determined qualitatively based on the relative intensities of the respective frequency spectra. However, the true hemodynamic dominance, i.e., difference in the ratio of arterial-to-venous volumetric flow rates corresponding to the relative intensities of the Doppler spectra, remains to be documented. In the presence of the dominance of the arterial flow signal the direction of the arterial and the venous flow should be noted. In the majority of patients flow is unidirectional, as in normal subjects. In a few patients, thought to be those with an advanced cirrhosis, bidirectional hepatopetal and hepatofugal flow can be seen. Similar qualitative observations were reported by others [16]. In this report the dominance of the arterial flow was termed "arterialization" and ascribed to patients with a diminished net portal venous flow. In some patients with portal hypertension the reversed

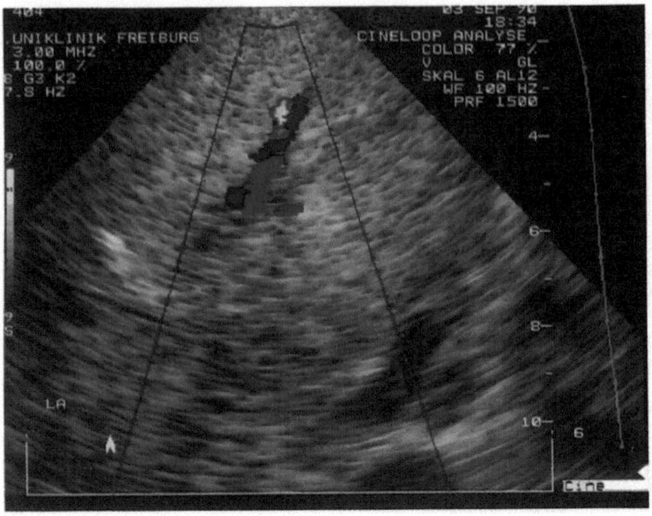

Fig. 9. Liver cirrhosis–portal flow reversal. The blood in the branches of the hepatic artery and the portal vein flows in opposite directions: hepatopetal arterial flow in *red*, hepatofugal venous flow in *blue*

flow in the portal vein, its large tributaries, or its efferent intrahepatic branches can be documented by CDFI, Fig. 9 [16].

CDFI greatly facilitates identification of collateral vessels with hepatofugal flow in the coronary (left gastric) veins of the stomach (Fig. 4), in the parumbilical vein (Fig. 5), the dilated short veins of the stomach walls, perisplenic and retroperitoneal vessels, superficial abdominal and gallbladder varicose veins [16]. In our experience some patients with liver cirrhosis complicated by a splenomegaly the splenic vein flow may increase four- to fivefold of the normal (normal value in our laboratory, 200–300 ml/min) potentially contributing to the severity of the portal hypertension. The flow in the hepatic venous circulation in patients with cirrhosis can also vary. In our experience in patients with a nearly normal calculated portal volumetric flow rate the maximum parenchymal hepatic vein flow velocity varies among individual vessels with maxima exceeding 30 cm/s (Fig. 10). These changes are, however, less prominent or are absent in patients with reduced portal flow. Flow acceleration is thought to be due to regional vein compressions secondary to regenerative nodules. However, flow accelerations in the central parenchymal branches of large hepatic veins are also occasionally observed in healthy subjects secondary to diaphragmatic hepatic compression during inspiration. To ascertain pathology flow accerelations should therefore also be documented in the peripheral branches of the hepatic veins. Due to the frequency of regional hepatic flow differences multiple segments of liver parenchyma should be evaluated. Regional flow acceleration can also be observed in the presence of space-occupying lesions in the liver. Therefore, liver metastasis should be considered even when they can not be seen on B-mode images.

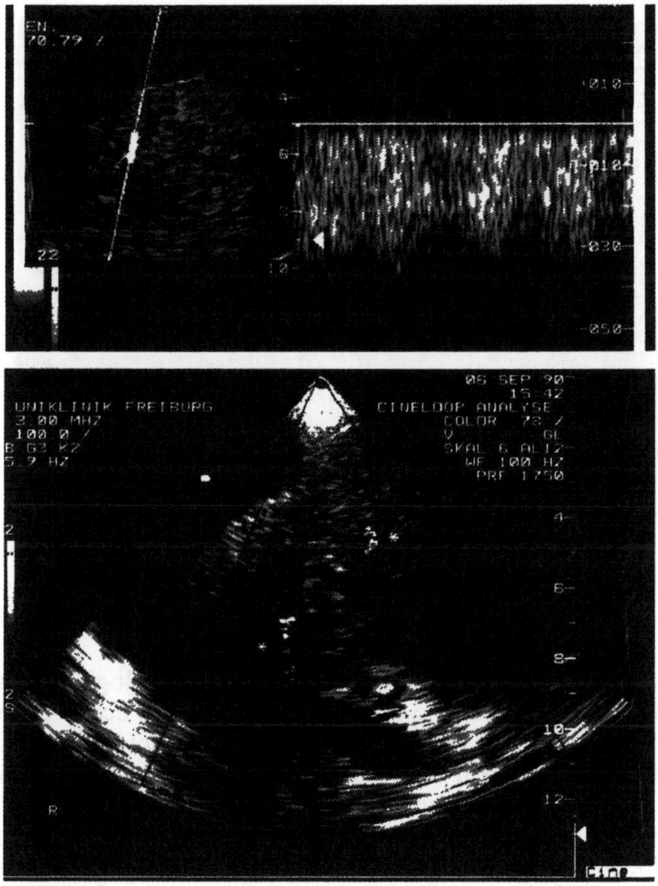

Fig. 10 a, b. Regional compression of a peripheral hepatic vein in liver cirrhosis. **a** $V_{max} > 30$ cm/s is demonstrated in the Doppler velocity tracing. **b** Flow velocity normalization down stream to the stenosis (*)

Portal Vein Thrombosis. A complete and incomplete portal vein thrombosis of various etiologies based on the absence of color-coded and Doppler signal is reliably detected on the CDFI examination [16]. Older thrombosis is often documented already on the B-mode image by direct visualization of the organized thrombus within the vessel lumen [1]. In about one-third of patients this fibrotic cord can be observed to be surrounded by numerous small collaterals (Fig. 11), the appearance of which is termed in duplex imaging "cavernous transformation" [17]. Occasionally, reconstitution of the portal vein or its branches with a partially patent lumen and residual flow can be documented. An acute portal vein thrombosis should always be excluded by duplex or CDFI because the hypoechoic fresh thrombus can easily be overlooked on the B-mode image.

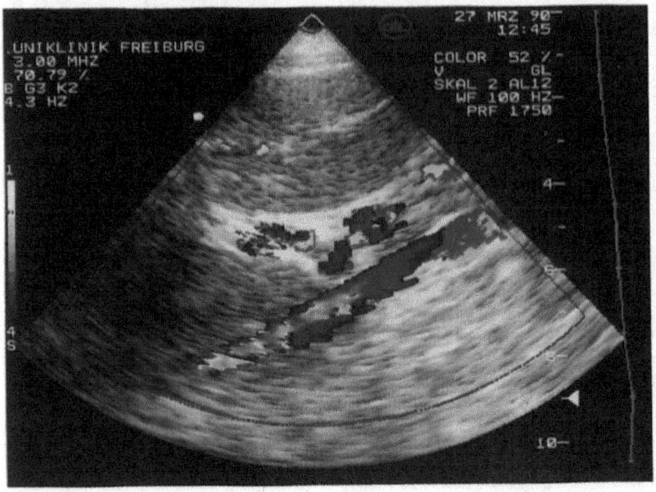

Fig. 11. Portal vein thrombosis. The small colored tortuous vessels ventral to the inferior vena cava correspond to the so-called cavernous transformation of the portal vein

Splenic Vein Thrombosis. A variety of pathologic conditions may accompany thrombosis of the splenic vein; this may occur either in isolation or in conjunction with trombosis involving other splanchnic veins [18]. The patency of the splenic vein should routinely be established in patients with suspected pathology of the pancreas and gastric vein varicosis [19]. Based on our experience the CDFI is useful in detecting thrombosis of the splenic vein. This is documented by the absence of the color-coded flow at the level of the body of the pancreas and by visualization of collaterals in the proximity of the hilum of the spleen. In patients with incomplete obliteration of the splenic vein the upstream blood flow velocity in the proximity of the splenic hilus is in our experience often reduced, usually to less than 15 cm/s. In contrast to patients with portal hypertension, in whom splenomegaly is a common finding, it appears to be less frequent in patients with an isolated incomplete thrombosis of the splenic vein.

Thrombosis of the Superior Mesenteric Vein. The occurrence of thrombosis in the superior mesenteric vein may be seen either in isolation or in conjunction with thrombosis of other splanchnic veins. It is often found in patients with an identifiable coagulopathy or in those with a previous history of thrombosis [20]. Experience in our laboratory shows that CDFI is useful in diagnosis of these thromboses. An occluded superior mesenteric vein usually appears as a hypoechoic cord on the B-mode image with absent Doppler and color-coded flow on the CDFI. The smaller intestinal veins accompanying the superior mesenteric vein often remain patent, maintaining a normal hepatopetal flow. However, the normally smooth curvilinear course of these vessels is often

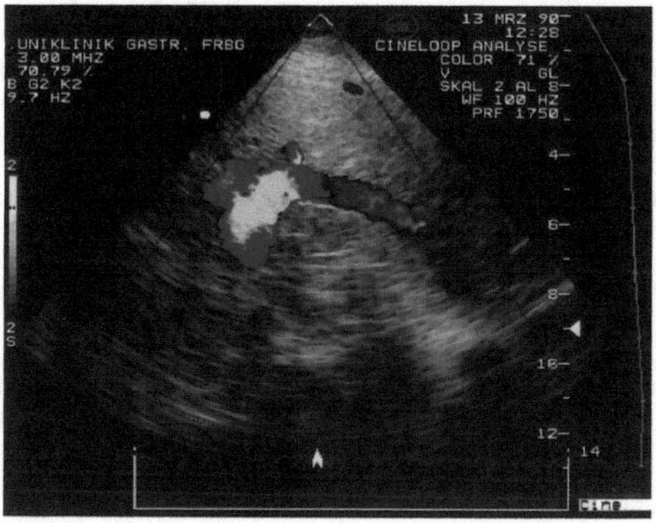

Fig. 12. Incomplete Budd-Chiari syndrome. Reversed flow direction in a hepatic vein with drainage of the blood via an irregular hepatic vein from the right to the left liver lobe is demonstrated

interrupted to follow a tortuous trajectory toward the hepatic porta. Close association between thrombosis of the superior mesenteric vein and pathology of the head of the pancreas mandates further diagnostic evaluations of this region when thrombosis of the mesenteric vein is detected.

Hepatic Venoocclusive Disease and Budd-Chiari Syndrome. The etiology of hepatic venooclusive disease and Budd-Chiari syndrome and the associated duplex findings have been reviewed [21, 22]. In patients with a venooclusive disease a reversed or bidirectional flow pattern and/or decrease in the volumetric flow rates in the portal vein can be documented [21]. Based on a limited experience in our laboratory and those of others [23], CDFI appears to be useful in the assessment of patients with Budd-Chiari syndrome. In these patients color-coded flow images reveal a reversed or turbulent blood flow, changes in flow pattern from phasic to continuous, or absence of flow in the hepatic veins (Fig. 12) and/or in the inferior vena cava. Slow blood flow in the portal vein [24] and in our experience an unusual tortuosity of the hepatic veins occasionally filled with echogenic material as well as a markedly increased flow velocity in the remaining still patent hepatic veins at the confluence with the inferior vena cava can often be recognized.

Tricuspid Insufficiency and Congestive Heart Failure. In patients with tricuspid incompetence or congestive heart failure the intrahepatic veins are typically dilated. The blood flow pattern in the portal vein suggestive of congestive heart failure includes monophasic pulsatile forward flow with maximum flow velocity during ventricular diastole, reversed flow during systole, biphasic pancyclic forward flow [25], and increased pulsatility [26]. An increased flow pulsatility in the portal vein [27] and decrease in the maximum forward systolic flow velocity or systolic flow reversal in the hepatic vein [28] appear to be associated with tricuspid insufficiency. Systolic backflow results, in some patients, in a pulsatile flow profile in the superior mesenteric and splenic veins. However, in contrast to patients with arterioportal shunts due to large hemangiomas or hepatocellular carcinoma the flow direction remains in most cases hepatopetal throughout the entire length of the cardiac cycle.

Portosystemic Shunts. Compared to simple duplex imaging CDFI allows a faster and more realible determination of patency of the portosystemic shunts. In addition, CDFI appears superior in assessing the patency of the distal splenorenal shunts and in evaluations of shunt anastomoses [29]. Based on our experience, a partial shunt obstruction should be suspected when the maximum flow velocity within the shunt lumen exceeds 200 cm/s (normal range established in our laboratory, 50–150 cm/s). However, quantitative Doppler measurements within the shunt lumen can be difficult at times due to the presence of local hematomas, deep and tortuous shunt position, or echogenitcity of the walls of the synthetic grafts. In these patients it is preferable to measure shunt flow upstream in the afferent vessel. Here, the blood flow pattern typically fluctuates and undulates with respiration; based on standards in our laboratory the expected minimum volumetric flow rate exceeds 500 ml/min. In our patients the reduced portal hypertension following a successful shunt implantation is reflected by disappearance of preoperatively documented flow in the paraumbilical vein. The exception to this observation appears to be the splenorenal shunt. Here, adequate shunt function appears to be consistent with flow reduction of at least 50% in the umbilical vein. Figures 6 and 7 represent some of the typical CDFI findings in patients with portosystemic shunts. Recently a nonoperative percutaneous palliative treatment of portal hypertension using a transjugular intrahepatic implantation of portocaval expandable stents has been developed [30]. Portal hemodynamics associated with these stents are characterized by a high flow, typically 1.5–2.0 l/min in the portal vein and maximum flow velocities of 100–200 cm/s within the stent (Fig. 13).

Upper Abdominal Arteries

The indications for a CDFI examination of the celiac trunk and its branches and the superior and inferior mesenteric arteries include those listed in section Portal System.

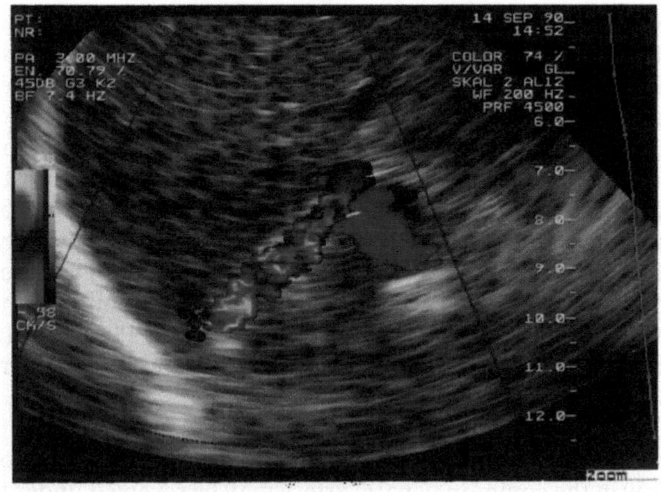

Fig. 13. Intrahepatic portocaval stent. The stent is visualized from right lateral by the highly turbulent flow (*green*) from the origin at the right portal branch (*red*) nearly up to the outflow into the hepatic vein

Examination Technique

The examination begins by locating the celiac trunk at its origin from the abdominal aorta. The longitudinal transducer orientation facilitates separation from the superior mesenteric artery. In a transverse transducer orientation the proximal segments of the common hepatic artery and the splenic artery can be visualized. The hepatic artery and the superior mesenteric artery can be imaged in their long axis left parallel to the portal vein and the superior mesenteric vein, respectively (see above). The ultrasonographer should be familiar with the common arterial variations of the upper abdominal arteries [31]. In particular, the common origin of the celiac trunk and the superior mesenteric artery in the celiacomesenterial trunk should be recognized. In this anomaly high blood flow with accompanying flow disturbance are frequently demonstrated on CDFI. Also, the anomalous origin of the right or common hepatic artery from the superior mesenteric artery can pose a differential diagnostic puzzle to an unsuspecting examiner. The knowledge of normal Doppler flow patterns is important to assess pathology [17].

Pathologic Findings

Celiac Trunk and Superior Mesenteric Artery. Stenoses of the celiac trunk and those up to approximately 5 cm from the origin in the superior mesenteric artery can be, in our experience, reliably confirmed or excluded by CDFI examination. A V_{max} greater than 300 cm/s has been advocated as a hallmark of a hemodynamically significant stenoses [32]. The increase in V_{max} can be less in long, complex lesions. In these patients CDFI typically reveals a highly turbulent flow pattern. Less pronounced turbulent color-coded flow pattern at the origin of the upper abominal arteries often represents complex flow at branch points and should not be confused with a poststenotic flow field. Although proximal stenoses are usually atherosclerotic, obstructions resulting from mechanical compression from carcinoma of the body of the pancreas or enlarged lymph nodes should also be considered. Detection of distal superior mesenteric artery stenoses by CDFI is, in our limited experience, less consistent. A pronounced downstream decrease in Pourcelot (resistive) index can raise a suspicion of an upstream stenosis. However, the sensitivity of this sign, when present, has not been established.

In an occlusion of the celiac trunk the arterial blood is supplied to the liver via the arcades of the pancreatic arteries and the gastroduodenal artery. Blood flow in the common hepatic artery may be reversed under these circumstances. Proximal occlusions of the superior mesenteric artery are documented by the absence of the color-coded signal on the CDFI. A concomitant compensatory flow increase with a prominent diastolic flow component in the inferior mesenteric artery and the celiac trunk has been observed in our laboratory in several patients with this rare pathology. Based on examinations in our laboratory the normal values for V_{max}, the resistive index (RI), and the volumetric flow rates (VFR) in the celiac trunk and the superior mesenteric artery are: $V_{max} = 100-200$ cm/s and $70-180$ cm/s; $RI = 0.6-0.75$ and over 0.80 (fasting); $VFR = 600 \pm 100$ ml/min and 500 ± 100 ml/min (fasting). However, the blood flow to the intestine is subject to a great number of intrinsic (e.g., metabolic, myogenic) and extrinsic regulatory factors (e.g., nervous system, hormones, drugs) [6, 33], thus making meaningful and reproducible measurements difficult. The ability to increase the blood flow in response to a meal has been evaluated as a means to determine the reactivity of the splanchnic arterial circulation [34, 35]. However, due to complexity of postprandial hemodynamics the clinical utility of this test remains uncertain [36].

Hepatic and Splenic Arteries. Duplex scanning [17, 37] and more recently CDFI [38] have been primarily advocated to establish patency of the hepatic artery in liver transplant recipients. Hepatic artery thrombosis represents the most common vascular complication after liver transplantation, requiring prompt retransplantation [38]. Splenic artery stenoses are of a lesser clinical importance, with the exception of when they are detected in patients with pancreas transplantation [39]. In these patients an increased (RI over 0.7) was highly correlated with episodes of rejection. Occasionally, an extremely tortuous course of the splenic artery may mimic the CDFI findings associated with splenic artery aneurysm. On the other hand, aneurysms, pseudoaneurysms, and mesenteric fistulas may simulate B-mode findings of a pseudocyst of the pancreas. Distinction is easily made, however by CDFI, where a circular blood flow in the presumed pseudocyst is documented (Fig. 14).

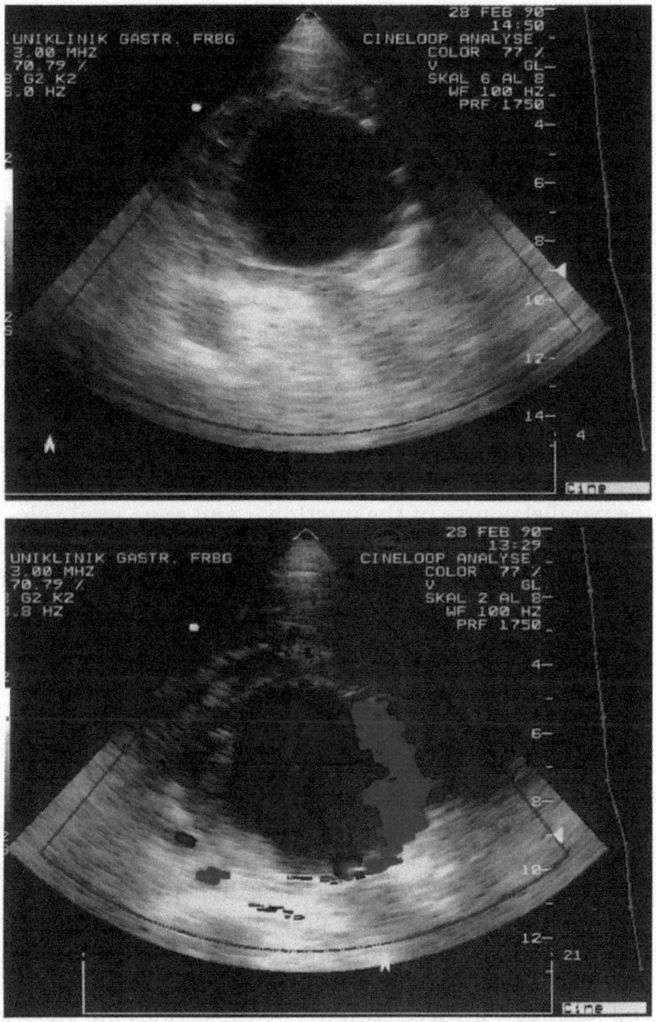

Fig. 14a, b. Pseudoaneurysm of the splenic artery. **a** In a patient with chronic pancreatitis a pseudocyst was diagnosed by B-mode sonography. After an episode of gastrointestinal bleeding no bleeding source was identified endoscopically. **b** CDFM examination revealed a circular pulsatile blood flow and confirmed the diagnosis of pseudoaneurysm

Renal Vasculature

The indications for a CDFI study of the renal vasculature are:

- renal artery stenosis
- renal artery occlusion
- renal vein thrombosis

- arteriovenous fistula
- acute renal failure
- monitoring of renal transplant

Examination Technique

The entire course of both renal arteries should be examined. In the majority of patients it is not feasible to examine their entire length from an anterior approach. To aid in the examination of the hilar segments of the renal arteries, its intrarenal posterior and anterior branches, and the segmental, interlobar, and arcuate arteries, the lateral approach is recommended with the patient in the contralateral decubitus position via the renal parenchyma [39]. To document pathology the proximal, central, and caudal segmental arteries along with the interlobar arteries which ascend perpendicularly to the renal capsula should be examined. The examination of the renal veins begins laterally. First, the parallel course of the interlobar and segmental veins to the corresponding arteries is evaluated. Subsequently, the course of renal veins is followed medially. The examiner should be familiar with common anatomic variants and typical Doppler flow findings [8, 40].

Pathologic Findings

Renal Artery Stenosis. A relatively rare, but treatable, cause of systemic hypertension is renal artery stenosis [41, 42]. Due its ability to detect reliably stenoses in the carotid arteries Doppler sonography and more recently CDFI have been proposed as a means to exclude hemodynamically significant renal artery stenosis in patients with systemic hypertension. A significant renal artery stenosis was reported to be present based on the decrease in volumetric flow rates, changes in spectral Doppler flow characteristics, increase in maximum flow velocity, ratio of peak renal artery to aortic velocities (for review see [40]), and changes in acceleration characteristics of the flow pulse, when the translumbar approach was used [43]. Conflicting data regarding the utility of Doppler sonographic methods in screening for renovascular systemic hypertension prevent unequivocal recommendations at this stage [44–46]. The reliability of detection renal artery stenosis appears to be determined primarily by the experience and skill of the operator as well as the pathologic substrate, i.e., location of the stenosis and the presence of accessory renal arteries. Based on our experience, proximal stenoses of the renal artery located close to the origin from the abdominal aorta can be reliably detected by the CDFI based on the peak velocity (in our laboratory, $V_{max} > 300$ m/s), the loss of the systolic window, and/or color-coded flow turbulence criteria (Fig. 15). However, the increase in V_{max} and in the intensity of the turbulence is highly dependent not only on the severity but also on the geometry of the lesions. For example, in our experience, the increase in V_{max} is often less, and the intensity of turbulence

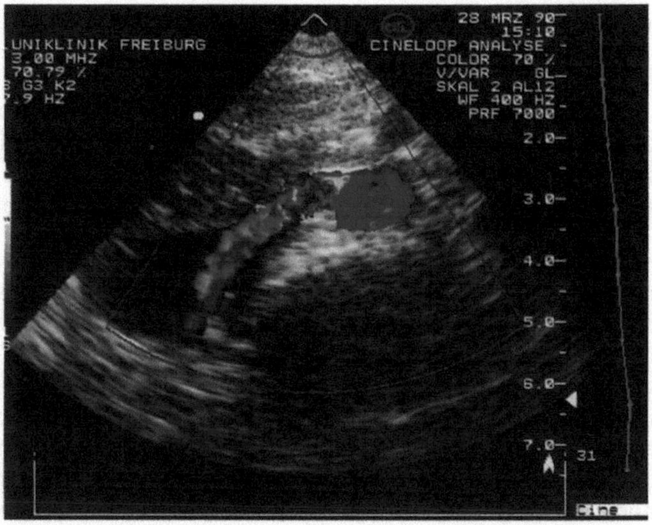

Fig. 15. Renal artery stenosis. Turbulent flow is visualized (*green*) at the origin of the right renal artery from the aorta (*red*) in a transverse view

is often higher in long, complex lesions. The detection of more distal lesions depends on the operator's ability to interrogate the distal course of the vessel. In some patients in whom a contiguous visualization of the renal arteries is not feasible a marked segmental reduction in the Pourcelot index (RI) may be indicative of an upstream hemodynamically significant stenosis.

Renal Artery Occlusion. Absence of renal blood flow may be secondary to thromboembolic or traumatic occlusion. Depending on the site of the embolization or thrombosis the renal artery blood flow ceases completely or segmentally. Acutely, the B-mode image may appear normal. Absence of flow should therefore be confirmed by lack of color filling of the renal artery and/or its branches. In patients with chronic renal failure and small fibrotic kidneys the renal artery blood flow may be very low, making a distinction to an occlusion difficult.

Renal Vein Occlusion. Doppler techniques have been proposed to assess patients with renal vein thrombosis [47]. Based on the reduction of a mean blood flow velocity in the renal vein to less than 17.0 cm/s, a moderate sensitivity and specificity were accomplished. Based on our experience in patients with chronic renal vein thrombosis, a low, predominantly systolic flow in the ipsilateral renal artery and a residual perfusion of the renal parenchyma are usually present. The flow in the renal vein compared to the flow in the inferior vena cava lacks pulsatility and respiratory variation. CDFI often reveals collaterals originating at the renal hilum, the left testicular, and ovarian veins. In acute thrombosis the B-mode image shows findings similar to those accompanying urinary tract obstruction with dilated renal veins.

Arteriovenous Fistulas. Arteriovenous renal parenchymal fistulas as a result of biopsy can be recognized on the CDFI based on localized flow turbulence at the site of the fistula, decrease in the RI, and increased flow velocities in the supplying artery and arterialization of the venous efferent limb [48].

Renal Transplant. Vascular causes of a posttransplant renal failure include renal artery or vein occlusion and stenosis, and renal vascular acute allograft rejection [40]. The examination technique and normal Doppler flow signals in the transplant's vasculature have been described [40]. Several sonographic criteria have been used successfully to identify transplant renal artery stenoses, including direct visualization of the stenotic lumen, perivascular artifacts around the stenotic lumen, spectral broadening, increase in peak flow velocity, and auditory impression of turbulence [49, 50]. Similarily, detection of arterial and venous thrombotic occlusions based on the absence of Doppler and/or color-coded flow signals (for review see [40, 50]) or on the presence of a retrograde arterial flow during diastole [51] have been highly reliable. The ability of duplex and CDFI imaging to detect an acute vascular rejection based mainly on an increase in the RI due to the presumably high impedance circulation in immunologically induced acute proliferative vasculitis has been proposed by several [52, 53] and refuted by other [54] investigators. Based on a large number of examinations, it appears that an increased RI indeed frequently accompanies episodes of acute vascular rejections; however, it can also be associated with other vascular and nonvascular graft complications [55]. Figure 16 is an example of an incomplete renal vein thrombosis in a transplant kidney.

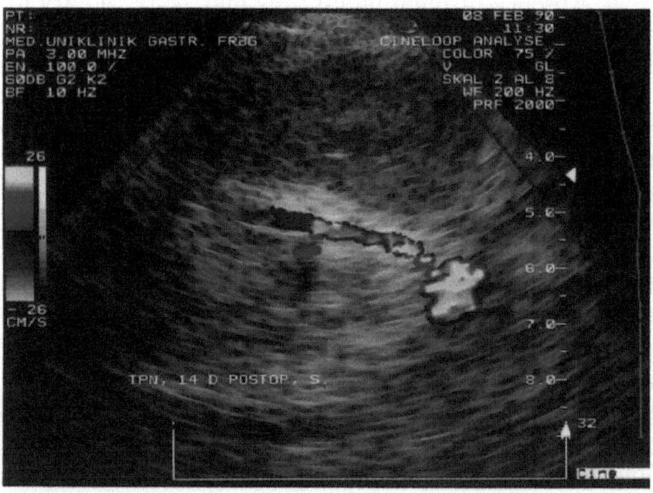

Fig. 16. Incomplete renal vein thrombosis. Acceleration of blood flow (*light blue*) corresponds to a partial thrombosis of the renal vein in a transplanted kidney

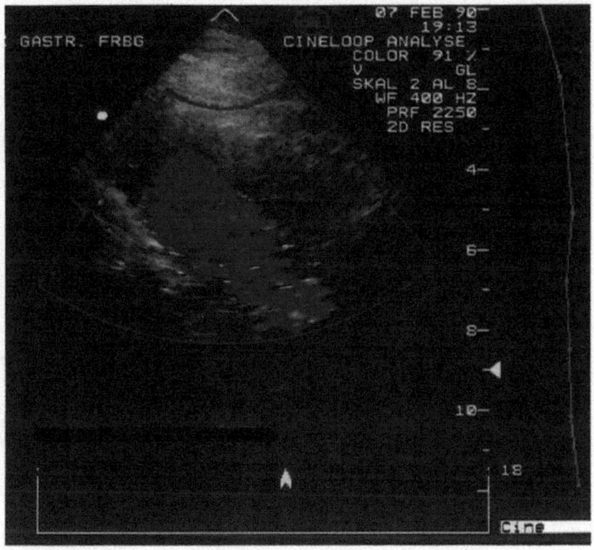

Fig. 17. Thrombosed abdominal aortic aneurysm. The patent lumen is visualized in *blue* whereas the thrombosed lumen is hypoechoic, and there is no flow signal

Abdominal Aorta, Vena Cava Inferior, and Blood Vessels of the Pelvis

Only limited experience on CDFI diagnostics of the abdominal aortic, inferior vena cava, and pelvic vascular pathology has been accumulated. Therefore, predictions of the clinical utility of CDFI in this vascular territory would be premature and speculative.

Based on our preliminary experience, CDFI is able to visualize both abdominal aortic aneurysms and dissections. In the former a clear distinction between the patent lumen with a disturbed flow pattern and hypoechoic intraluminal thrombus has been possible (Fig. 17). In the latter the patent or thrombosed true and false lumen have also been successfully identified. In these patients blood flow in the true and the false lumen can be documented to be out of phase.

Similarily, our early experience suggests that CDFI might become useful in the identification if abdominal aortic stenoses and occlusions based on an increase in flow velocity (normal values in our laboratory, 70–140 cm/s), loss of the systolic window, and turbulent color-coded flow pattern. However, in patients with high-grade subtotal stenosis and occlusions the upstream flow velocity can be low and the flow pattern bidirectional, whereas downstream typically no flow signal can be recorded. In the presence of distal aortic occlusions CDFI frequently identifies the presence of dilated collaterals with a prominent diastolic flow.

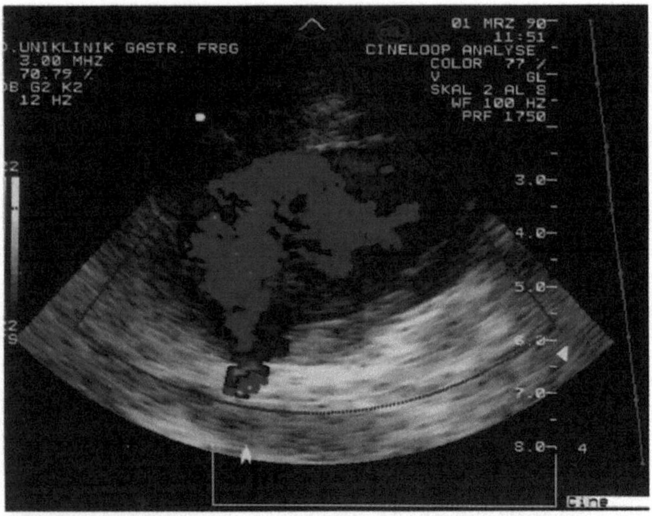

Fig. 18. False aneurysm of the right external iliac artery at endsystole. Following a routine coronary angiography in this patient a pulsatile tumor in the right groin was shown by CDFM to be a pseudoaneurysm. An arteriovenous fistula was excluded because the blood left the pseudoaneurysm during diastole by the same path as it entered during systole

Based on initial experience stenoses, occlusions, and aneurysms (Fig. 18) of the pelvic arteries, similarily to other vascular territories, can also be documented by CDFI. A consistent and accurate quantitation of the grade of the stenosis however, appears, more difficult due to the often suboptimal incidence angle of the Doppler beam. The value of CDFI in diagnosis of patients with a suspected vasculogenic impotence is currently under evaluation [56, 57].

Thrombosis of the inferior vena cava and of the common and external iliac veins has been successfully visualized by CDFI in several patients studied in our laboratory (Fig. 19). The thrombosis of the inferior vena cava was visualized as a hypoechoic cord with an absent color-coded intraluminal flow. Upstream to the thrombosis the blood flow is slow; downstream it shows reduced respiratory variation and loss of the normal pulsatile pattern. The examination of patients with thrombosis of the inferior vena cava should include the search for the presence of collateral circulation. One can often document the epigastric inferior and the ascending lumbar veins with the cranially directed blood flow and the testicular or ovarian veins with the caudally directed blood flow. The large lumbar veins may occasionally be confused with the persistent left inferior vena cava which drains the left lower extremity. In patients with a suspected thrombosis of the inferior vena cava a comparison of the findings with the contralateral vessel often aids the diagnosis. Stenoses of the inferior vena have been recognized in a few patients based on an increased flow velocity (up to $V_{max} = 150$ cm/s) and the presence of turbulence in the color-coded flow

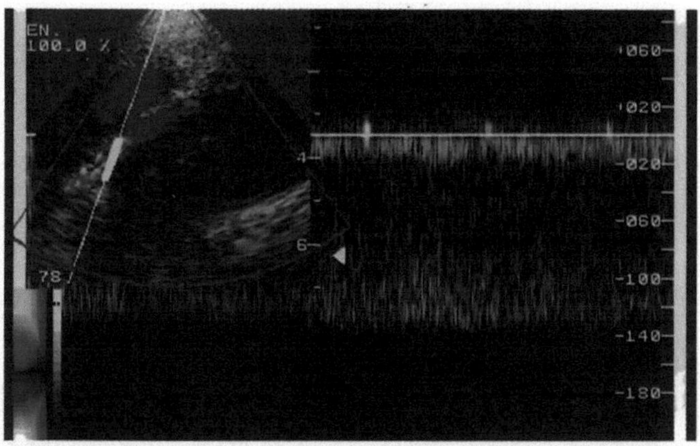

Fig. 19. Distal thrombosis of the inferior vena cava. The blood of the left renal vein is drained via the partially thrombosed cranial portion of the inferior vena cava

map. However, an undue pressure of the transducer can simulate stenosis and must be avoided.

Although the initial experience is clearly promising, the results of ongoing clinical evaluations must be awaited to determine the role of CDFI in diagnosis of patients with vascular pathology in these important vascular territories.

Tumor Vessels

An increase in flow velocities due to arteriovenous anastomoses in patients with hepatocellular carcinoma [58, 59] and hypernephromas [40] have been documented by duplex sonography. More recently, several vascular patterns on CDFI have been proposed to indicate the presence of hepatocellular carcinoma [60].

Based on our experience, presently available CDFI technology does not allow a systematic evaluation of vessels less than 1 mm in diameter. However, it is these vessels which are felt to be pathologic. These small vessels may be detected indirectly on CDFI images based on regional flow accelerations in parenchymal veins. However, these findings are nonspecific and liver cirrhosis must also be excluded. It appears that pathologic parenchymatous arteries are either of low resistance and low Pourcelot index (Fig. 20) or high resistance and high Pourcelot index (Fig. 21) vessels. The former appear to be associated with the presence of arteriovenous anastomoses. In addition, in patients with a hepatocellular carcinoma high diastolic flow in a large afferent artery and retrograde, frequently pulsatile flow in the portal vein can be interpreted as indirect signs of multiple parenchymal arteriovenous shunts. In patients with liver hemangiomas the flow within the arteriovenous shunts is usually slow

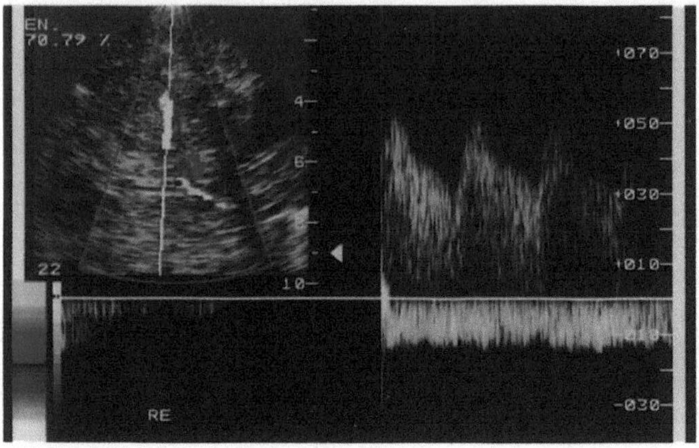

Fig. 20. Intrahepatic artery with a high diastolic blood flow. In a patient with hepatocellular carcinoma retrograde hepatofugal flow in the accompanying branch of the portal vein was observed, thus indicating an arterioportal shunting as demonstrated by Doppler

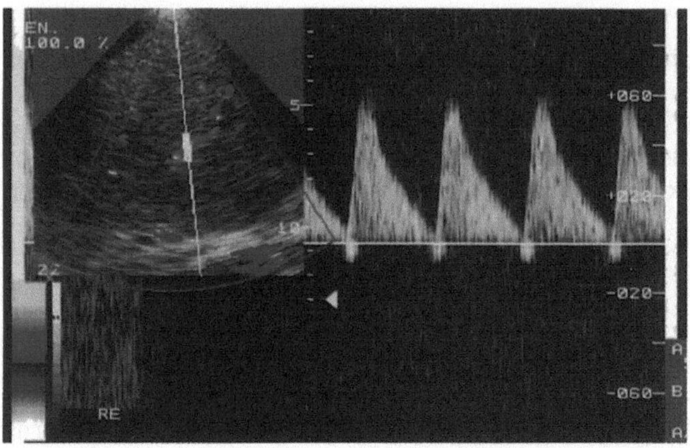

Fig. 21. Intrahepatic artery with a diminished diastolic blood flow. In a patient with hepatocellular carcinoma the enddiastolic flow velocity normally accounting for approximately 30% of the systolic flow velocity was absent due to increased peripheral resistance

and frequently evade Doppler sonographic detection, with the exception of giant hemangiomas associated with a prominent diastolic flow in the supplying artery.

Summary

CDFI is a novel ultrasound method for assessing the intraabdominal and pelvic vascular pathology. The method represents an extension of the well established principles of duplex sonography to a planar color-coded flow mapping. Although new and presently subject to extensive ongoing clinical evaluations, CDFI technology holds great promise for the noninvasive diagnosis of diseases involving the intraabdominal vasculature. Compared to duplex imaging the identification of the intraabdominal and pelvic vessels is faster and more complete, the visualization of the large tortuous as well as the small parenchymal vessels is facilitated and greatly enhanced. The initial clinical results indicate that CDFI will likely become the leading sonographic modality in the nonivasive diagnosis of a variety of gastrointestinal, renal, and pelvic vascular disorders. However, to determine its diagnostic value in each of the individual fields of potential applications the results of a more extensive clinical experience in a greater number of medical centers must be awaited. Improved flow sensitivity and quantitative flow mapping should further enhance the relevance of CDFI in the abdominal and pelvic vascular diagnosis in the future.

References

1. Goldberg BB (ed) (1984) Abdominal ultrasonography. Wiley, New York
2. Foley WD, Erickson SJ (1991) Color Doppler flow imaging. AJR 156:3–13
3. Taylor KJW, Burns PN, Wells PNT (eds) (1988) Clinical applications of Doppler ultrasound. Raven, New York
4. Gill RW (1985) Measurement of blood flow by ultrasound: accuracy and sources of error. Ultrasound Med Biol 11:625–641
5. Wilson LS, Dadd MJ, Gill RW (1990) Automatic vessel tracking and measurement for Doppler studies. Ultrasound Med Biol 16:645–652
6. Kvietys PR, Barrowman JA, Granger DN (eds) (1987) Pathophysiology of splanchnic circulation. CRC, Boca Raton
7. Landwehr P, Schindler R, Heirich U, Doelken W, Krahe T, Lackner K (1991) Quantification of vascular stenosis with color Doppler flow imaging: in vitro investigations. Radiology 178:701–704
8. Lusza G (1963) X-ray anatomy of the vascular system. Lippincott, Philadelphia
9. Sherlock S (1989) Diseases of the liver and biliary system. Blackwell, Oxford, Chap 10
10. Grant EG, Tessler FN, Gomes AS, Holmes CL, Perrella RR, Duerinckx AJ, Busuttil RW (1990) Color Doppler imaging of portosystemic shunts. AJR 154:393–397
11. Taylor KJW, Burns PN (1985) Duplex Doppler scanning in the pelvis and abdomen. Ultrasound Med Biol 11:643–658
12. Ohnishi K, Saito M, Nakayama T, Iida S, Nomura F, Koen H, Oduda K (1985) Portal venous hemodynamics in chronic liver disease: effects of posture change and excercise. Radiology 155:757–761
13. Moriyasu F, Nishida O, Ban N, Nakamura T, Sakai M, Miyake T, Uchido H (1986) "Congestion index" of the portal vein. AJR 146:735–739
14. Bolondi L, Gandolfi L, Arienti V, Caletti GC, Corcioni E, Gasbarrini G, Labo G (1982) Ultrasonography in the diagnosis of portal hypertension: diminished response of portal vessels to respiration. Radiology 142:167–172
15. Lafortune M, Marleau D, Breton G, Viallet A, Lavoie P, Huet P-M (1984) Portal venous system measurements in portal hypertension. Radiology 151:27–30
16. Ralls PW (1990) Color Doppler sonography of the hepatic artery and portal venous system. AJR 155:517–525
17. Taylor KJW (1988) Gastrointestinal Doppler ultrasound. In: Taylor KJW, Burns PN, Wells PNT (eds) Clinical applications of Doppler ultrasound. Raven, New York, pp 162–200

18. Vogelzang RL, Gore RM, Anschuetz SL, Blei AT (1988) Thrombosis of the splanchnic veins: CT diagnosis. AJR 150:93–96
19. Marn CS, Glazer GM, Williams DM, Francis IR (1990) CT-angiographic correlation of collateral venous pathways in isolated splenic vein occlusion: new observations. Radiolgoy 175:375–380
20. Harward TRS, Green D, Bergan JJ (1989) Mesenteric vein thrombosis. J Vasc Surg 9:328–333
21. Brown BP, Abu-Yousef M, Farmer R, LaBrecque D, Gingrich R (1990) Doppler sonography: a noninvasive method for evaluation of hepatic venoocclusive disease. AJR 154:721–724
22. Stanley P (1989) Budd-Chiari syndrome. Radiology 170:625–627
23. Grant EG, Perrella R, Tessler FN, Louis J, Busuttil R (1989) Budd-Chiari syndrome: the results of duplex and color Doppler imaging. AJR 152:377–381
24. Hosoki T, Kuroda C, Tokunoga K, Marukawa T, Masuike M, Kozuka T (1989) Hepatic venous outflow obstruction: evaluation with pulsed Duplex sonography. Radiology 170:733–737
25. Duerinckx AJ, Grant EG, Perella RR, Szeto A, Tessler FN (1990) The pulsatile portal vein in cases of congestive heart failure: correlation of duplex Doppler findings with right atrial pressure. Radiology 176:655–658
26. Hosoki T, Arisawa J, Marukawa T, Tokunaga K, Kuroda C, Kozuka T, Nakano S (1990) Portal blood flow in congestive heart failure: pulsed duplex sonographic findings. Radiology 174:733–736
27. Abu-Yousef MM, Milam SG, Farmer RM (1990) Pulsatile portal vein flow: a sign of tricuspid regurgitation on duplex Doppler sonography. AJR 155:785–788
28. Abu-Yousef MM (1991) Duplex Doppler sonography of the hepatic vein in tricuspid regurgitation. AJR 156:79–83
29. Grant EG, Tessler FN, Gomes AS, Holmes CL, Perella RR, Duerinckx AJ, Busuttil RW (1990) Color Doppler imaging of portosystemic shunts. AJR 154:393–397
30. Richter GM, Noeldge G, Palmaz JC, Roessle M, Slegerstetter V, Franke M, Gerok W, Wenz W, Farthman (1990) Transjugular intrahepatic portocaval stent-shunt: preliminary clinical results. Radiology 174:1027–1030
31. Wenz W (1972) Abdominale Angiographie. Springer, Berlin Heidelberg New York
32. Seitz K, Kubale R (/1988) Duplexsonographie der abdominalen und retroperitonealen Gefaesse. VCH, Weinheim, FRG
33. Granger DN, Richardson PDI, Kvietys PR, Mortillaro NA (1980) Intestinal blood flow. Gastroenterology 78:837–863
34. Qamar MI, Read AE, Skidmore R, Evans JM, Wells PNT (1984) Transcutaneous Doppler ultrasound measurement of the superior mesenteric artery blood flow in man. Gut 27:100–105
35. Moneta GL, Taylor DC, Helton WS, Mulholland MW, Strandness DE Jr (1988) Duplex ultrasound measurements of postprandial intestinal blood flow: effect of meal composition. Gastroenterology 95:1294–1301
36. Taylor GA (1990) Blood flow in the superior mesenteric artery: estimation with Doppler US. Radiology 174:15–16
37. Flint EW, Sumkin JH, Zajko AB, Bowen A (1988) Duplex sonography of hepatic artery thrombosis after liver transplantation. AJR 151:481–483
38. Hall TR, Diarmid S, Grant EG, Boechat M, Busuttil RW (1990) False-negative Duplex Doppler studies in children with hepatic artery thrombosis after liver transplantation. AJR 154:573–575
39. Isikoff MB, Hill MC (1980) Sonography of the renal arteries: left lateral decubitus position. AJR 134:1177–1179
40. Rigsby CM, Burns PN, Taylor KJW (1988) Renal duplex sonography. In: Taylor KJW, Burns PN, Wells PNT (eds) Clinical applications of Doppler ultrasound. Raven, New York, pp 201–245
41. Tegtmeyer CJ, Sos TA (1986) Techniques of renal angioplasty. Radiology 161:577–586

42. McNeil BJ, Varady PD, Burrows BA, Adelstein SJ (1975) Measures of clinical efficacy: cost-effectiveness calculations in the diagnosis and treatment of hypertensive renovascular disease. N Engl J Med 293:216–221
43. Handa N, Fukunaga R, Etani H, Yoneda S, Kimura K, Kamada T (1988) Efficacy of echo-Doppler examination for the evaluation of renovascular disease. Ultrasound Med Biol 14:1–5
44. Hansen KJ, Tribble RW, Reavis SW, Canzanello VJ, Craven TE, Plonk GW, Dean RH (1990) Renal duplex sonography: evaluation of clinical utility. J Vasc Surg 12:227–236
45. Berland LL, Koslin B, Routh WD, Keller FS (1990) Renal artery stenosis: prospective evaluation of diagnosis with color duplex US compared with angiography. Radiology 174:421–423
46. Dresberg AL, Pauschter DM, Lammert GK, Hale JC, Troy RB, Novick AC, Nally JV Jr, Welevreden AM (1990) Renal artery stenosis: evaluation with color Doppler flow imaging. Radiology 177:749–753
47. Avashi PS, Greene ER, Scholler C, Fowler CR (1983) Noninvasive diagnosis of renal vein thrombosis by ultrasonic echo-doppler flowmetry. Kidney Int 23:882–887
48. Middleton WD, Kellman GM, Melson GL, Madrazo BL (1989) Postbiopsy renal transplant arteriovenous fistulas: color Doppler US characteristics. Radiology 171:253–257
49. Taylor KJW, Morse SS, Rigsby CM, Bia M, Schiff M (1987) Vascular complications in renal allografts: detection with duplex Doppler US. Radiology 162:31–38
50. Grenier N, Douws C, Morel D, Ferriere J-M, Guillou ML, Potaux L, Broussin J (1991) Detection of vascular complications in renal allografts with color Doppler flow imaging. Radiology 178:217–223
51. Kaveggia LP, Perella RR, Grant EG, Tessler FN, Rosenthal JT, Wilkinson A, Danovitch GM (1990) Duplex Doppler sonography in renal allografts: the significance of reversed flow in diastole. AJ 155:295–298
52. Rifkin MD, Needleman L, Pasto ME, Kurtz AB, Foy PM, McGlynn E, Canico C, Baltarovich OH, Pennell RG, Goldberg BB (1987) Evaluation of renal transplant rejection by duplex Doppler examination: value of the resistive index. AJR 148:759–762
53. Rigsby CM, Burns PN, Weltin GG, Chen B, Bia M, Taylor KJW (1987) Doppler signal quantitation in renal allografts: comparison in normal and rejecting transplants, with pathologic correlation. Radiology 162:39–42
54. Drake DG, Day DL, Letourneau JG, Alford BA, Sibley RK, Mauer SM, Bunchman TE (1990) Doppler evaluation of renal transplants in children: a prospective study with histopathologic correlation. AJR 154:785–787
55. Warshauer DM, Taylor KJW, Bia MJ, Marks WH, Weltin GG, Rigsby CM, True LD, Lorber MI (1988) Unusual causes of increased vascular impedance in renal transplants: duplex Doppler evaluation. Radiology 169:367–370
56. Paushter DM (1989) Role of duplex sonography in the evaluation of sexual impotence. AJR 153:1161–1163
57. Schwartz AN, Wang KY, Mack LA, Lowe M, Berger RE, Cyr DR, Feldman M (1989) Evaluation of normal erectile function with color flow Doppler sonography. AJR 153:1155–1160
58. Taylor KJW, Ramos I, Morse SS, Fortune KL, Hammers L, Taylor CR (1987) Focal liver masses: differential diagnosis with pulsed Doppler US. Radiology 164:643–647
59. Taylor KJW, Ramos I, Carter D, Morse SS, Snower D, Fortune KL (1988) Correlation of Doppler US tumor signals with neovascular morphologic features. Radiology 166:57–62
60. Tanaka S, Kitamura T, Fujita M, Nakanishi K, Okuda S (1990) Color Doppler flow imaging of liver tumors. AJR 154:509–514

Chapter 11

Color Doppler Flow Imaging of the Peripheral Vascular System

G. L. MONETA, J. CASTER, C. CUMMINGS, and J. M. PORTER

Introduction

Angiography and contrast venography have traditionally served as the definitive means of evaluating the peripheral arterial and venous systems. In fact, the tremendous burgeoning of vascular surgical and endovascular catheter-based techniques has in large part been dependent upon widespread availability of high-quality, low-risk angiography and venography. However, noninvasive techniques assessing physiologic parameters of pressure and flow are assuming ever-increasing importance in the care and evaluation of patients with vascular disease. In particular, color Doppler flow imaging (CDFI) is capable of becoming of paramount importance in the noninvasive evaluation of arterial and venous disease. The purpose of this chapter is to provide a comprehensive update of the current status of CDFI as applied to peripheral arteries and veins.

Lower Extremity Arterial Color Doppler Flow Imaging

Equipment/Personnel

CDFI systems, while not absolutely necessary for imaging of peripheral arteries, do facilitate such examinations by aiding in more rapid identification of the arteries and thereby decreasing the overall time required for examination. CDFI systems are particularly useful in examining the iliac arteries, the region of the popliteal trifurcation, and the tibial vessels (Fig. 1). CDFI can also be used to measure the length of arterial occlusions and to identify the precise site of reconstitution of an occluded vessel (Fig. 2). Other than in characterizing occluded arteries, however, we do not at present use CDFI changes alone to grade peripheral arterial stenosis. Subocclusive levels of stenosis are determined by velocity waveform analysis (see below).

Optimal lower extremity arterial CDFI requires a selection of ultrasound transducers. For examining iliac vessels a 2- or 3-MHz probe is usually required. Infrainguinal vessels can usually be examined with a 5-MHz probe. Occasionally, however, transducers of lower MHz are needed to examine superficial femoral arteries in patients with large legs; conversely, distal tibial vessels are occasionally best examined with higher MHz probes.

We consider the use of angle-corrected velocity recordings to be a practical necessity in peripheral artery CDFI. Varying patient body habitus and depth of vessels from the skin surface make it impossible with current technology to insonate all portions of all peripheral arteries at the same angle in all patients.

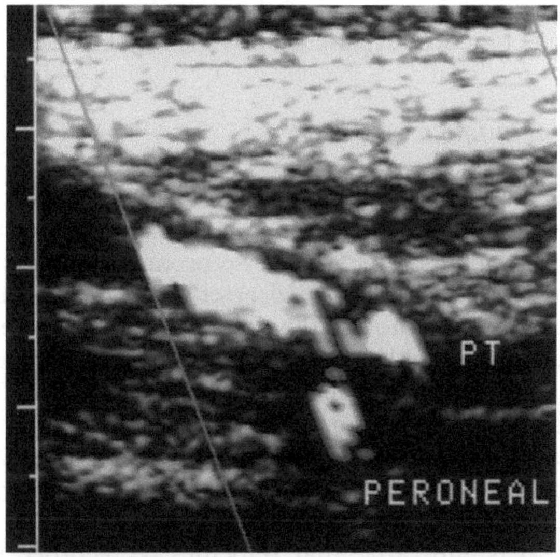

Fig. 1. CDFI facilitates identification of tibial arteries. Here, the tibial peroneal trunk is seen to divide into the peroneal and posterior tibial (*PT*) arteries

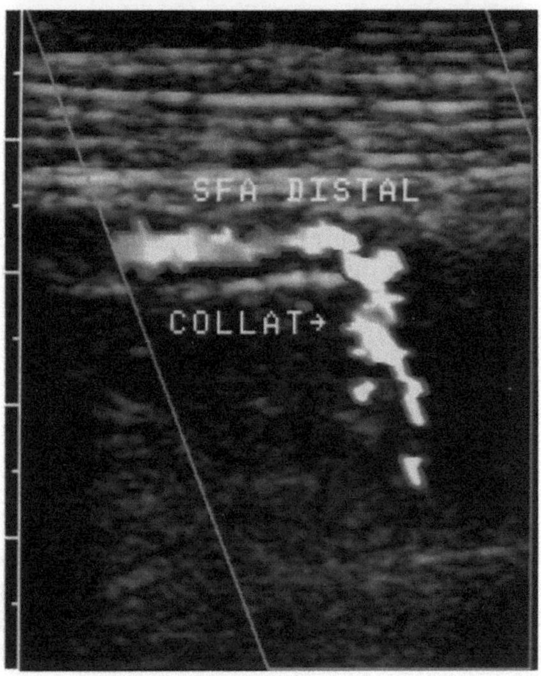

Fig. 2. Occluded arteries frequently reconstitute at sites of major collateral vessels. CDFI demonstrates distal reconstitution of a proximally occluded superficial femoral artery (*SFA*) at the takeoff of a major collateral artery (*collat*)

While 60° is generally agreed to be the ideal angle of insonation [1], angles between 30° and 70° are practical for the examination of peripheral arteries and are sufficiently accurate for clinical purposes [2, 3].

It is important to realize that lower extremity arterial CDFI is a highly technician-dependent procedure. Regardless of the sophistication of the equipment employed, experienced, highly trained vascular technologists are absolutely required to obtain accurate results. Such technologists should have a thorough knowledge of abdominal and extremity arterial anatomy and extensive practical experience in CDFI techniques.

Technique

Lower extremity arterial CDFI should include evaluation of the distal aorta, common and external iliac arteries, common femoral arteries, origin of the profunda femoris artery, the proximal, middle and distal superficial femoral artery, and the popliteal artery. In our laboratory the examination also includes evaluation of the tibial arteries from the level of the popliteal trifurcation to the ankle. If possible, patients should be examined after a fast of 8–12 h to reduce abdominal gas. With the exception of the popliteal artery, where the patient may be examined either in the prone or lateral position, the entire examination is performed with the patient supine. Tibial artery examinations may in some cases be facilitated by beginning near the ankle and following individual vessels proximally.

It is imperative that velocities be recorded routinely from several sites in each vessel and from any site where a flow disturbance is identified. Areas of high peak systolic velocity suggestive of a hemodynamically significant stenosis and areas of marked velocity decrease indicating a more proximal stenosis or occlusion should especially be noted.

Lower extremity CDFI is made more efficient by prior physical examination and determination of segmental Doppler pressures and exercise testing. We do not recommend peripheral arterial CDFI if the physical examination, segmental Doppler pressures, and postexercise Doppler evaluations are normal. Abnormalities in the physical examination and/or segmental Doppler pressures can, however, direct the technologist to examine specific areas more intensely with the CDFI system. Currently, in our laboratory, a complete examination of the arterial systems of both legs using CDFI with angle correction requires approximately 1–1.5 h.

Velocity Patterns/Classifications of Stenosis

CDFI-derived velocity waveforms obtained from normal peripheral arteries are triphasic, with a short reverse flow component at the end of systole. End-diastolic flow is near zero because of the high end-organ resistance of the extremity circulation (Fig. 3). The triphasic waveform is maintained through-

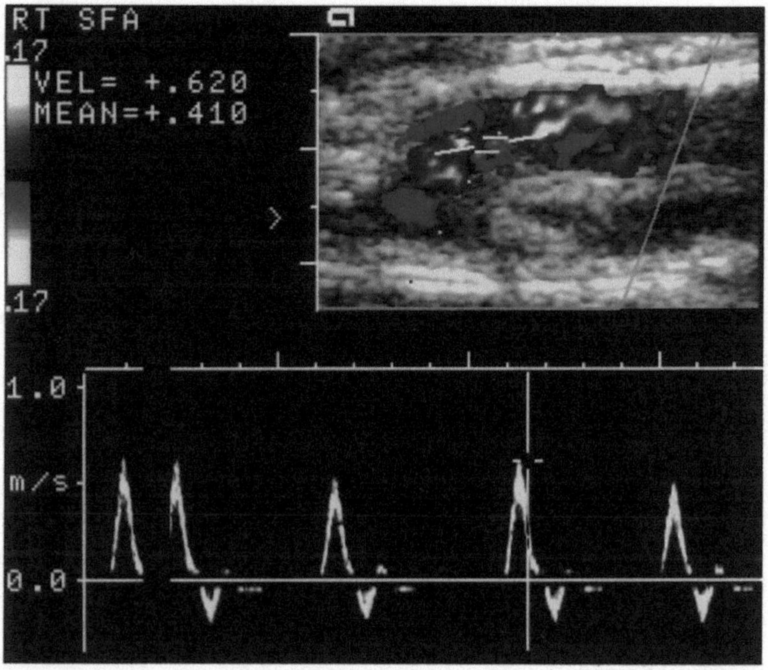

Fig. 3. Waveforms from a normal superficial femoral artery (*SFA*) demonstrate the expected triphasic configuration with absence of flow at end-diastole

out the length of the leg [4]. However, peak systolic velocities decrease steadily from the level of the iliac arteries to the tibial vessels.

Information regarding the hemodynamic significance of individual lesions in vessels proximal to the popliteal trifurcation is obtained from analysis of CDFI-determined velocity waveforms. Of particular importance is determination of the presence or absence of the reverse flow component and determination of the magnitude of peak systolic velocity. Markedly depressed peak systolic velocities and/or the absence of a reverse flow component imply a pressure and flow-reducing lesion proximal to the site of examination. Conversely, elevated peak systolic velocities (greater than 100% of the immediate proximal arterial segment) suggest a stenosis greater than 50% at that examination site [5].

The original CDFI-derived criteria for noninvasive quantification of peripheral artery stenosis developed at the University of Washington also utilized spectral broadening in an attempt to discriminate between lesions of 1%–19% versus those of 20%–49% angiographic stenosis [5] (Table 1). Unfortunately, assessment of spectral broadening requires reference to a set of standards developed at a constant Doppler angle [6]. The maintenance of a constant Doppler angle however, is, not possible in clinical peripheral artery CDFI, and we therefore find analysis of varying degrees of spectral broadening impractical in routine lower extremity arterial CDFI.

Table 1. University of Washington spectral criteria for CDFI-derived stenosis in peripheral arteries

Stenosis	Criteria
0%	Normal waveforms and velocities
1% –19%	Normal waveforms and velocities. Spectal broadening is present primarily on the downslope of the systolic portion of the curve.
20% –49%	Normal waveform with marked spectral broadening. There is a 30% or more increase in peak systolic velocity.
50% –99%	Reverse flow component is lost from the waveform. Peak systolic velocity is increased 100% or more.
Occluded	No flow can be detected at multiple sites in an adequately visualized arterial segment.

Table 2. Peak systolic velocities in normal arteries

Artery	Peak systolic velocity (cm/s)
External iliac	119.3 ± 21.7
Common femoral	114.1 ± 21.9
Superficial femoral	
Proximal	90.8 ± 13.6
Distal	93.6 ± 14.1
Popliteal	68.8 ± 13.5

Blood flow velocities and waveforms may vary somewhat with cardiac output, arterial inflow, and outflow resistance. It is also important to note the relatively high standard deviations of CDFI-determined mean peak systolic velocities obtained from arteries of normal volunteers (Table 2). Each patient undergoing peripheral artery CDFI must therefore, to some extent, be considered individually when noninvasively quantifying peripheral arterial stenoses. Waveforms must be carefully analyzed immediately proximal and distal to an area suspicious for a hemodynamically significant lesion. Velocity artifacts may be produced by misplacement of the sample volume or as a result of respiratory motion or tortuosity of the vessel under examination. These errors can usually be avoided by attempting to obtain the most normal-appearing velocity waveform at each site of examination. Reproducible high-velocity signals in a well-visualized vessel cannot of course be ignored and indicate a significant localized arterial stenosis (Fig. 4). A small prospective analysis of the classification system detailed in Table 1, has shown peripheral artery CDFI to have a sensitivity of 82%, a specificity of 92%, a positive predictive value of 80%, and a negative predictive value of 93% in the determination of a hemodynamically, significant stenosis (greater than 50% diameter reduction) in the ileo-femoral-popliteal system [7].

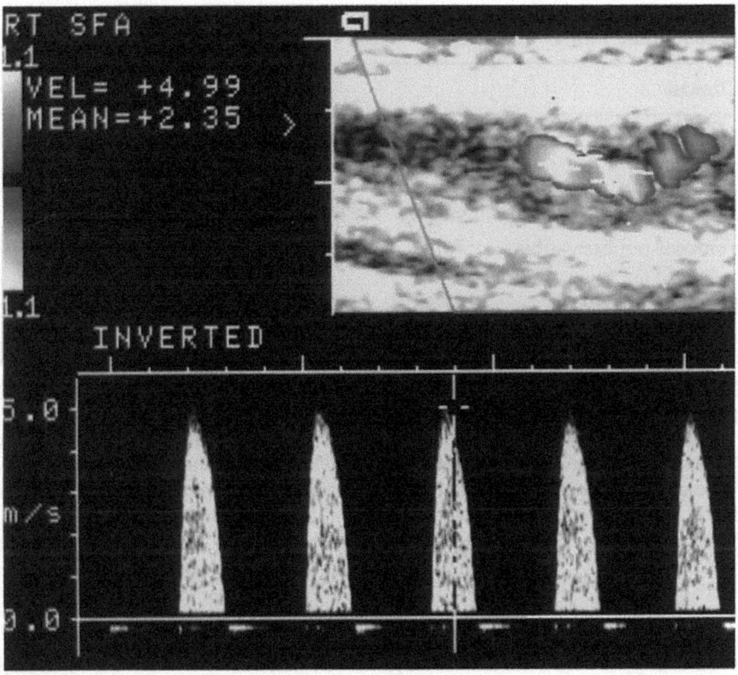

Fig. 4. A peak systolic velocity of about 500 cm/s combined with absence of the end-systole reverse flow component indicates the presence of a >50% stenosis at the site of examination

At the Oregon Health Sciences University we have recently completed a larger prospective, blinded study evaluating the accuracy of lower extremity arterial CDFI. A total of 150 consecutive patients (286 limbs) underwent peripheral artery CDFI, segmental Doppler pressures, and arteriography prior to a contemplated arterial reconstruction. Doppler pressures and clinical data were used to classify limbs as normal, aortoiliac disease, femoral-popliteal-tibial disease, or multilevel inflow-outflow disease to ascertain whether the clinically determined disease category affected the accuracy of peripheral arterial CDFI. In this study, 99% of 2036 arterial segments from the popliteal artery cephalad were visualized. Using CDFI 95% of anterior tibial and posterior tibial artery segments were also successfully visualized; however, only about 80% of peroneal artery segments were able to be identified with CDFI. In vessels proximal to the tibials, lower extremity arterial CDFI was evaluated for its ability to detect an angiographic stenosis greater than 50% and to distinguish stenosis from occlusion. In the tibials, lower extremity arterial CDFI was assessed for its ability to predict continuous tibial artery patency from the level of the popliteal trifurcation to the ankle. The results of the comparison of lower extremity arterial CDFI with angiography are summarized in Table 3. In addition, CDFI successfully distinguished stenosis from occlusion 97% of the time in the ileo-femoral-popliteal systems. These results confirm that lower

Table 3. Accuracy of arterial CDFI in comparison to angiography in 150 preoperative patients (286 limbs) with lower extremity arteriosclerosis

Clinical			Location			
Classification [a] (no. limbs)	Iliac [b]	SFA [b]	Popliteal [b]	AT [c]	PT [c]	Per [c]
Aortoiliac ($n=44$)	94/100	77/99	No lesions	100/100	100/100	90/73
Femoral-popliteal-tibial ($n=117$)	71/99	92/98	67/99	96/73	100/75	80/76
Multilevel ($n=45$)	92/98	78/97	69/98	81/85	89/71	79/57
All ($n=286$) [d]	89/99	87/98	67/99	93/75	97/74	–

SFA, Superficial femoral artery; AT, anterior tibial artery; PT, posterior tibial artery; Per, peroneal artery.

[a] Clinical classification based on physical examination and segmental Doppler pressures.

[b] Sensitivity/specificity for detecting $>50\%$ stenosis and/or occlusion.

[c] Sensitivity/specificity for detecting continuous patency from popliteal trifurcation to the ankle.

[d] Includes limbs clinically felt to be normal.

extremity arterial CDFI is highly accurate in comparison to angiography. In addition, they suggest peripheral arterial CDFI to be of roughly equal accuracy in all commonly encountered, clinically determined patterns of lower extremity arteriosclerosis.

It is important to note, however, the rather poor specificity of the technique in evaluating continuous patency of the peroneal artery as compared to the anterior and posterior tibial arteries (Table 3). In addition, the ability of lower extremity arterial CDFI to determine the quality of individual arterial segments, and therefore their suitability for arterial grafting, has not been fully evaluated. It is our impression that while CDFI can accurately determine the patency of ileo-femoral-popliteal and tibial arteries, determination of the most suitable vessel for a distal arterial anastomosis may not be possible currently with sufficient accuracy using CDFI alone. Such considerations are, of course, of considerble importance in the evaluation of the ultimate routine clinical utility of lower extremity arterial CDFI.

Clinical Applications

Peripheral artery CDFI and angiography are complimentary tests. Current clinical applications of lower extremity arterial CDFI can be considered within the context of evaluation of preoperative patients and as a means of follow-up of surgical and endovascular reconstructions.

Nonoperative Patients

Lower extremity arterial CDFI is useful as a screening modality prior to angiography. Arterial CDFI is used to define precisely the location and character of various occlusive lesions. CFDI can therefore be used as a noninvasive means for determining the extent of an operative or endovascular procedure required to relieve patient symptoms. Jägger and Bollinger from Zürich, Switzerland, first suggested this approach in 1986 when they reported a series of patients who underwent percutaneous transluminal angioplasty without preceding diagnostic angiography [8]. Their patients were selected for angioplasty on the basis of symptoms, physical examinations, and peripheral artery CDFI studies.

Cossman et al. reported in 1989 on a series of 61 patients in whom CDFI was used to map the ileo-femoral and femoral-popliteal arterial segments [9]. In these patients, who were being evaluated for possible excimer laser angioplasty, a blinded comparison with angiography showed CDFI to have a sensitivity of 99% and a specificity of 87% in detecting stenoses of 50% or greater. In 94% of patients with occlusions, peripheral artery CDFI accurately predicted the site and length of occlusion.

More recently, Edwards et al. reported the University of Washington experience in the use of lower extremity arterial CDFI in selection of patients for transluminal angioplasty [10]. In 110 cases angiography for lower extremity ischemia was preceded by CDFI. Based on this system, 50 cases (45%) were scheduled for transluminal angioplasty at the time of initial angiography. Transluminal angioplasty was actually performed in 47 of the 50 cases (94%). These data, and those of Cossman et al. [9], strongly suggest that CDFI of lower extremities allows the accurate detection of lesions amenable to catheter-based endovascular techniques.

In a recent paper, also from the University of Washington group, Kohler et al. has considered the complex question of whether CDFI can be used in lieu of angiography prior to lower extremity surgical reconstructions [11]. In this study several prominent vascular surgeons were asked to formulate operative plans for 29 patients with lower extremity occlusive disease. The information provided on each patient included a brief clinical history and the patient's Doppler-determined ankle pressures. Information describing the location and extent of arterial lesions was derived both from angiography and CDFI in each patient. This information was indicated on anatomic line drawings. The surgeons, however, were not informed as to the source of the information concerning arterial stenosis. Analysis of the data indicated that clinical decisions made by individual surgeons were very similar regardless of whether the information concerning arterial stenosis was obtained from CDFI or angiography.

In summary, with regard to preoperative patients, it appears that CDFI of peripheral arteries is highly accurate in comparison to angiography for the detection of peripheral arterial lesions. The clinical utility of this modality, however, remains to be determined, and significant questions remain. Will lower extremity arterial CDFI actually permit surgical reconstructions without preoperative angiography? Under what circumstances is peripheral arterial CDFI likely to be most useful? Will it be best applied in patients with moderate claudication, who are likely to have single-level disease, or will it also prove useful in patients with very short distance claudication or rest pain who usually have more extensive vascular occlusive disease? What combination of patient symptoms, Doppler systolic pressures, and peripheral arterial CDFI results will best predict angiographic findings? These and other questions are obvious areas for future research in the determination of the clinical utility of lower arterial extremity CDFI in the preoperative patient.

Operated Patients

CDFI appears ideal for monitoring the results of surgical and endovascular lower extremity arterial reconstructions. In many vascular laboratories CDFI is now the method of choice for postoperative monitoring of lower extremity vein graft arterial reconstructions (Fig. 5). It appears that either serially falling vein graft peak systolic velocities or a peak systolic velocity below 45 cm/s at any point in the vein graft are markers of impending vein graft failure [12]. In many laboratories measurements of graft flow velocities have proven to be more sensitive in predicting lower extremity vein graft failure than changes in ankle/brachial systolic blood pressure ratios [13, 14].

In our vascular laboratory patients with lower extremity autogenous vein grafts are monitored with CDFI-determined graft flow velocities every 3 months for the 1st year after a lower extremity vein graft reconstruction and

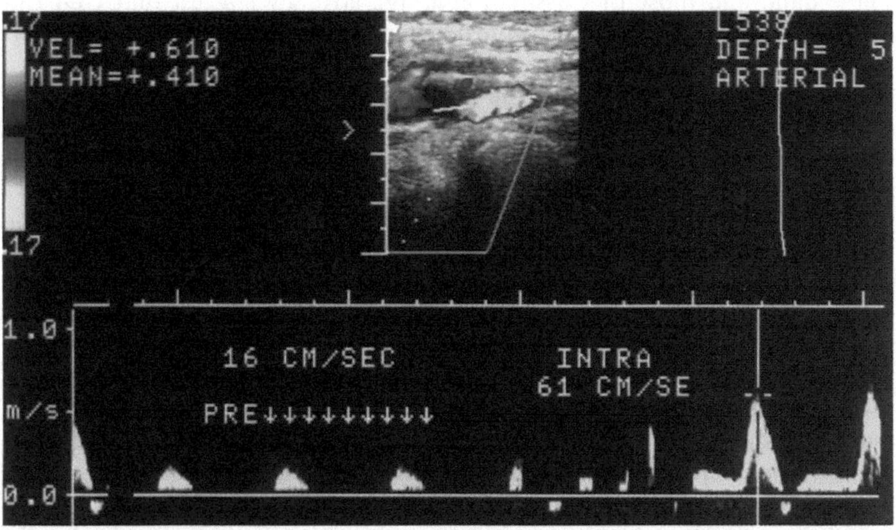

Fig. 5. CDFI of a failing femoral-popliteal vein graft. A peak systolic velocity of 16 cm/s is clearly below the critical level of 45 cm/s and suggests impending vein graft thrombosis. The step-up in velocity from 16 cm/s to 61 cm/s indicates an intrinsic graft stenosis as the etiology of this patient's failing graft

every 6 months thereafter. At each examination graft flow velocities are obtained from the proximal, middle, and distal aspects of the vein graft. If a graft flow velocity below 45 cm/s is detected, CDFI is then used to examine the entire vein graft in an effort to identify a flow-limiting anastomotic or intrinsic graft stenosis. If no lesion is found within the graft itself, the arteries proximal and distal to the vein graft are examined. Frequently it is progression of occlusive disease involving the arterial inflow or outflow of the vein graft that is responsible for a low graft flow velocity. Correction of flow-reducing lesions prior to graft failure is of utmost importance as actual thrombosis of vein grafts results in poor ultimate salvage of the graft [15, 16].

The principal disadvantage of monitoring graft flow velocities is the uncertain specificity of the test. It is unknown exactly which lesions detected by abnormal graft flow velocities will eventually result in thrombosis of the graft. Reducing unnecessary vein graft revisions will require greater knowledge of precisely which lesions lead to vein graft thrombosis. For instance, low graft flow velocities are commonly found in dilated vein grafts. Such grafts, however, do not appear to be at an increased risk of thrombosis [14].

Peripheral artery CDFI may also be used directly to monitor sites of endovascular arterial reconstructions. Failure of velocity waveforms to improve at sites of endovascular treatment has been suggested to be associated with short-term clinical failure of an endovascular procedure [17]. At a recent meeting of the Association of Veterans Affairs surgeons, Kinney et al. reported their results using peripheral arterial CDFI to monitor the precise sites of dilatation in patients undergoing percutaneous balloon angioplasty. They noted that when the immediate postprocedure CDFI scan suggested a residual stenosis of greater than 50%, even if associated with a technically satisfactory angiographic result, the incidence of clinical and/or hemodynamic failure at 3 months was 45% and at 1 year was 88%. Conversely, if the CDFI examination revealed a residual stenosis of less than 50%, clinical and/or hemodynamic failure rates at 3 and 18 months were only 7%, and 20%, respectively [18].

Localized dissections and associated intimal flaps at angioplasty sites may in many cases preclude angiography as a means of accurately determining the technical success of angioplasty. The functional result of balloon dilatation, as termed by CDFI, is perhaps a more accurate predictor of long-term hemodynamic and clinical success following percutaneous transluminal angioplasty. Other investigators, however, have reached different conclusions. Sacks et al. examined 22 patients with pre- and post-angioplasty CDFI [19]. Lesions greater than 50% were shown by CDFI in eight patients after transluminal angioplasty. However, only two of these patients were felt to have a residual hemodynamic stenosis as determined by immediately postangioplasty arteriography, pulse volume recordings, ankle-brachial indices, or symptom relief.

Unfortunately, comparison of the Kinney and Sacks studies is not appropriate. Sacks employed nonhemodynamic criteria (i.e., relief of symptoms) as a marker of success. In addition, follow-up in the Sacks study was very short, an average of 4 days. The proper conclusion from examination of these two studies is that CDFI and angiography give disparate results in the immediately postangioplasty period. Residual high-flow velocities may predict poor intermediate and long-term success, as suggested by Kinney et al., but not evaluated in the Sacks et al. report.

Lower Extremity Venous Color Doppler Flow Imaging

Acute Deep Venous Thrombosis

CDFI may be used in the evaluation of acute and chronic venous problems. Currently the greatest interest is in the diagnosis of deep venous thrombosis. In only about 50% of cases is this diagnosis correctly made based on the history and physical examination alone [20]. Venography, while serving as the gold standard for the diagnosis of acute venous thrombosis, is impractical in evaluation of the large number of patients who present with acute leg symptoms which may be consistent with deep venous thrombosis. An accurate, noninvasive method to detect deep venous thrombosis is therefore critical. Of all the methods available for the noninvasive evaluation of acute deep venous thrombosis CDFI is the most exciting. Although in widespread clinical use for only about 5 years, venous CDFI is rapidly becoming the noninvasive method of choice in the initial evaluation of patients with possible deep venous thrombosis.

Technique/Criteria for Diagnosis

Ileo-femoral-popliteal System

Unlike CDFI of the peripheral arterial system, that of the venous system to detect acute deep venous thrombosis is comparatively easy to learn. The patient is placed supine with the legs slightly externally rotated and lowered 10°–20°. The venous system is examined from the vena cava down to and including the tibial veins. (The popliteal vein is examined with the patient in the lateral position.) Probes of 2- or 3 MHz are required for examination of abdominal veins. These veins are visualized in a longitudinal fashion and assessed with the Doppler for flow, response to a Valsalva maneuver, and variation of flow with respiration. The common femoral and superficial femoral veins are examined both transversely and in longitudinal section with a 5-MHz probe. These veins are also assessed for spontaneous flow, Valsalva response, and variation of flow with respiration. It is usually also possible to observe functioning venous valves when the extremity veins are normal. The transducer is then used to apply gentle pressure to the veins in an attempt to coapt their walls. Failure of the vein walls to coapt with gentle probe pressure indicates an intraluminal thrombus. Scanning is then continued down to the level of the popliteal vein. The examination is most difficult in the region of the

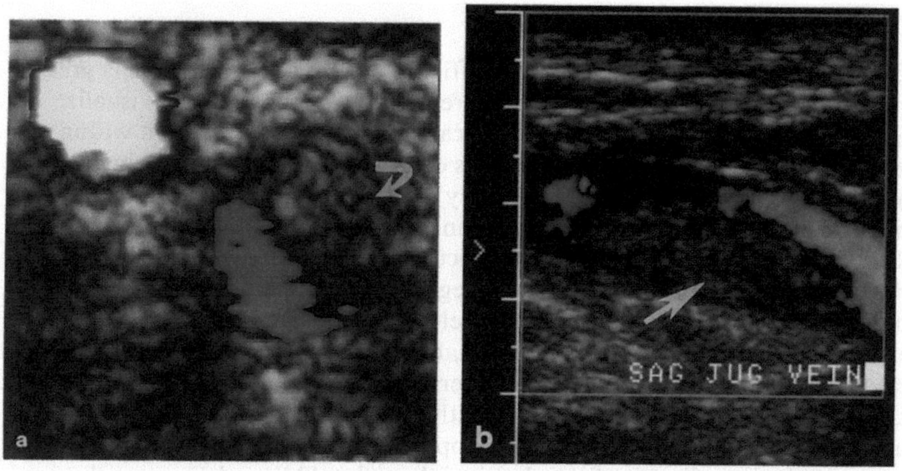

Fig. 6. CDFI venous examinations demonstrate partially occluding venous thrombi in the common femoral vein (**a**) and the internal jugular (*JUG*) vein (**b**)

adductor canal. Confusing echoes are produced by overlapping fascial plains and the superficial femoral vein is difficult to compress in this location.

It is important to note that actual visualization of thrombus is not necessary to make the diagnosis of deep venous thrombosis. Failure of the vein walls to coapt with probe pressure is diagnostic for the presence of a venous thrombus even if it is not visible on the B-mode portion of the examination. In such cases lack of color filling of the vein lumen confirms the presence of a venous thrombosis (Fig. 6). The criteria for diagnosis of deep venous thrombosis by CDFI are as follows:
- Noncompressible vein wall
- No spontaneous flow
- No color filling of vein/lumen
- No flow variation with respiration
- Visible thrombus
- Venous distension
- Visualized collateral vessels

When thrombi are visible, their echo characteristics may help determine their age and thereby help distinguish acute from chronic venous occlusion. Although many thrombi are of indeterminate age, a smooth homogeneous thrombus, perhaps with a free-floating component, suggests a more acute process. An irregular, adherent, heterogeneous echo pattern implies that the thrombosis is more chronic. Chronic thrombi are also frequently associated with venous dilatation and numerous collateral vessels. Such findings are usually not present in the early stages of acute deep venous thrombosis in patients with previously normal extremities [21].

Tibial Veins

Examination of tibial veins is more difficult, and in our experience is greatly enhanced by the use of CDFI. It is frequently not possible to visualize all segments of all tibial veins [22, 23]. Because of their deep position within the leg visualization of the peroneal veins is often the most difficult. Using CDFI, however, we have found it relatively easy to examine the posterior and anterior tibial veins, although others have found the anterior tibial veins relatively difficult to examine. Like the suprageniculate veins, tibial veins are best examined with the patient supine and the legs slightly dependent. It is frequently best to begin the examination at the level of the ankle and to proceed cephalad examining the veins in both longitudinal and cross-sections and assessing for evidence of flow using both the Doppler and the presence of color. The veins are also assessed for compressibility with probe pressure. Unfortunately, failure of tibial veins to compress with application of pressure does not always appear to be an indicator of intraluminal thrombus [24], underscoring the need to confirm the absence of flow using the Doppler and/or color flow portion of the examination. The difficulty of visualizing these vessels along with their deep anatomic locations makes it impossible accurately to detect coaption of the vein wall in all cases. Some clinicians feel that augmentation of tibial venous flow with distal compression is a useful adjunct to tibial vein CDFI examination [25]. Increased flow with distal pressure is suggested as a criteria for patency of tibial veins. We have not, however, routinely employed such maneuvers in our vascular laboratory.

Accuracy

Venous CDFI appears to be highly accurate in the detection of deep venous thrombosis from the level of the popliteal vein cephalad. Early reports describing the utility of this technique are almost evangelistic in tone. Huge numbers of patients were described as having undergone venous CDFI, excellent sensitivities and specificities were claimed, but few venograms were done for objective correlation. Fortunately, there are now several well-conducted prospective trials correlating the accuracy of venous CDFI with venography in the detection of acute deep vein thrombosis. These trials uniformly indicate that in the ileo-femoro-popliteal veins, venous CDFI, when compared to venography, has sensitivities and specificities exceeding 90% in the detection of acute deep venous thrombosis [26–29]. The overall accuracy of venous CDFI in the detection of acute deep venous thombosis therefore appears to be significantly superior to any other noninvasive modality routinely available [30].

Errors in the diagnosis of acute deep venous thrombosis in the thigh veins appear to be related primarily to two factors. Difficulty in compressing the superficial femoral vein at the level of the abductor canal can in some cases lead to a diagnosis of venous thrombosis when, in fact the vein is actually patent. Secondly, chronic thrombi may be misinterpreted as being acute [31].

It would seem that these errors could be avoided in most cases by using the Doppler and color flow portion of the examination to confirm that an incompressible vein is truly not patent and by paying close attention to the echo characteristics of the thrombus and the presence or absence of significant collateral vessels (see above).

There is currently no consensus as to the accuracy of venous CDFI in the detection of isolated calf vein thrombosis. Little venographic information on the ability of venous CDFI to detect tibial vein thrombosis is available. Analysis of available data suggests, however, that the overall accuracy of venous CDFI in detection of tibial vein thrombi is significantly less than in the detection of venous thrombosis involving the thigh veins [25, 26, 29]. It is our opinion, that, although tibial vein venous CDFI appears promising, available data do not currently support its use as the sole means for diagnosis of isolated tibial vein thrombosis.

Clinical Applications

We feel that venous duplex scanning is sufficiently accurate in the detection of above-knee deep venous thrombi that inpatient anticoagulation with heparin followed by outpatient warfarin therapy is justified when the examination is positive. A negative examination should lead to analysis of other causes of the patient's lower extremity symptoms. With a properly performed venous CDFI examination we do not feel that a confirming venogram is necessary prior to the institution of anticoagulant therapy.

The situation is more vexing with respect to the detection and/or treatment of isolated calf vein thrombosis. The treatment of such patients is controversial. While isolated calf vein thrombosis appears rarely to lead to symptomatic pulmonary embolism [31], it is clear that tibial vein thrombus may propagate into the popliteal and superficial femoral veins. At the present time one can justify treating isolated calf vein thrombosis either conservatively, i.e., by aspirin and walking, or with full-dose anticoagulation. If the physician feels it necessary to treat calf vein thrombosis with anticoagulation, we feel that the risk of anticoagulation in selected patients versus the proven accuracy of CDFI in detection of isolated infrapopliteal venous thrombosis does not justify anticoagulation therapy based on CDFI alone. Higher risk patients therefore require confirmatory venography prior to anticoagulant therapy for isolated calf vein thrombosis. If the physician chooses to institute anticoagulant therapy only for propagation of thrombus into the popliteal or superficial femoral venous systems, we feel the patient should be entered into a surveillance program. Venous CDFI scans can then be repeated very 2–3 days for a total of three examinations and the patient treated conservatively if the above-knee veins remain free of thrombus.

Chronic Venous Insufficiency

CDFI is now being advocated as an ideal method for quantitating venous reflex in patients with chronic venous insufficiency. Of all the minimally invasive and noninvasive methods available for quantitating venous reflex in patients with chronic venous insufficiency, only CDFI can both detect the presence or absence of reflux at specific anatomically defined sites and also provide a quantitative measurement of the severity of reflux. Some feel that precise identification of sites of venous reflux may allow treatment directed at the source of reflux in patients with severe sequelae of chronic venous insufficiency.

Technique

Quantitation of venous reflux with CDFI is best performed with the patient upright and placing essentially no weight on the leg undergoing examination. Because Valsalva maneuver and manual compression produces inconsistent and nonreproducible degrees of venous reflux, pneumatic cuffs connected to an automatic inflation and deflation device are used to produce standard compressive forces on the leg [32]. Cuffs of 24 cm used for the thigh, 12 cm for the calf, and 7 cm for the foot. B-mode scanning is used to identify the vein to be examined. The site to be examined is then insonated with the CDFI probe just cephalad to the cuff. The cuff is then inflated to a predetermined level (80 mm Hg for thigh cuffs, 100 mm Hg for calf cuffs, and 120 mm Hg for foot cuffs). The cuffs are inflated in 3 s and deflated in less than 0.5 s. Venous reflux in response to cuff decompression is recorded with the CDFI scanner. The peak velocity of venous reflux as determined with the Doppler portion of the CDFI scanner is then used along with the B-mode determined diameter of the vein under examination to calculate the volume of reflux at peak reflux flow. Using this technique, Nicolaides and associates from London have measured retrograde flow at peak reflux in the major axial veins of the lower extremities (short and long saphenous veins, superficial femoral vein, popliteal vein). In a study of 47 patients with chronic venous insufficiency this group then determined that the skin changes of lipodermatosclerosis and/or ulceration do not occur until the sum of peak reflux flows in all axial veins is greater than 10 ml/s. In addition, it does not appear that the source of reflux is important. The incidence of lipodermatosclerosis and/or ulceration is significantly higher when the total peak reflux exceeds 10 ml/s regardless of whether the reflux is primarily in the superficial or deep veins [33].

Clinical Applications

At this time, the clinical implications of CDFI quantification of venous reflux are somewhat speculative. The technique may perhaps allow more precise identification of the sites of venous reflux and therefore aid in the selection of patients for venous valvular reconstruction and/or transplantation. The technique may also be used in the selection of patients with chronic venous disease who may benefit from selective stripping and avulsion of superficial veins. Patients whose superficial venous system contributes significantly to their total volume of reflux can be identified with CDFI and offered removal of their superficial venous systems.

Conclusion

It is now well established that CDFI can be used to study both the arteries and veins of the extremities. The use of this technique is now well accepted in the detection of acute deep venous thrombosis and in the serial monitoring of lower extremity arterial vein grafts. Other clinical applications of extremity CDFI, as described above, will obviously require further prospective clinical evaluation. The future, however, is exciting. The day may be upon us when noninvasive physiological techniques such as CDFI examinations will provide the bulk of information required intelligently to treat patients with arterial and venous disease of the extremities.

References

1. Beach KW, Lawrence R, Phillips DJ, Primozich J, Strandness DE Jr (1989) The systolic velocity criterion for diagnosing significant internal carotid artery stenosies. J Vas Tech 13:65–68
2. Rizzo RJ, Sandager G, Astleford P, Payne K, Peterson Kennedy L, Flinn WR, Yao JST (1990) Mesenteric flow velocity variations as a function of angle of insonation. J Vasc Surg 11:688–694
3. Thiele BL, Strandness DE Jr (1987) Duplex scanning and ultrasonic arteriography in the detection of carotid disease. In: Kempczinski RF, and Yao JST (eds) Practical noninvasive vascular diagnosis, 2nd eds. Yearbook Medical, Chicago, pp 339–363
4. Jäger KA, Phillips DJ, Martin RL, Hanson C, Roederer GO, Langlois YE, Ricketts HJ, Strandness DE Jr (1985) Noninvasive mapping of lower limb arterial lesions. Ultrasound Med Biol 11:515–521
5. Jäger KA, Ricketts HJ, Strandness DE Jr (1985) Duplex scanning for evaluation of lower limb arterial diseases. Jn: Bernstein EF (ed) Noninvasive diagnostic techniques in vascular disease. Mosby, St. Louis, pp 619–631
6. Beach KW, Strandness DE Jr (1985) Carotid artery velocity waveform analysis. In: Bernstein EF (ed) Noninvasive diagnostic techniques in vascular surgery. Mosby, St. Louis, pp 409–422
7. Kohler TR, Nance DR, Cramer MM, Vandenburghe N, Strandness DE Jr (1987) Duplex scanning for diagnosis of aortoiliac and femoropopliteal disease: a prospective study. Circulation 76:1074–1080
8. Jäger KA, Johl H, Seifert H, Bollinger A (1986) Perkutane transluminale Angioplastie (PTA) ohne vorangehende diagnostische Arteriographie. VASA [Suppl 15]:24
9. Cossman DV, Ellison JE, Wagner WH, Carroll RM, Treiman RL, Foran URF, Levin PM, Cohen JL (1989) Comparison of contrast arteriography to arterial mapping with color-flow duplex imaging in the lower extremities. J Vasc Surg 10:522–529
10. Edwards JM, Coldwell DM, Goldman ML, Strandness DE Jr (1991) The role of duplex scanning in the selection of patients for transluminal angioplasty. J Vasc Surg 13:69–74
11. Kohler TR, Andros G, Porter JM, Clowes A, Goldstone J, Johansen K, Raker E, Nance DR, Strandness DE Jr (1990) Can duplex scanning replace arteriography for lower extremity arterial disease? Ann Vasc Surg 4:280–287

12. Bandyk DF, Cates RF, Towne JB (1985) A low flow velocity predicts failure of femoropopliteal and femorotibial bypass grafts. Surgery 98:799–809
13. Bandyk DF (1986) Postoperative surveillance of femorodistal grafts: the application of echo-Doppler (duplex) ultrasonic scanning. In: Bergan JJ, Yao JST (eds) Reoperative arterial surgery. Grune and Stratton, Orlando, pp 59–79
14. Mills JL, Harris EF, Taylor LM Jr, Beckett WC, Porter JM (1990) The impact of routine surveillance of distal bypass grafts with duplex scanning: a study of 379 reversed vein grafts. J Vasc Surg 12:379–389
15. Whittmore AD, Clowes AW, Couch NP, Mannick JA (1981) Secondary femoropopliteal reconstruction. Ann Surg 193:35–42
16. Belkin M, Donaldson MC, Polak JF (1990) Observations on the use of thrombolytic agents for thrombotic occlusion of infrainguinal grafts. J Vasc Surg 11:289–296
17. Moneta GL, Strandness DE Jr (1987) Peripheral arterial duplex scanning. JCU 15:645–651
18. Kinney E, Bandyk D, McWissen M, Lauza D, Bergamini T, Lipchick E, Towne J (1991) Monitoring functional patency of percutaneous transluminal angioplasty. Arch Surg 126:734–747
19. Sacks D, Robinson ML, Marinelli DL, Perlmutter GS (1990) Evaluation of peripheral arteries with duplex ultrasound after angioplasty. Radiology 176:39–44
20. Cranley JJ, Canos AJ, Sull WJ (1976) The diagnosis of deep venous thrombosis: fallibility of clinical signs and symptoms. Arch Surg 111:34–36
21. Hobson RW, Mintz BL, Jamil Z, Breitbart GB (1991) Current status of duplex ultrasonography in the diagnosis of acute deep venous thrombosis. In: Bergan JJ, Yao JST (eds) Venous disorders. Saunders, Philadelphia, pp 55–62
22. Vogel P, Laing FC, Jeffrey RB Jr, Wing VW (1987) Deep venous thrombosis of the lower extremity: US evaluation. Radiology 163:747–751
23. Cronan JJ, Dorfman GS, Scola FH, Schepps B, Alexander J (1987) Deep venous thrombosis: US assessment using vein compression. Radiology 162:191–194
24. Polak JR, Culter SS, O'Leary DH (1989) Deep veins of the calf: assessment with color Doppler flow imaging. Radiology 171:481–485
25. Wright DJ, Sheard AD, McPharlin M, Ernst CB (1990) Pitfalls in lower extremity venous duplex scanning. J Vasc Surg 11:675–679
26. Lensing AWA, Prandoni P, Brandjes D, Huisman PM, Vigo M, Tomasella G, Krekt J, tenCate JW, Huisman MV, Büller HP (1989) Detection of deep vein thrombosis by real-time B-mode ultrasonography. N Engl J Med 320:342–345
27. Danzat MM, Laroche J-P, Charras C, Blin B, Domingo-Faye M-M, Sainte-Luce P, Domergue A, Lopez F-M, Jaubon C (1986) Real-time B-mode ultrasonography for better specificity in the noninvasive diagnosis of deep venous thrombosis. J Ultrasound Med 5:625–631
28. Killewich LA, Bedford GR, Beach KW, Strandness DE Jr (1989) Diagnosis of deep venous thrombosis: a prospective study comparing duplex scanning to contrast venography. Circulation 79:810–814
29. Comerota AJ, Katz ML, Greenwald LL, Leefmaus E, Czeredarczuk M, White JV (1990) Venous duplex imaging: should it replace hemodynamic tests for deep venous thrombosis? J Vasc Surg 11:53–61
30. Moneta GL, Strandness DE Jr (1989) Basic data concerning noninvasive vascular testing. Ann Vasc Surg 3:190–193
31. Markel AL, Manzo RA, Bergelin RO, Strandness DE Jr (1991) Pattern and distribution of thrombi in acute venous thrombosis. Presentation before the Third Annual Meeting of the American Venous Forum, Fort Lauderdale, Florida, USA, February 20–22
32. van Bemmelen PS, Bedford G, Beach K, Strandness DE Jr (1989) Quantitative segmental evaluation of venous valvular reflux with duplex ultrasound scanning. J Vasc Surg 10:425–431
33. Vasdelois SN, Clarke GM, Nicolaides IV (1989) Quantification of venous reflux by means of duplex scanning. J Vasc Surg 10:670–677

Part V

Chapter 12

Magnetic Resonance Arteriography: Initial Clinical Results

P. LANZER

Introduction

Definite diagnosis of vascular diseases traditionally relies on detection of filling defects on contrast-enhanced radiograms [1]. Some of the limitations of this luminographic approach, such as risk of complications, exposure to ionizing radiation, and inadequate spatial definition of lesion morphology, are well recognized. Others, such as frequent underestimation of the intramural extent and severity of the disease, the lack of tissue characterization, and the virtual absence of information regarding flow and vessel wall functin have only recently received due attention [2, 3]. However, despite these limitations, at the outset of the 1990s, radiographic arteriography remains the definite means to assess patients with vascular disease.

A better understanding of the biology of vascular diseases [4] and increasing numbers of patients participating in prevention programs [5] or undergoing endovascular catheter interventions [6, 7] have simulated research and development of modern vascular imaging technology. The objectives are to design methods allowing noninvasive, quantitative, and tissue-specific definition of the vascular morphology and function. Vascular ultrasound and magnetic resonance imaging (MRI) represent by far the most promising attempts to achieve these goals. The versatility of MRI allows, depending on the imaging parameters selected, either the definition of the vessel wall or of its lumen. Definition of the vessel wall allows, at least theoretically, the assessment of wall morphology [8], its funciton [9], and tissue composition [10]. Definition of the vessel lumen allows a qualitative arteriographic [11] and quantitative flow [12] characterization of the blood flow. This chapter reviews the present status of clinical applications of magnetic resonance arteriography (MRA).

Hitorical Overview and General Comments

Defintion of flow by MRI proceeded historically in several steps of development: definition [13, 14] and characterization [15, 16] of the interdependence between the spin motion and MR signal; development of nonimaging MR methods to measure flow [17]; rediscovery of the effects of flow in the conventional MR tomography [18, 19]; definition [20] and realization [11] of the concept of MRA and velocimetry [12]. The principles of MRA are outlined in chapters 6–8 (this volume).

MRA shares some of the limitations of the standard radiographic approach. In particular, the vessel wall is not directly visualized, and its tissue composition is not characterized. However, the distinct advantages of MRA include:

a) the noninvasive approach – skin penetration or injection of contrast media are not required;
b) freedom from known harmful side effects – the patient's discomfort is minimized;
c) the possibility of serial applications; and
d) measurement of blood flow.

On the other hand, compared with standard radiographic arteriography, MRA at this stage provides a highly inconsistent image quality and a lower spatial resolution. The inconsistent image quality in MRA appears to relate predominantly to its inability to retrieve an adequate MR signal from vascular regions with a disturbed flow and to its susceptibility to artifacts. Although it appears that the occurrence of the flow artifacts can be reduced by shortening of TE/TR time intervals [21], the question of their elimination remains open. The potentials of an artifact-free definition of vessels using the echo planar MR technology is currently being evaluated [22, 23].

As discussed in Chaps. 6–8, two-dimensional Fourier transform (2DFT) and 3DFT "time-of-flight" (TOF) or phase-based strategies can be implemented in MRA. At the time of writing, a systematic comparison of the methods is not available, and optimal protocols for imaging of the different vascular beds have not yet been established. The distinct disadvantages of the indirect phase-based [24] as well as indirect TOF [25] method are the necessity to acquire two projections, with different flow characteristics and the need for postacquisition subtraction reconstruction. Although spin dephasing represents a common problem to any MRA method, its detrimental effect in phase-based MRA may be more critical. MR signal loss due to spin saturation represents a particular difficulty in 3DFT TOF MRA. From the limited available evidence it appears at present that the acquisition of three-dimensional data sets using sequential two-dimensional [26, 27] and thin-section three-dimensional [28, 29] TOF methods with subsequent maximum intensity projection (MIP) [30] to calculate the planar arteriograms are the most efficient strategies for clinical MRA.

In majority of the MRA methods arterial and venous blood flow can be imaged simultaneously or successively, depending on the selected imaging protocol. To allow a clearer vascular definition the latter approach using presaturation pulses [31] appears, however, preferable. Figures 1 and 2 demonstrate the differences between the simultaneous and successive arterious and venous blood flow MRA.

The vessel contrast in MRA is determined by the relative magnitudes of the enhanced blood flow related intraluminal signal and of the supressed extraluminal background signal. It is usually the magnitude and the completeness of

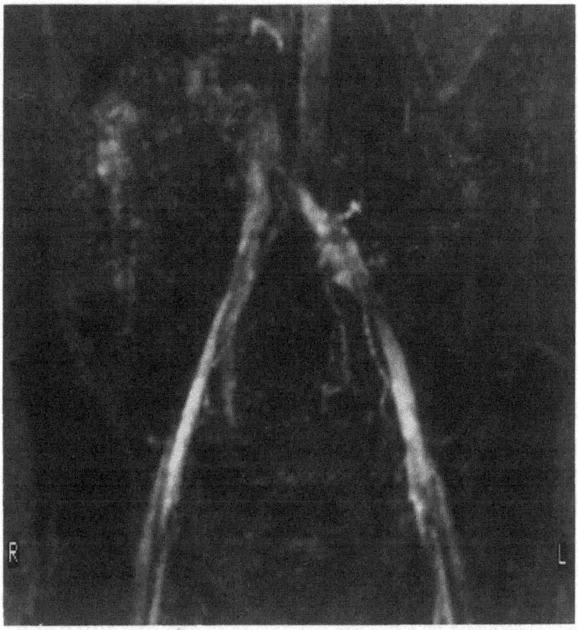

Fig. 1. Combined aortoiliac arteriogram and venogram obtained using sequential 2DFT time of flight (TOF) angiography with coronal plane orientation

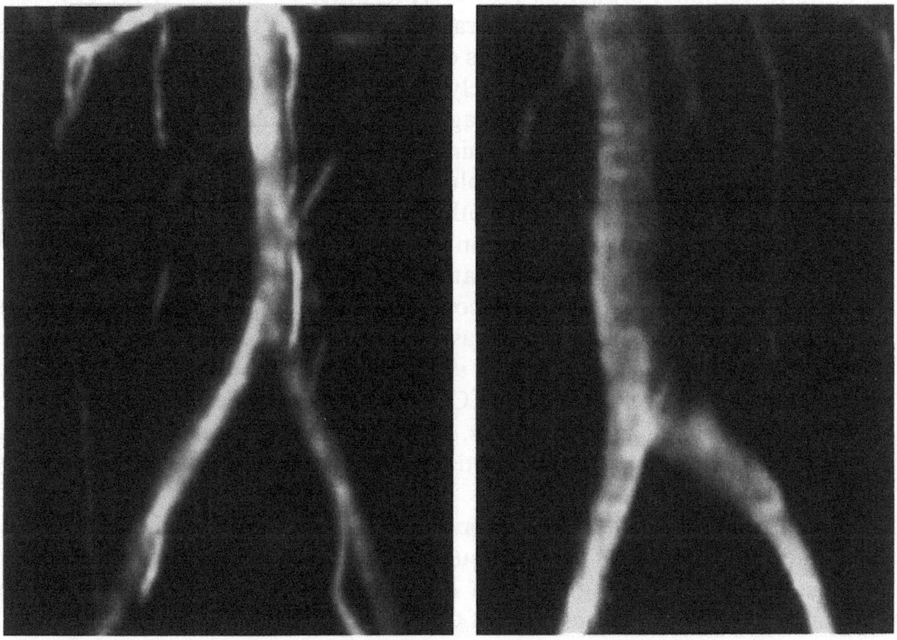

Fig. 2. Selective aortoiliac 2DFT TOF arteriogram (*left*) and venogram (*right*) obtained using transverse plane orientation and flow presaturation

the intraluminal MR signal which determine the overall quality of the MR arteriograms. Compared with the standard radiographic cut films, the quality of MRA films is lower and depending on the presence of flow, instrument and/or method related artifacts more variable. In addition, small branches and morphologic intraluminal features such as venous valves are not resolved in MRA.

Vascular Anatomy

Blood Vessels of the Head and Neck

Diagnostic MR arteriograms of normal arterial intracranial and extracranial vasculature were obtained in healthy volunteers using the time-averaged three-dimensional phase-based [32] and TOF [33, 34] approach. Typically all the major extracranial arteries supplying the brain as well as the four cerebral arteries forming the circle of Willis are well visualized. Among the primary and secondary branches the ramifications from the pars petrosa and supraclinoidalis of the internal carotid and the unpaired meningeal and the paired branches of the vertebral arteries are usually not depicted by MRA. In the neck the trunks of the common, internal, and external carotid and the vertebral arteries are typically well defined. The external carotid artery is usually depicted up to its terminal branches. However, its anterior and posterior branches or their anastomoses with the intracranial branches of the internal carotid are not visualized by MRA. The origins of the carotid and the vertebral arteries from the aortic arch are not routinely included in the MR field of view, and only recently the possibility of imaging these vascular segments has been explored [35]. The imaging time required to define the intracranial vasculature varies, depending on the imaging volume and the MRA method, between 15 and 40 min. The imaging time for both extracranial vessels is less than 15 min. The frequently used image resolution is 1 mm^3. Figures 3–5 are examples of the depiction of cerebrovascular anatomy in healthy subjects.

Due to the complexity of flow secondary to the conduit's geometry and its inside morphology (see Chap. 2) phase dispersion and consecutive signal loss can be a problem even in healthy subjects. Thus for example, the carotid bifurcation at the level of C3–4 or C5–6 with its complex flow fields [36] is prone to flow-related MRA artifacts. In addition, projection calculation using the MIP algorithm may cause additional distortion of the vessel's geometry [37].

Visualization of the veins of the brain using 3DFT TOF methods is incomplete due to saturation of slow venous; however, using contrast enhancement with gadopentetate dimeglumine a routine visualization of the transverse, sagittal, and sigmoid sinuses, vein of Galen, internal cerebral and thalamostriate veins, as well as lateral mesencephalic, jugular and cortical veins became possible [38].

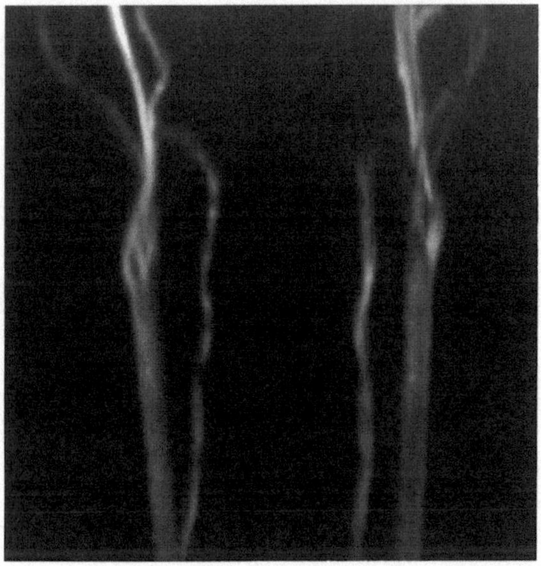

Fig. 3. Extracranial sequential 2DFT TOF arteriogram made in a healthy volunteer

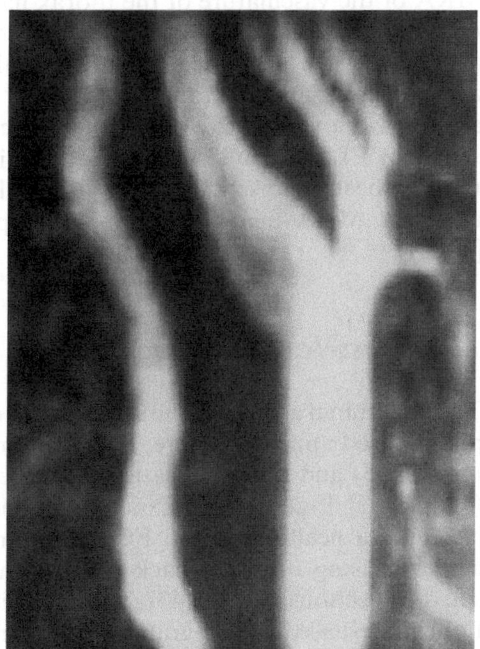

Fig. 4. Extracranial 3DFT TOF arteriogram made in a healthy volunteer. (Courtesy Dr. T. Masaryk, Cleveland Clinic, Ohio)

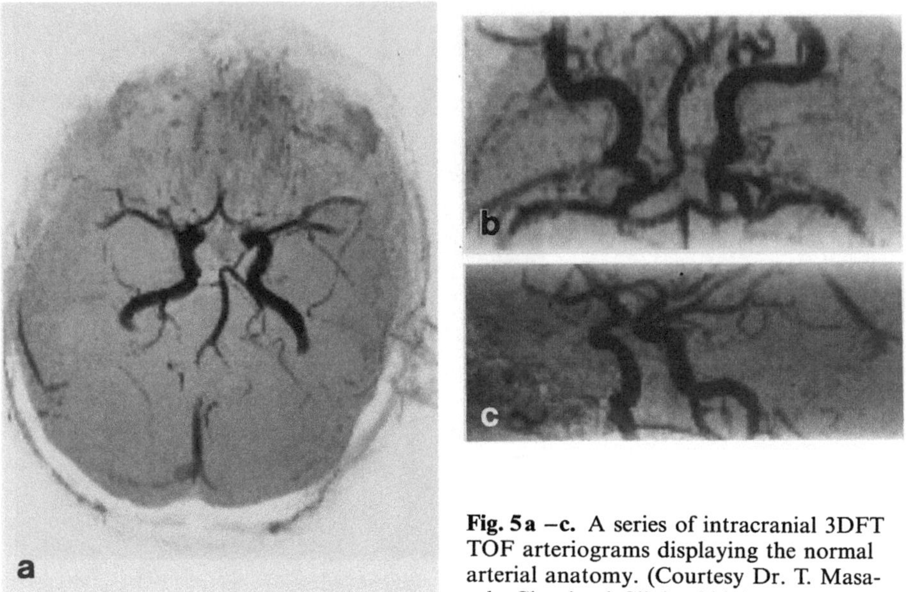

Fig. 5a–c. A series of intracranial 3DFT TOF arteriograms displaying the normal arterial anatomy. (Courtesy Dr. T. Masaryk, Cleveland Clinic, Ohio)

Blood Vessels of the Thorax

MRA of the vasculature of the thorax has been hampered by the presence of breathing and motional artifacts. Recently, however, diagnostic quality images have been generated using the 3DFT TOF MRA [35]. In these MR arteriograms the aortic arch and its three branches as well as smaller thoracal arteries were visualized. Initial experience with multiple-shot imaging using snapshot FLASH pulse sequence in the imaging of thoracic vessels has also recently been reported [39]. Imaging of the coronary arteries using presently available MRA methods has not yet produced clinically relevant results [40, 41].

Blood Vessels of the Abdomen

The abdominal aorta and the iliac arteries can be imaged using the projectional phase-based approach. Here, the optimum results are achieved using triggering (Fig. 6) and synchronization with the flow pulse in the abdominal aorta (Fig. 7) [42]. Figure 8 is an example of a biplane phase-based aortoiliac arteriogram in a healthy subject. Further improvement in image quality has been achieved using a back-to-back acquisition of flow sensitive and flow-compensated projections (Fig. 9) [43]. The acquisition time for a triggered two-projection phase-based arteriogram is 25–30 min. High-quality aortoiliac arteriograms can also be obtained using the sequential 2DFT TOF with MIP recon-

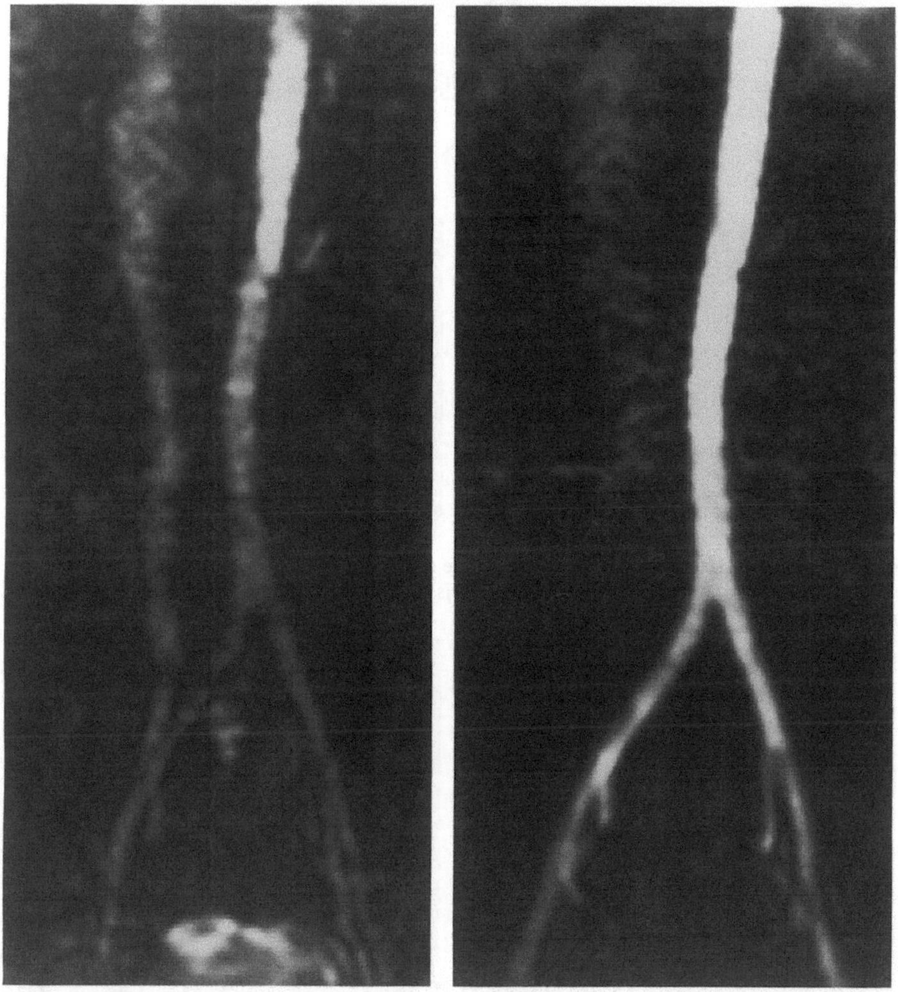

Fig. 6. Aortoiliac projective phase-sensitive untriggered (*left*) and triggered (*right*) flow adjusted gradient (FLAG) phase-based angiograms in anterior-posterior projection

struction (Fig. 10) [44]. The acquisition time for the complete three-dimensional data set is approximately 15 min. The anatomic definition of the aortoiliac segment using the phase-based approach is comparable to that using TOF. In both methods the abdominal aorta and the common, external, and internal iliac ateries are usually clearly defined. The visceral branches of the abdominal aorta and the peripheral branches of the internal iliac artery are inconsistently visualized. However, in contrast, a diagnostic image quality of the renal arteries was reported by others using the 2DFT TOF MIP approach [45]. Recently, using the 3DFT TOF MRA the visceral branches of the abdominal aorta

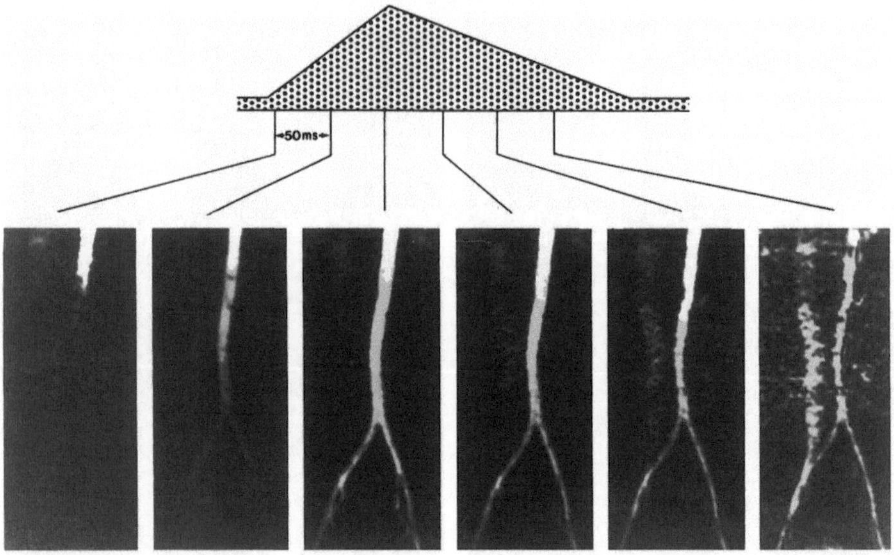

Fig. 7. A series of six projective phase-sensitive aortoiliac FLAG angiograms. Shown is the effect of timing of data acquisition during the rising and the falling phase of the systolic flow pulse (*upper tracing*)

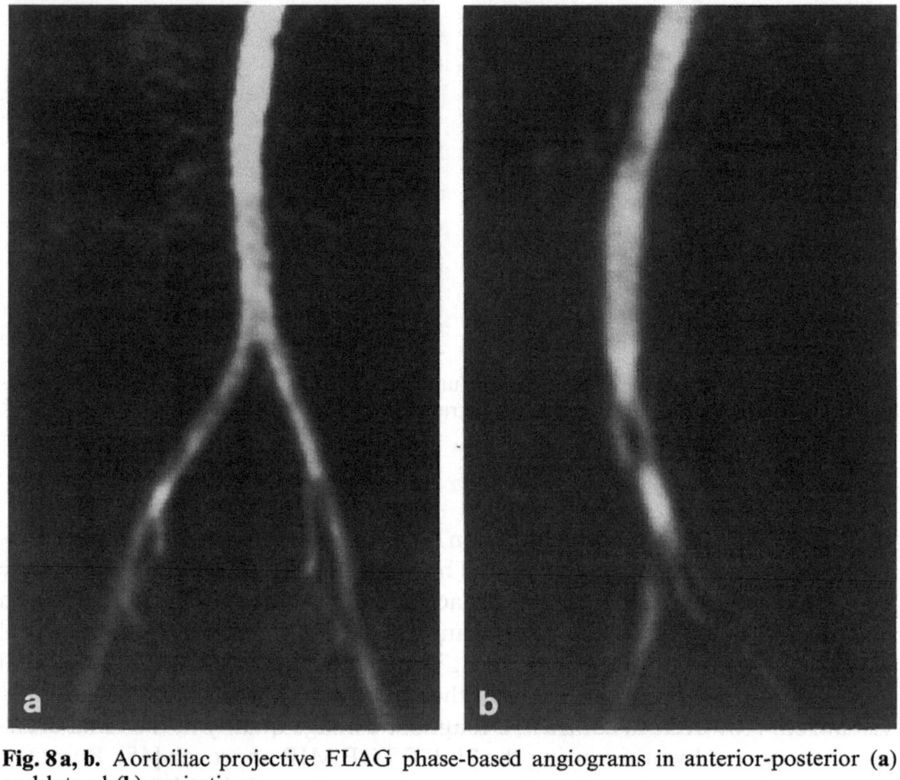

Fig. 8 a, b. Aortoiliac projective FLAG phase-based angiograms in anterior-posterior (**a**) and lateral (**b**) projections

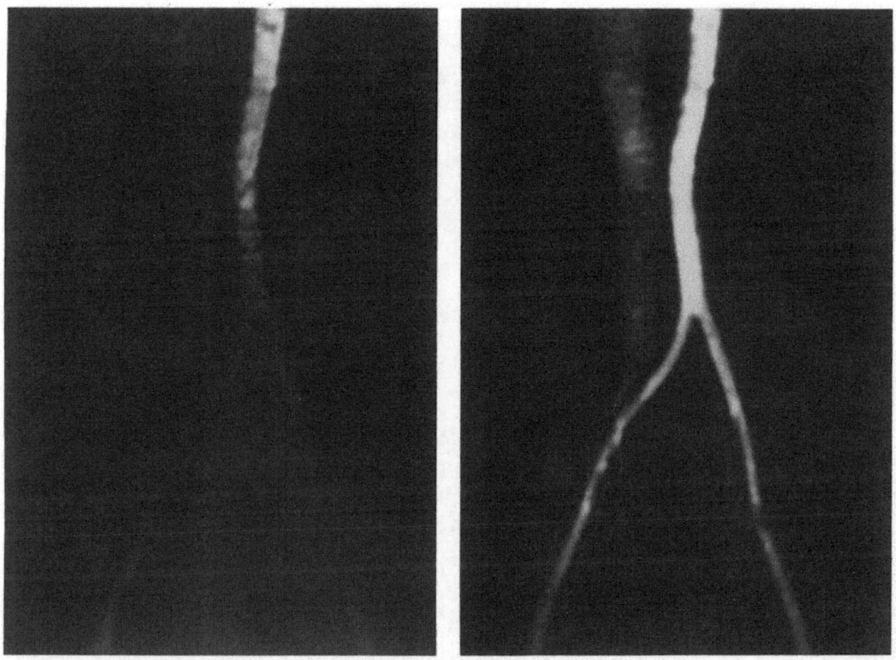

Fig. 9. FLAG (*left*) and rapid sequential echo (RSE) (*right*) aortoiliac phase-based arteriograms in a healthy subject. The improvement in the vessel contrast at identical window settings with RSE is clearly evident

including the celiac trunk, the renal arteries, and the superior mesenteric arteries have been successfully visualized [39].

Compared to the arteries the venous system may be a better target for MRA due to slow flow velocities, high volumetric flow rates and lower propensity to flow disturbances. Initial results indicate that high-quality venograms of the iliac veins, inferior vena cava [46], and splanchnic veins [47] can be obtained with 2DFT TOF.

An optimum strategy for aortoiliac and iliocaval MRA has not yet been defined. However, the TOF methods may be preferable due to shorter imaging times and ability to provide multiple projections from a single three-dimensional data set.

Blood Vessels of the Lower Extremities

The femoropopliteal arterial segment can be successfully visualized using phase-based [48] or TOF method [44]. Similar to the aortoiliac MRA triggering and synchronization with the femoral artery flow pulse are required with the former approach. With both methods definition of the common, superfi-

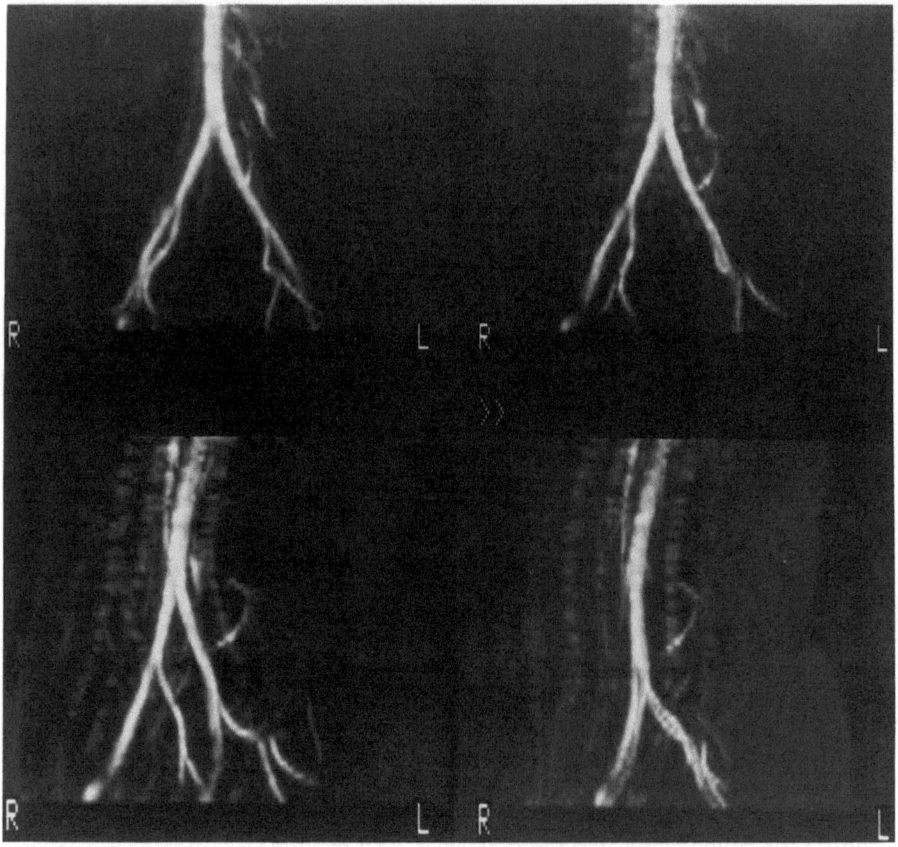

Fig. 10. Sequential 2DFT TOF aortoiliac angiograms in anterior-posterior (*upper left*) LAO 30° (*upper right*), LAO 60° (*lower left*) and LAO 90° (*lower right*) projections

cial, and deep femoral artery is usually adequate. The medial and lateral circumflex femoral arteries and the perforating branches of the deep femoral artery are not consistently defined. Similarily, the superficial external pudendal arteries, superficial circumflex iliac artery, and geniculate branches of the popliteal artery are typically not well seen. The ability of MRA methods to define the lower leg arteries distal to the trifurcation has not yet been systematically evaluated. Figures 10–12 are examples of 2DFT TOF peripheral MR arteriograms. Figure 13 depicts the relationship between the peripheral flow pulse and the intraarterial blood flow signal in a peripheral TOF MR arteriogram.

The superficial and the deep veins above the knee have been successfully visualized using the 2DFT TOF approach [46]. In this study, however, definition of the veins below the knee was suboptimal. In addition, the perforating veins and venous valves are not visualized with this method (Fig. 14).

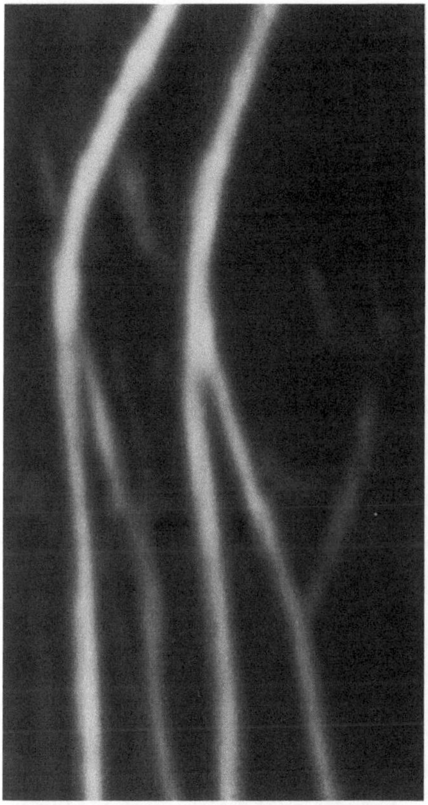

Fig. 12. Origin of the deep femoral arteries in a sequential 2DFT arteriogram

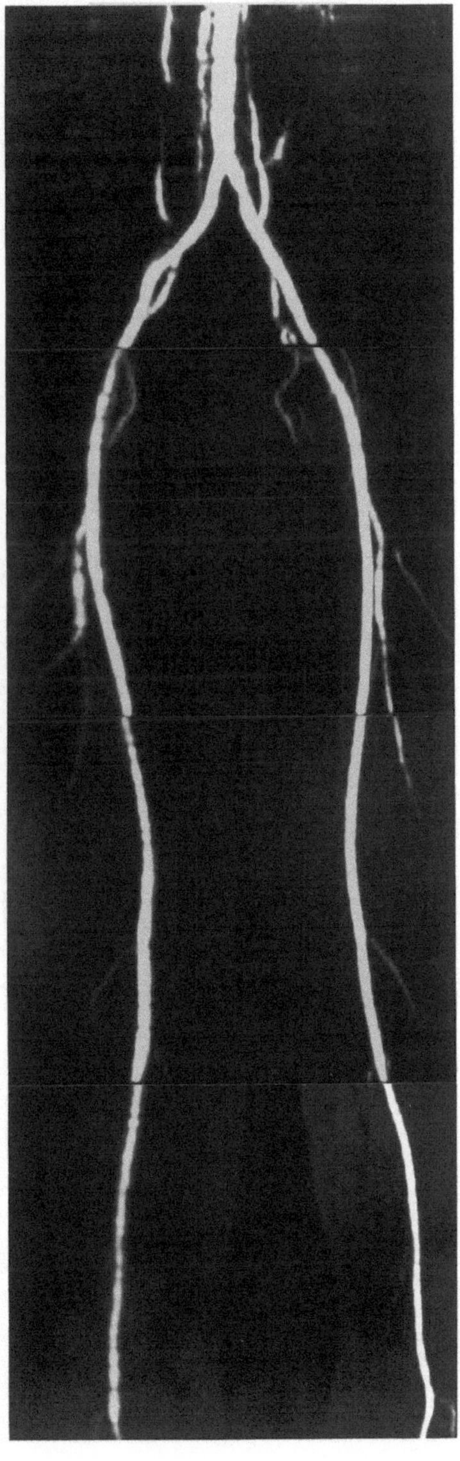

Fig. 11. Peripheral sequential 2DFT TOF angiogram made in a healthy subject

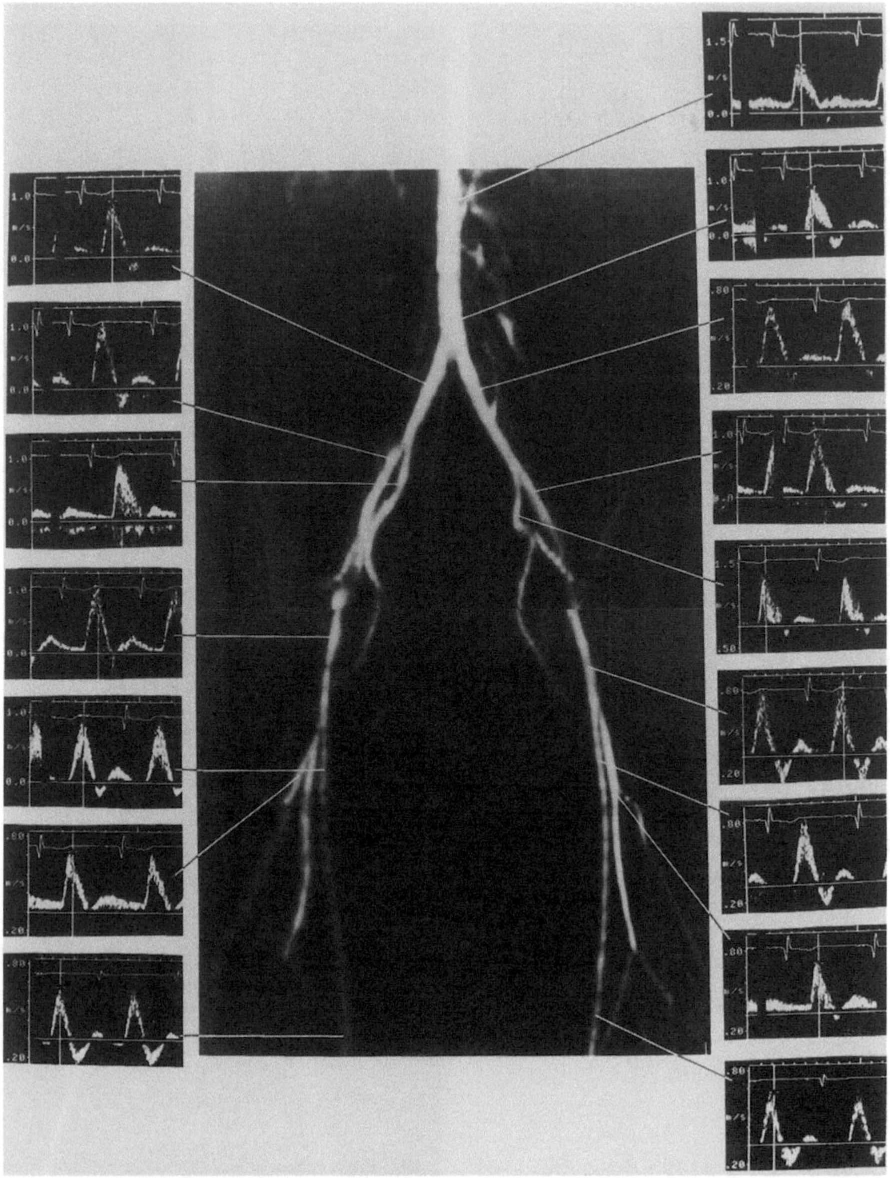

Fig. 13. Aortoiliac 2DFT TOF arteriogram with the corresponding Doppler frequency spectra in the main peripheral arteries in a healthy subject. Despite of the marked differences in the shape of the flow pulse and the flow velocity characteristics the measured differences in the average intravascular signal intensities were not significantly different

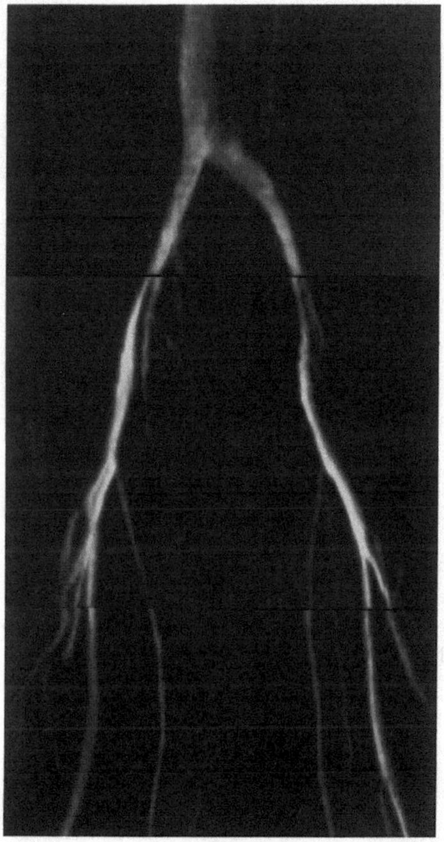

Fig. 14. Peripheral 2DFT TOF MR venogram made in a normal volunteer

Vascular Pathology

Experience with MR arteriography in patients with documented vascular pathology is at present limited to small numbers of patients studied at a small number of institutions. Although these numbers are growing large prospective clinical trials and a better MRA technology will be needed to determine the value of MRA in vascular diagnostics. A brief summary of the present status of clinical MRA is provided in this section.

Blood Vessels of the Head and Neck

Comparison of standard radiographic methods and MRA for defining extracranial carotid artery pathology was reported in two studies [49, 50]. When 3DFT TOF was compared with intraarterial digital subtraction angiography

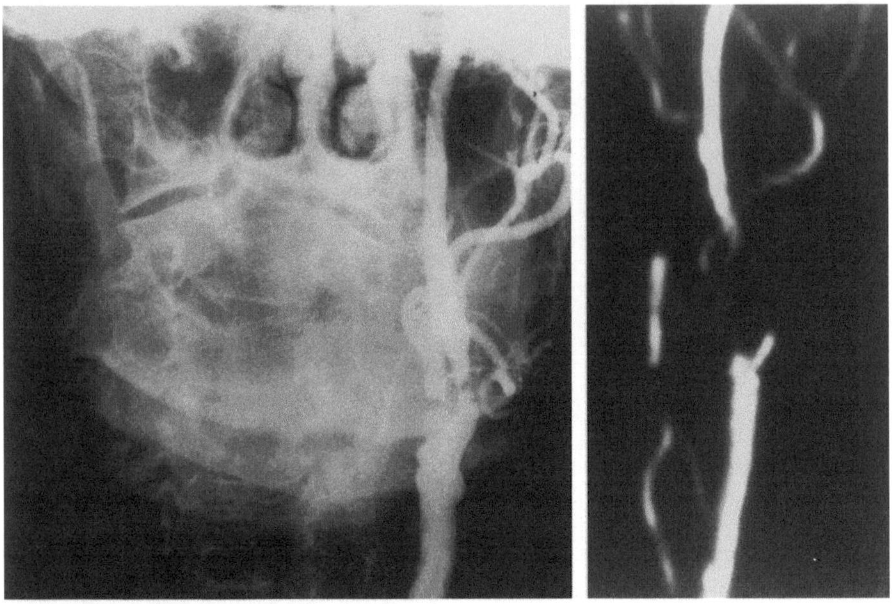

Fig. 15. Radiographic (*left*) and 2DFT TOF (*right*) angiograms showing severe stenoses in the left internal and external carotid arteries

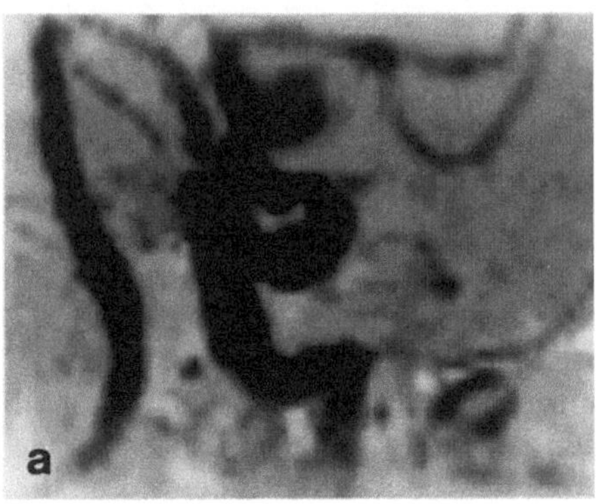

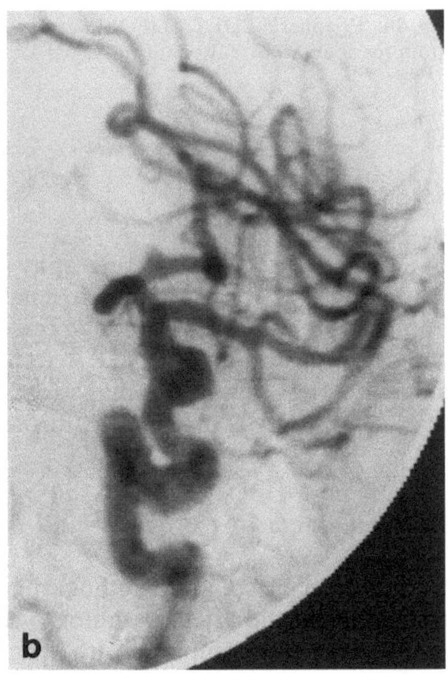

Fig. 16. 3DFT TOF (**a**) intracranial arteriogram in a patient with a large pituitary adenoma documented on T1-weighted spin-echo scans (not shown) and left internal carotid artery aneurysm arising from the supraclinoid segment. The aneurysm was subsequently confirmed by conventional arteriography (**b**). (Courtesy Dr. T. Masaryk, Cleveland Clinic, Ohio)

(DSA) in 24 carotid bifurcations, a high agreement in definition of stenoses of all severity grades as well as occlusions was reported in 22 cases depicted on MRA [49]. A lower agreement was noted when sequential 2DFT TOF was compared with the standard intraarterial contrast angiography in 94 carotid arteries [50]. In this study, the highest correlations between the methods were observed in patients with severe stenoses and the lowest in those with carotid artery occlusions. Ability to demonstrate extracranial vascular pathology was also provided using the three-dimensional phase-based approach [51]. However, in this study no independent confirmation of vascular pathology was provided. Emerging usefulness of MRA in definition of intracranial vascular abnormalities was documented in 40 patients, in 27 of whom a comparison with intraarterial DSA was available [52]. In 72% of MRA images which were considered diagnostic aneurysms, intracerebral artery occlusions, vascular malformations, and tumors were successfully demonstrated. Figures 15–17 are examples of MRA in patients with cerebrovascular pathology.

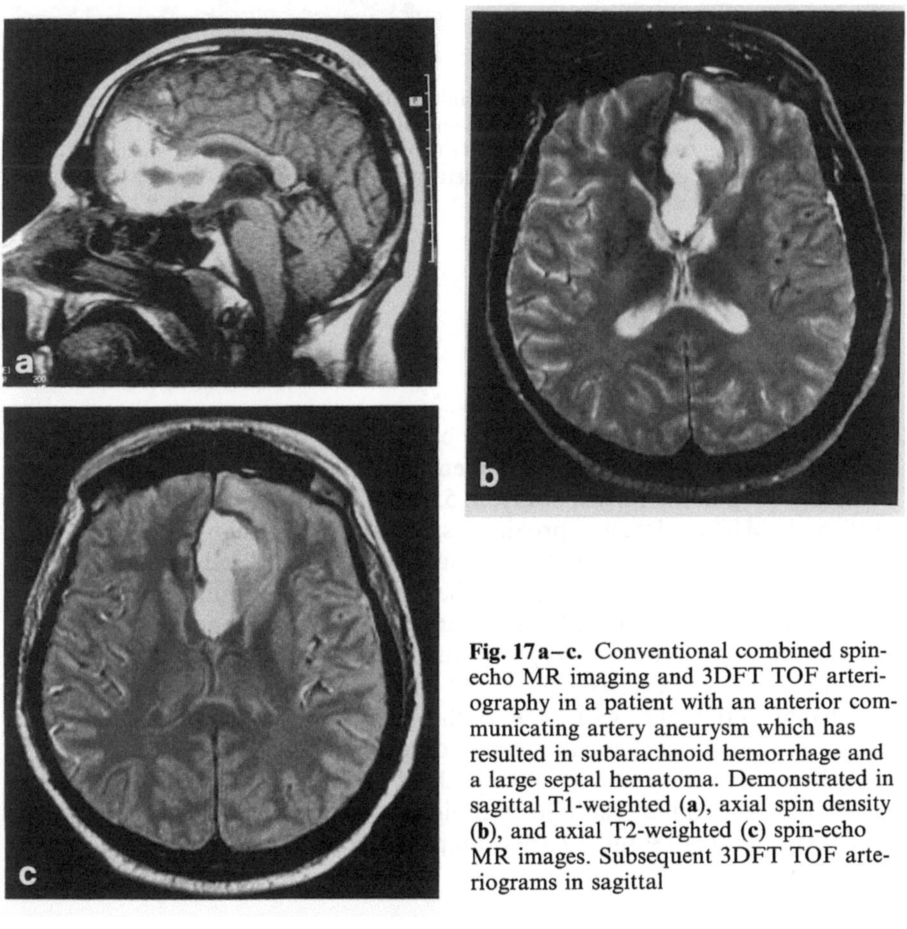

Fig. 17a–c. Conventional combined spin-echo MR imaging and 3DFT TOF arteriography in a patient with an anterior communicating artery aneurysm which has resulted in subarachnoid hemorrhage and a large septal hematoma. Demonstrated in sagittal T1-weighted (**a**), axial spin density (**b**), and axial T2-weighted (**c**) spin-echo MR images. Subsequent 3DFT TOF arteriograms in sagittal

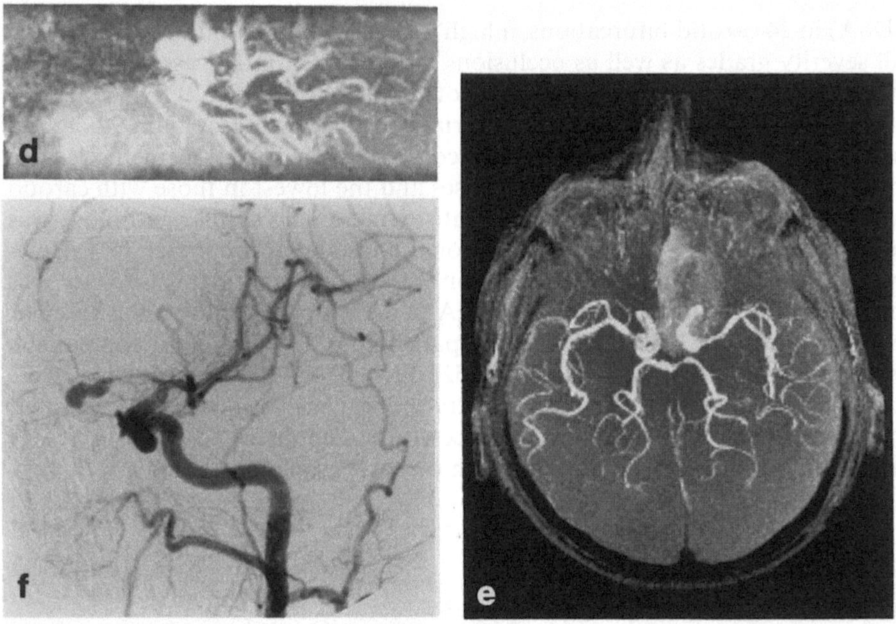

Fig. 17 d–f. (**d**) and coronal (**e**) projections fail to demonstrate the aneurysm because it is obscured by the surrounding methemoglobin within the septal hematoma and the interhemispheric fissure. Conventional arteriogram (**f**) demonstrates not only the aneurysm at the junction of the A1 and A2 segments of the left carotid artery but also the spasm involving the left anterior cerebral artery

Blood Vessels of the Abdomen

Initial results suggest that MRA can be useful in diagnosis of patients with renal artery and abdominal aorta stenoses [53] and those with portal vein pathology [54]. In the former study, in 55 renal arteries the sensitivity of 2DFT TOF with MIP to detect a proximal stenosis greater than 50% was 100%. However, distal stenoses could not be successfully visualized, and several mild to moderate lesions were overrated. The grading of abdominal aortic stenoses was correct in 88% compared to standard abdominal aortography. The results of the latter study indicate that portal vein patency as well as its abnormalities, such as occlusions, reversed flow, and collateral circulation, can be reliably detected by 2DFT TOF with MIP. In this study comparisons with duplex flow measurements, surgical and autopsy findings were available in 28 out of 30 and in all 30 patients, respectively. Potentials of 2DFT phase-based MRA to document intraabdominal vascular pathology was documented only in case reports [55]. Figures 18 and 19 are examples of abdominal aortic pathology as defined by MRA.

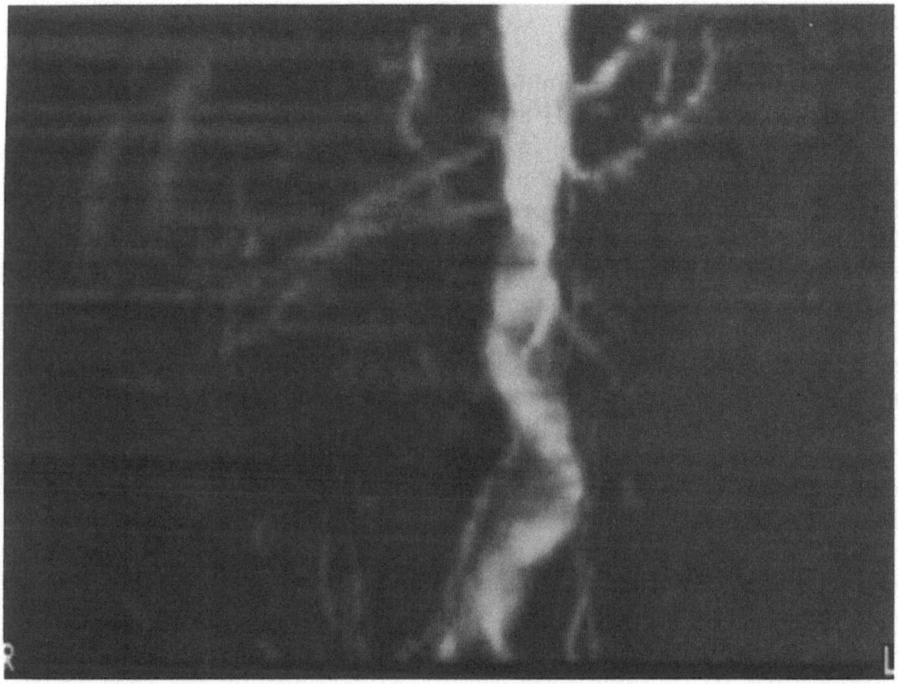

Fig. 18. 2DFT TOF arteriogram showing aneurysmic dilatation of the abdominal aorta distal to the renal arteries

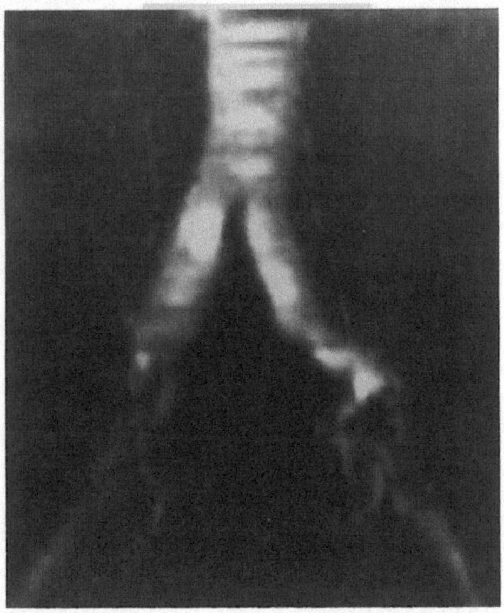

Fig. 19. 2DFT TOF arteriogram showing an abdominal aneurysm extending into both common iliac arteries and severe lesions at the origin of the internal iliac arteries

Blood Vessels of the Lower Extremities

Comparisons between the definition of the peripheral vascular pathology by sequential 2DFT TOF with MIP and the standard aortography and peripheral arteriography were performed in 12 patients [56]. The results showed that MRA agreed with conventional arteriography in grading significant (greater than 50% diameter reduction) or nonsignificant (less than 50% diameter reduction) in 100 out of 140 (71%) lesions. This degree of disagreement along with a suboptimal definition of lesion morphology accounted for major differences when the angiographic findings based choices of appropriate revascularization interventions were considered. Here, MRA agreed with the standard methods in only 5 out of 21 (24%) interventions. Figures 20–22 are examples of peripheral artery abnormalities depicted by MRA.

The results of our own unpublished observations in 415 arterial segments show that 2DFT TOF with MIP tends to overestimate low-grade lesions, whereas it may more frequently underestimate high grade stenoses. In addition, the agreement on lesions severity varied considerably between the individual peripheral arteries, this being the highest ($r=0.85$) in the superficial femoral arteries and the lowest in the profunda femoris ($r=0.28$). Figure 23 is an example of correlations in the definition of the severity of stenoses by both arteriographic methods in the abdominal aorta.

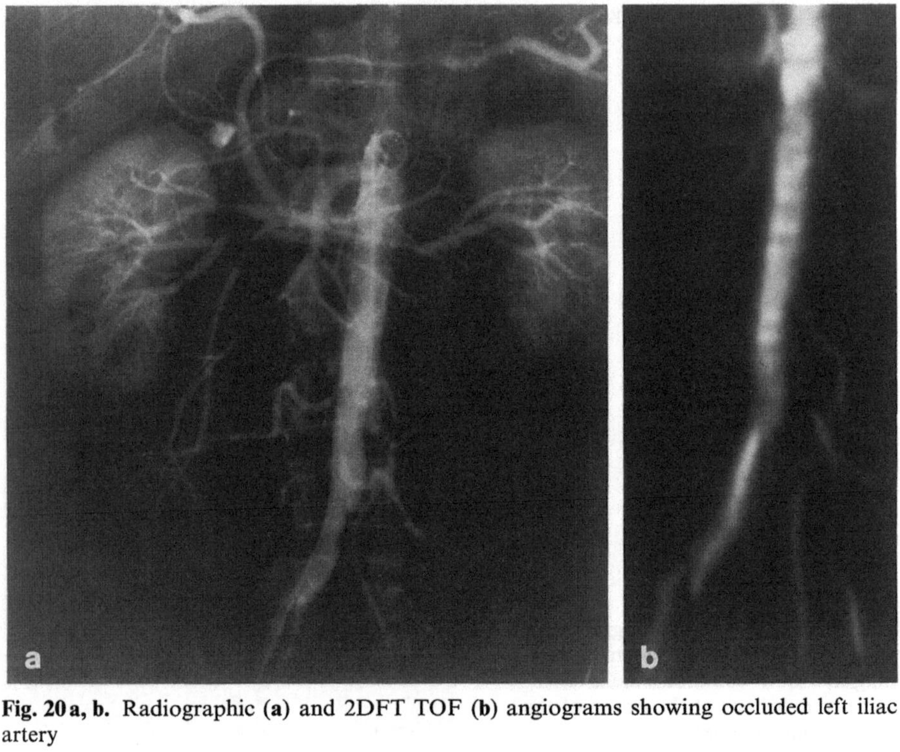

Fig. 20 a, b. Radiographic (**a**) and 2DFT TOF (**b**) angiograms showing occluded left iliac artery

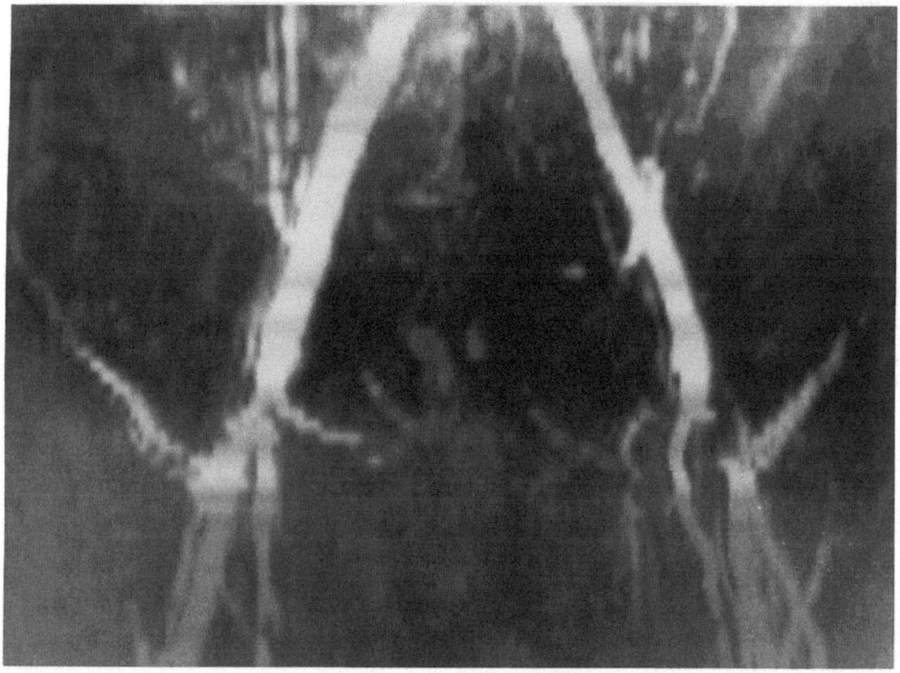

Fig. 21. Sequential 2DFT TOF showing a subtotal occlusion of the left distal graft anastomosis and severe stenosis of the right distal graft anastomosis in a patient with a patent aortobifemoral bypass graft

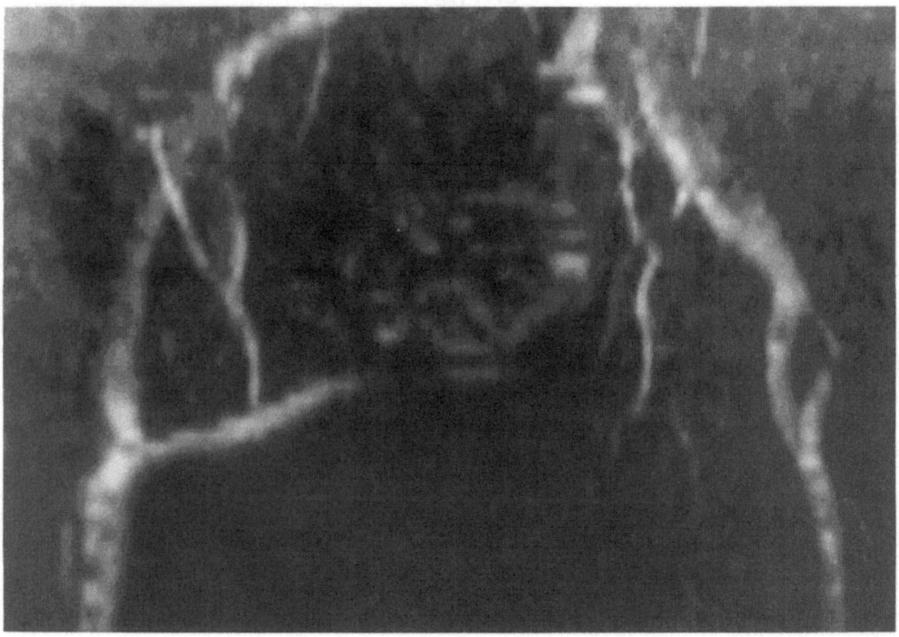

Fig. 22. 2DFT TOF arteriogram showing partial occlusion of a femorofemoral bypass graft

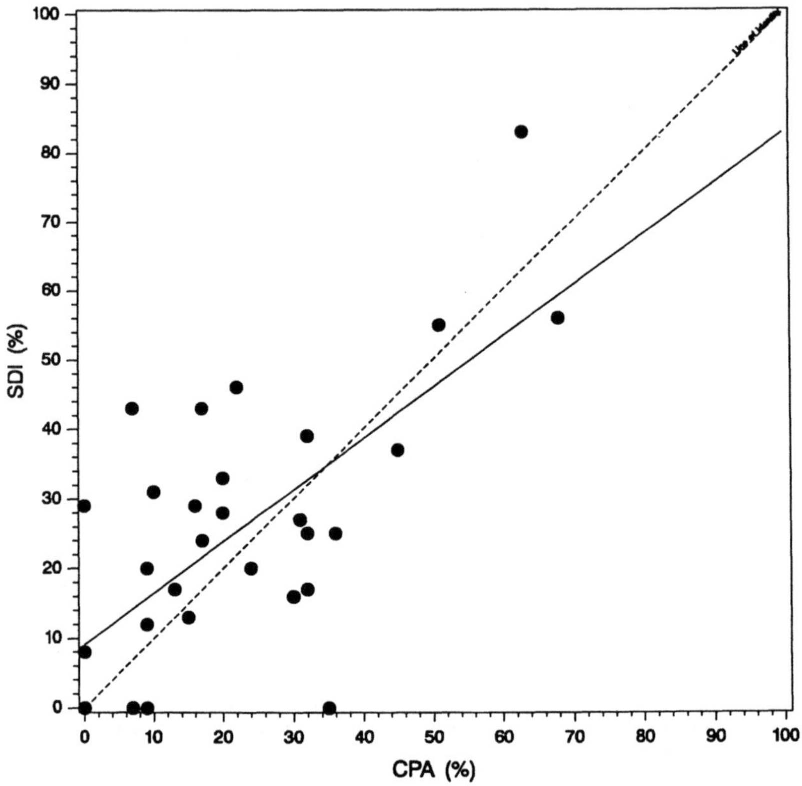

Fig. 23. Measurements of the severity of stenoses (% reduction in the vessel diameter) in the abdominal aorta using conventional aortography (*CPA*) and 2DFT TOF (*SDI*) arteriography. The line of identity (*stippled*) and the correlation line (*solid*) show that 2DFT TOF tends to overestimate abdominal aortic lesions <50%, whereas lesions >50% are more frequently underestimated

The ability of 2DFT TOF imaging [57–59] and venography [46] to detect deep vein thrombosis has been documented in a limited number of patients. Thus, for example, among 26 patients with confirmatory venography the sensitivity and specificity of MR imaging was 100% and 92.9%, respectively [59]. In our own experience in five patients with documented deep vein thrombosis only occluding thrombi above the knee were, with one exception, reliably visualized by 2DFT TOF venography [46]. Figures 24, 25 are examples of deep vein thrombosis documented by MR venography.

Summary

MRA and venography is a new, evolving modality for assessing patients with vascular diseases. The method is noninvasive, free of known harmful side

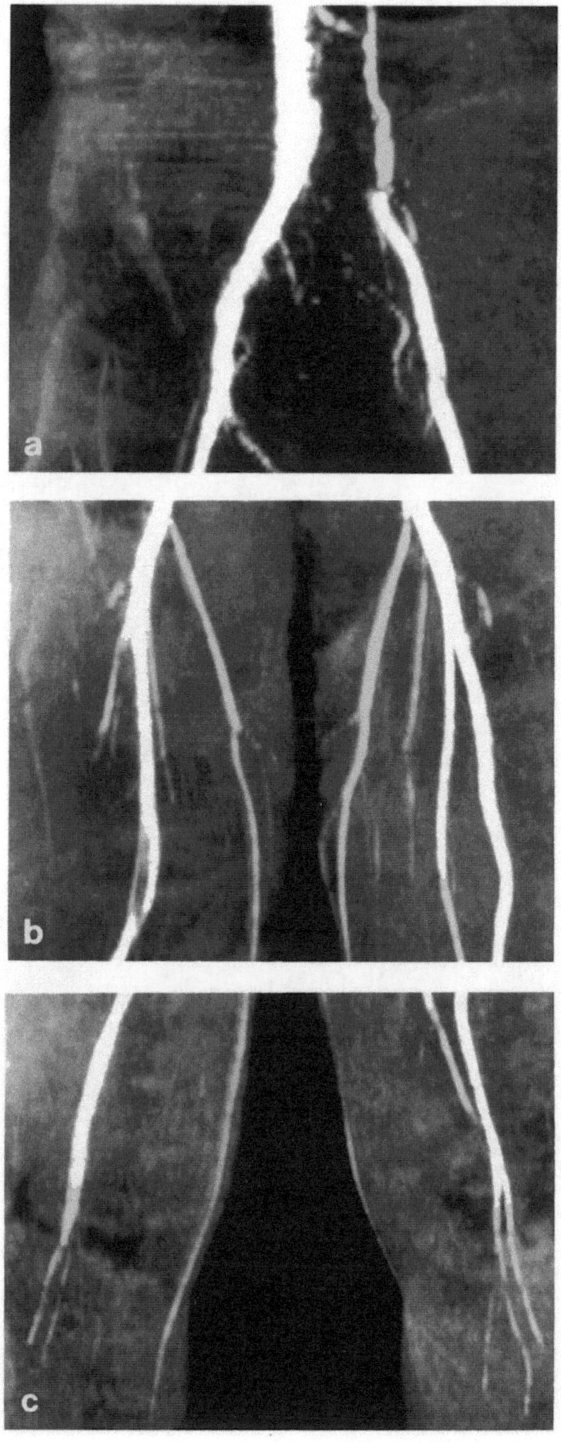

Fig. 24 a–c. Sequential 2DFT TOF venograms showing chronic occlusion of the left common iliac vein

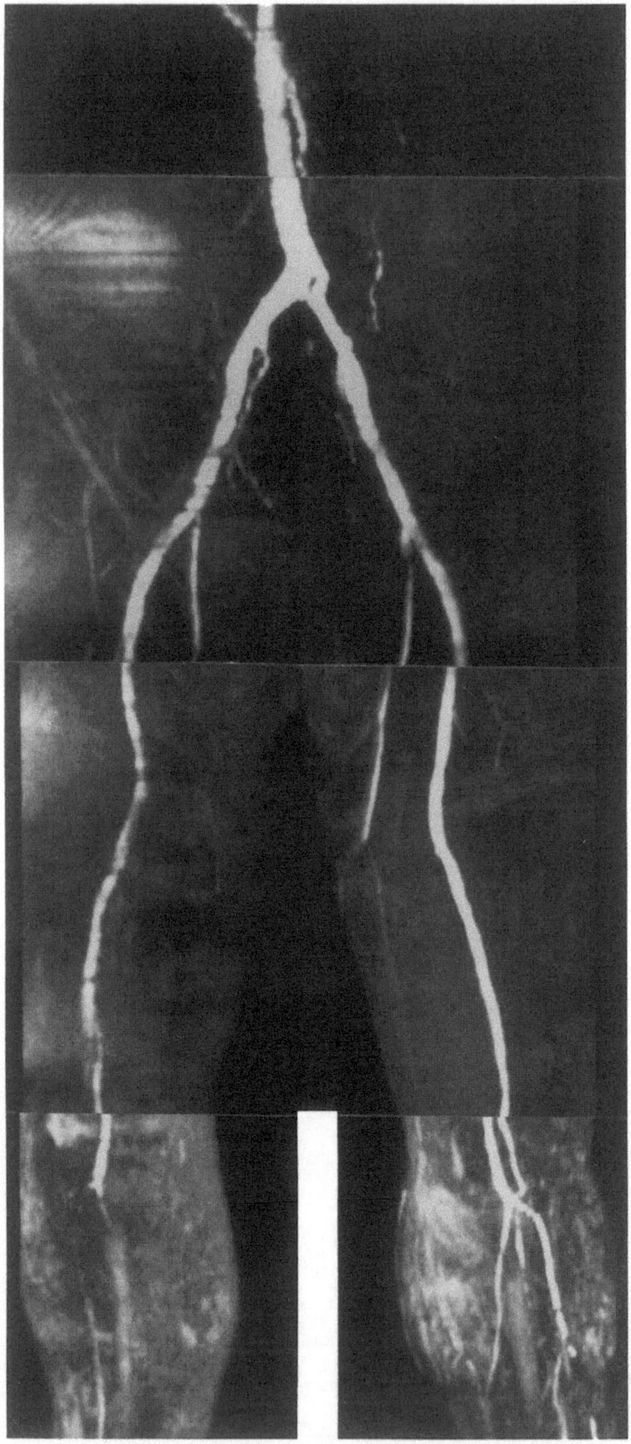

Fig. 25. Sequential 2DFT TOF venogram showing occlusions of the veins of the right lower leg

effects, and well tolerated by patients. Initial clinical results suggest that, although the MRA techniques hold a substantial future promise particularly in defining intracranial vascular pathology in conjunction with standard MRI, further technical improvements and a greater clinical experience will be necessary before these methods may become useful in clinical diagnostics.

References

1. Abrams HL (ed) (1983) Abrams angiography, vol I–III. Little, Brown, Boston
2. Pagnoli P (1989) Ultrasonic evaluation of arterial intima and media thickness: development and validation of methodology. In: Glagov S, Newman WP III, Schafer SA (eds) Pathobiology of the human atherosclerotic plaque. Springer, Berlin Heidelberg New York, pp 705–732
3. Glagov S, Weisenberg E, Zarins CK, Stankunavicius R, Kolettis GJ (1987) Compensatory enlargement of human atherosclerotic coronary arteries. N Engl J Med 316:1371–1375
4. Ryan US (ed) (1988) Endothelial cells, vol I–III. CRC, Boca Raton
5. Anonymous (1986) Dyslipoproteinemia in North America: The Lipid Research Clinics Program Prevalence Study Circulation [Suppl II] 73:1
6. Topol EJ (ed) (1990) Textbook of interventional cardiology. Saunders, Philadelphia
7. Dondelinger RF, Rossi P, Kurdziel JC, Wallace S (eds) (1990) Interventional radiology. Thieme, Stuttgart
8. Herfkens RJ, Higgins CB, Hricak H, Lipton MJ, Crooks LE, Sheldon PE, Kaufman L (1983) Nuclear magnetic resonance imaging of atherosclerotic disease. Radiology 148:161–166
9. Bogren HG, Mohiaddin RH, Klipstein RK, Firmin DN, Underwood RS, Rees SR, Longmore DB (1989) The function of the aorta in ischemic heart disease: a magnetic resonance and angiographic study of aortic compliance and blood flow patterns. Am Heart J 118:234–247
10. Pearlman JD (1986) Nuclear magnetic resonance spectral signatures of liquid crystals in human atheroma as basis for multi-dimensional digital imaging of atherosclerosis. Thesis, School of Engineering, University of Virginia
11. van Weeden J, Meuli RA, Edelman RR, Frank LR, Brady TJ, Rosen BR (1985) Projective imaging of pulsatile flow with magnetic resonance. Science 230:946–948
12. Bryant DJ, Payne JA, Firmin DN, Longmore DB (1984) Measurements of flow with NMR using a gradient pulse and phase difference technique. J Comput Assist Tomogr 8:588–593
13. Suryan G (1951) Nuclear resonance in flowing liquids. Proc Indian Acad Sci [A] 33:107–111
14. Carr HY, Purcell EM (1954) Effects of diffusion on free precession in nuclear magnetic resonance experiments. Phys Rev 94:630–638
15. Singer JR (1959) Blood flow rates by nuclear magnetic resonance measurements. Science 130:1652–1653
16. Hahn EL (1960) Detection of sea-water motion by nuclear precession. J Geophys Res 65:776–777
17. Battocletti JH (1985) Blood flow measurements by NMR. Crit Rev Biomed Eng 13:311–367
18. Hinshaw WS, Bottomley PA, Holland GN (1977) Radiogaphic thin-section image of the human wrist by nuclear magnetic resonance. Nature 270:722–723
19. Waluch V, Bradley WG (1984) NMR even echo rephasing in slow laminar flow. J Comput Assist Tomogr 8:594–598
20. Macovski A (1982) Selective projection imaging: applications to radiography and NMR. IEEE Trans Med Imag MI-1:42–47

21. Schmalrock P, Chun Y, Chakeres DW, Kohli J, Pelc NJ (1990) Volume MR angiography: methods to achieve very short echo times. Radiology 175:861–865
22. Firmin DN, Klipstein RH, Hounsfield GL, Paley MP, Longmore DB (1989) Echo-planar high-resolution flow velocity mapping. Magn Res Med 12:316–327
23. Guilfoyle DN, Gibbs P, Ordidge RJ, Mansfield P (1991) Real-time flow measurements using echo-planar imaging. Magn Res Med 18:1–8
24. Dumoulin CL, Hart HR (1986) Magnetic resonance angiography. Radiology 161:717–720
25. Dixon WT, Du LN, Faul DD, Gado M, Rossnick S (1986) Projection angiograms of blood labeled by adiabatic fast passage. Magn Res Med 3:454–462
26. Gullberg GT, Wherli FW, Shimakawa A, Simons MA (1987) MR vascular imaging with a fast gradient refocusing pulse sequence and reformatted images from transaxial sections. Radiology 165:241–246
27. Keller PJ, Drayer BP, Fram EK, Williams KD, Dumoulin CL, Souza SP (1989) MR angiography with two-dimensional acquisition and three-dimensional display. Radiology 173:527–532
28. Laub GA, Kaider WA (1988) MR angiography with gradient motion refocusing. J Comput Assist Tomogr 12:377–382
29. Ruggieri PM, Laub GA, Masaryk TJ, Modic MT (1989) Intracranial circulation: pulse-sequence considerations in three-dimensional (volume) MR angiography. Radiology 171:785–791
30. Rossnick S, Laub G, Braekle G, Bachus R, Kennedy D, Nelson A, Dzik S, Starewicz P (1986) Three-dimensional display of blood vessels in MRI. In: Proceedings of the IEEE Computer Cardiology Conference. Institute of Electrical and Electronic Engineers, New York, pp 193–196
31. Felmlee JP, Ehman RL (1987) Spatial presaturation: a method for suppressing flow artifacts and improving depiction of vascular anatomy in MR imaging. Radiology 164:559–564
32. Pernicone JR, Siebert JE, Potchen EJ, Pera A, Dumoulin CL, Souza SP (1990) Three-dimensional phase-contrast MR angiography in the head and neck: preliminary report. AJNR 11:457–466
33. Masaryk TJ, Ross JS, Modic MT, Lenz GW, Haacke EM (1988) Carotid bifurcation: MR imaging. Radiology 166:461–466
34. Masaryk TJ, Modic MT, Ross JS, Ruggieri PM, Laub GA, Lenz GW, Haacke EM, Selman WR, Witznitzer M, Harik SI (1989) Intracranial circulation: preliminary clinical results with three dimensional (volume) MR angiography. Radiology 171:793–799
35. Lewin JS, Laub G, Hausmann R (1991) Three-dimensional time-of-flight MR angiogrpahy: applications in the abdomen and thorax. Radiology 179:261–264
36. Ku DN, Giddens DP, Phillips DJ, Strandness DE Jr (1985) Hemodynamics of the normal human carotid bifurcation: in vitro and in vivo studies. Ultrasound Med Biol 11:13–26
37. Anderson CM, Saloner D, Tsuruda JS, Shapeero LG, Lee RE (1990) Artifacts in maximum-intensity-projection display of MR angiograms. AJR 154:623–629
38. Chakeres DW, Schmalbrock P, Brogan M, Yuan C, Cohen L (1991) Normal venous anatomy of the brain: demonstration with gadopentetate dimeglumine in enhanced 3-D MR angiography. AJR 156:161–172
39. Matthaei D, Haasse A, Henrich D, Duehmke (1990) Cardiac and vascular imaging with an MR snapshot technique. Radiology 177:527–532
40. Wang SJ, Hu BS, Macovski A, Nishimura DG (1991) Coronary angiography using fast selective inversion recovery. Magn Res Med 18:417–423
41. Cho ZH, Mun CW, Friedenberg RM (1991) NMR angiography of coronary vessels with 2-D planar image scanning. Magn Res Med 20:134–143
42. Lanzer P, McKibbin W, Bohning D, Thorn B, Gross G, Cranney G, Nanda N, Pohost G (1990) Aortoiliac imaging by projective phase sensitive MR angiography: effects of triggering and timing of data acquisition on image quality. Magn Res Imag 8:107–116

43. Lanzer P, Bohning D, Groen J, Gross G, Nanda N, Pohost G (1990) Aortoiliac and femoropopliteal phase-based NMR angiography: a comparison between FLAG and RSE. Magn Res Med 15:372–385
44. Lanzer P, Pinheiro L, Thorn B, Pohost G (1991) Periphere Spinaustausch-Kernspinresonanz-Angiographie: Untersuchungen zum Effekt des Blutflusses auf Kontrast. Z Kardiol 80:37–43
45. Edelman RR, Wentz KU, Mattle H, Zhao B, Liu C, Kim D, Laub G (1989) Projection arteriography and venography: initial clinical results with MR. Radiology 172:351–357
46. Lanzer P, Gross GM, Keller FS, Pohost GM (1991) Sequential 2D inflow venography: initial clinical observations. Magn Res Med 19:470–476
47. Edelman RR, Zhao B, Liu C, Wentz KU, Mattle HP, Finn JP, McArdle C (1989) MR angiography and dynamic flow evaluation of the portal venous system. AJR 153:755–760
48. Lanzer P, Gross G, Nanda N, Pohost G (1990) Timing of data acquisition determines image quality in femoropopliteal phase-sensitive MR angiography. Angiology 41:817–824
49. Masaryk TJ, Modic MT, Ruggieri PM, Ross JS, Laub G, Lenz GW, Tkach JA, Haacke EM, Selman WR, Harik SI (1989) Three dimensional (volume) gradient echo imaging of the carotid bifurcation: preliminary clinical experience. Radiology 171:801–806
50. Litt AW, Eidelman EM, Pinto RS, Riles TS, McLachlan SJ, Schwartzenberg S, Weinreb JC, Kricheff II (1991) Diagnosis of carotid artery stenosis: comparison of 2DFT time-of-flight MR angiography with contrast angiography in 50 patients. AJR 156:611–616
51. Pernicone JR, Siebert JE, Potchen EJ, Pera A, Dumoulin CL, Souza SP (1990) Three-dimensional phase-contrast MR angiography in the head and neck: preliminary report. AJR 155:167–176
52. Masaryk TJ, Modic MT, Ross JS, Ruggieri PM, Laub GA, Lenz GW, Haacke EM, Selman WR, Wiznitzer M, Harik SI (1989) Intracranial circulation: preliminary clinical results with three-dimensional (volume) MR angiography. Radiology 171:793–799
53. Kim D, Edelman RR, Kent KC, Porter DH, Skillman JJ (1990) Abdominal aorta and renal artery stenosis: evaluation with MR angiography. Radiology 174:727–731
54. Finn JP, Edelman RR, Jenkins RL, Lewis WD, Longmaid HE, Kane RA, Stokes KR, Mattle HP, Clouse ME (1991) Liver transplantation: MR angiography with surgical validation. Radiology 179:265–269
55. Edelman RR, Wentz KU, Mattle H, Zhao B, Liu C, Kim D, Laub G (1989) Projection arteriography and venography: initial clinical results with MR. Radiology 172:351–357
56. Mulligan S, Matsuda T, Lanzer P, Gross GM, Routh WD, Keller FS, Koslin DB, Berland LL, Fields MD, Doyle M, Cranney GB, Lee JY, Pohost GM (1991) Peripheral arterial occlusive disease: prospective comparison of MR angiography and color duplex US with conventional angiography. Radiology 178:695–700
57. Spritzer CE, Sussman SK, Blinder RA, Saeed M, Herfkens RJ (1988) Deep venous thrombosis evaluation with limited-flip-angle, gradient-refocused MR imaging: preliminary experience. Radiology 166:371–375
58. Totterman S, Francis CW, Foster TH, Brenner B, Marder VJ, Bryant RG (1990) Diagnosis of femoropopliteal venous thrombosis with MR imaging: a comparison of four MR pulse sequences. AJR 154:175–178
59. Spritzer CE, Sostman HD, Wilkes DC, Coleman RE (1990) Deep venous thrombosis: experience with gradient-echo MR imaging in 66 patients. Radiology 177:235–241

Chapter 13

Quantitative In Vivo Blood Flow Measurements with Magnetic Resonance Imaging

S. E. MAIER and P. BOESIGER

Introduction

Doppler ultrasound (US) is still considered the gold standard for in vivo blood flow measurements [1]. However, within past years several new magnetic resonance (MR) imaging methods for flow quantification evolved. All MR modalities essentially rely on the strong influence of the movement of spins on the amplitude and phase of the MR imaging signal. The phase-modulation method [3] has gained the greatest importance for flow quantification. The linear relationship between the phase of the MR imaging signal and the velocity of the moving spins permits a straightforward quantification of instantaneous flow velocity and direction in every pixel over the entire cross section of the vessel. The flow sensitivity of the phase mapping sequences can be adjusted widely to suit physiological velocities ranging from slow cerebrospinal fluid flow to accelerated flow in a stenosed artery.

Fast gradient-echo (GRE) sequences enable short repetition times (TRs) on the order of 20 ms and less by means of a time-saving reversal of the readout gradient instead of the 180° radio-frequency (RF) echopulse applied in spin-echo sequences. This and electrocardiographic triggering allows for a temporal sufficiently resolved quantitative analysis of repetitive pulsatile flow over the entire cardiac cycle. In comparison to other established methods such as thermodilution, electromagnetic flowmetry and Doppler US, the new MR imaging modality is neither invasive nor limited in its application by access windows. These are some of the reasons why MR velocity mapping is becoming an important method for flow quantification in clinical diagnosis [4]. The clinical relevance of quantitative flow measurements should further increase once the normal reference values for various vessels are established.

Previously, in vivo flow measurements by means of MR imaging were compared with other methods. Lipton et al. [5] measured the blood flow in the ascending and descending aorta with MR and standard thermodilution. Matsuda et al. [6] compared the MR flow measurements in the abdominal aorta with those achieved by a single-gated Doppler US. However, these authors utilized time of flight MR methods which are not sensitive to flow direction and provide no spatial resolution, thus preventing comparisons between the cross-sectional velocity profiles. These drawbacks can be overcome when phase-based MR methods are employed, allowing the measurement of the instantaneous two-dimensional velocity distribution within vessels. The results have been confirmed using multigated Doppler US methods [7, 8]. This allows the acquisition of one-dimensional velocity profiles and under the assumption of an axially symmetric velocity distribution [9] flow rates can be calculated and compared with MR volumetric flow.

Initially, the quantification of the flow rate with phase-based MR velocity mapping was validated with continuous and pulsatile flow on phantoms [10, 11]. Based on the results of these initial studies it became obvious that particularly for in vivo examinations the accuracy of the MR flow determination strongly depends upon a precise phase correction, which is applied to the data after the examination. The phase correction includes the subtraction of motion independent phase errors due to different causes and the exact determination of the zero-velocity baseline.

Established Methods for Noninvasive Flow Quantification

Plethysmography and Doppler US are well-known noninvasive flow quantification methods. Only the latter can be considered sufficiently accurate providing a semiquantitative estimation of blood flow. The analysis at the spectrum of the reflected Doppler waves typically allows only the determination of mean velocity and direction, hence velocity mapping or accurate flow rate determination is not feasible except under invasive conditions [12]. To acquire one- or two-dimensional velocity profiles it is necessary to use a pulsed multigated Doppler US instrument [13]. The velocity information is sampled at multiple sites along a focused US beam. The repeated acquisition of the same volume elements at a rate equal to the pulse rate gives an accurate estimation of the motion-induced Doppler frequency shifts. The scan rate should be at least twice as high as the maximal Doppler frequency shifts to avoid aliasing. With current designs utilizing a pulse rates of several kHz it is possible to measure physiological blood flow velocities up to a voxel size of 1 mm^3 at a temporal resolution of 20 ms. Employing the semiannular integration technique [9] and assuming an axially symmetric velocity distribution, volume flow rate can be estimated with a high accuracy. The multigated US is however, very sensitive to misalignments of the Doppler beam and to the measurements vessel diameter.

Flow Quantification by Phase-Based MR Imaging

Physical Principles

For the precession frequency ω of a spin system excited by a RF pulse at the time t_0 the following general relation applies

$$\omega(\vec{r}, t) = \gamma B(\vec{r}, t) = \gamma B_0 + \gamma \vec{G}(t)\vec{r}(t), \tag{1}$$

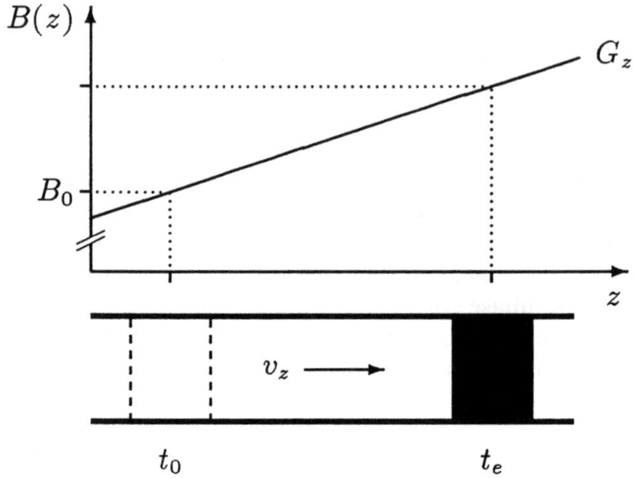

Fig. 1. Principle of MR phase velocity mapping. Movement of a bolus at constant velocity v_z along a constant magnetic field gradient G_z

where $\vec{r}$ is the spin position, t the time, γ the gyromagnetic ratio, B the local magnetic field, B_0 the main magnetic field and $\vec{G}$ a magnetic field gradient. At the instant t_e the phase Φ of the complex MR response signal of the spin system is equal to

$$\Phi = \int_{t_0}^{t_e} \omega(\vec{r}, t)\, dt = \gamma B_0 (t_e - t_0) + \gamma \int_{t_0}^{t_e} \vec{G}(t)\, \vec{r}(t)\, dt. \tag{2}$$

The signals of stationary spins and spins moving along a magnetic field gradient as shown in Fig. 1 differ in phase. This effect is also present in two-dimensional Fourier transform MR imaging with GRE sequences as depicted in Fig. 2. The common approach is to measure the phase shift due to flow in the direction of the slice-encoding gradient. For a constant velocity v orthogonal to the imaging plane the proportion of the phase shift $\Delta\Phi$ is obtained with the first order moment of the slice-encoding gradient G_s,

$$\Delta\Phi = \gamma v \int_{t_0}^{t_e} G_s(t)\,(t - t_0)\, dt. \tag{3}$$

For optimal results the first moment of the gradient and, hence, velocity sensitivity should match the velocities under examination. In order to measure in-plane velocities in the direction of the phase-encoding or readout gradient the standard imaging sequence must be modified. The modification is necessary because in the conventional acquisition schemes the first-order moments are not constant but dependent on the phase-encoding step or sample position during readout.

Application and In Vivo Validation

By means of standard GRE sequences in which the echoes are induced by a reversal of the readout gradient (Fig. 2) quantification of pulsatile flow in

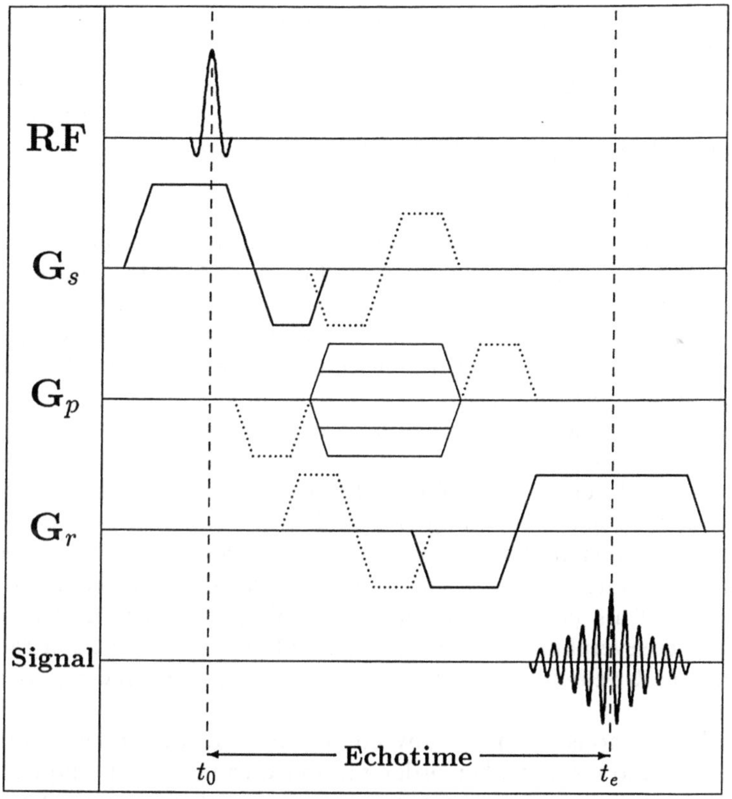

Fig. 2. Timing of a gradient echo (GRE) pulse sequence. G_s, slice-selection gradient; G_p, phase-encoding gradient; G_r readout (frequency encoding) gradient. *Dotted lines* indicate possible additions to the gradient design for compensation of the velocity-dependent phase shift (*GMN*)

sections perpendicular to the vessel axis is possible. Echo times on the order of 10–15 ms and an electrocardiogram-synchronized acquisitions allow for a temporal resolution of the crossectional flow velocity pattern at 20-ms intervals. To gain sufficient signal when imaging single planes at TRs down to 20 ms a rather large slice excitation thickness between 10 and 20 mm is preferable. For vessels sections without side branches and a straight course perpendicular to the imaging plane this is not a limitation, because with a pulse wave propagation in the human artery system at 5–10 m/s [14] and accelerations in the order of 10 m/s^2 velocities in the main flow direction can be considered to be constant over this distance. Depending upon the vessel size a lateral resolution between 1.0 and 2.0 mm and an acquisition matrix set to 256 by 256 is appropriate. If only the volume flow rate is analyzed a reduced resolution at only 128 phase encoding steps will still give accurate results at a reduced total imaging time. Image artifacts and mismapping of the phase due to in-plane flow, as it occurs in vessel bends and bifurcations, are less severe at a larger field of view (FOV).

The flip angle of the excitation pulse α should be adjusted to gain the maximum signal within the vessel. According to Ernst and Anderson [15] the optimal value α_E for a steady state at a chosen TR is given by

$$\alpha_E = \arccos\left(e^{-T_R/T_1}\right), \tag{4}$$

where T_1 is the longitudinal relaxation time (appr. 1000 ms for blood at 1.5 T). Within one cardiac cycle the excited bolus of blood is completely washed out and replaced by unsaturated spins and thus except for layers proximate to the vessel wall a steady state is never attained. This allows a considerably higher excitation angle, and a value twice the calculated optimal value proved to be useful in most situations.

The pulsatile flow is measured over the entire cardiac cycle at time intervals equivalent to the TRs. In contrast to ultrasonographic methods two-dimensional MR velocity mapping with GRE sequences is not possible in real time. With each phase-encoding step and within one heart-beat interval only one line in the spatial frequency domain of the images is acquired. The total imaging time is given by the number of cardiac cycles required for the acquisition of all averages and phase-encoding steps, i.e., under normal circumstances several minutes. In volunteers at rest the cardiac contraction is regular and MR velocity mapping is possible and reproducible to a great extent (Fig. 3).

Under nonstationary flow conditions a further phase shift is caused by acceleration and higher-order motion during the TE (typically 10–15 ms). With accelerations in a physiological range of ± 10 m/s^2, the estimated error in velocity phase mapping does not exceed a few percent and thus in general can be disregarded. The volume flow rate over the complete cardiac cycle can be determined accurately, because the net acceleration is zero.

A significant phase misregistration may occur in presence of an in-plane flow in the frequency-encoding direction (Fig. 4). Sequences with phase correction of the upper echo plane allow correct estimation of the flow rate, however, simultaneous compensation of all sampling positions is not feasible and a slight distortion of the velocity profiles in the frequency encoding direction will in presence of an in-plane flow still occur.

As stated before it is possible to calculate the velocity sensitivity of the sequence with the first-order moment of the gradient. Especially if asymmetric RF pulses are applied, the instant of excitation is not well defined, and a more accurate estimation of the velocity sensitivity with an error of even less than 1% can be achieved in vitro. As the phase information is derived from the real and imaginary part of the complex MR signal the values become cyclically represented in a range of $\pm\pi$. Thus, similar to the multigated US method, velocities inducing phase values which exceed this range are ambiguous. This effect is referred to as phase aliasing or phase wrapping. Although it may be feasible to recover this lost phase information to prevent phase wrap a low-motion sensitivity of the sequence is preferable. On the other hand errors in the

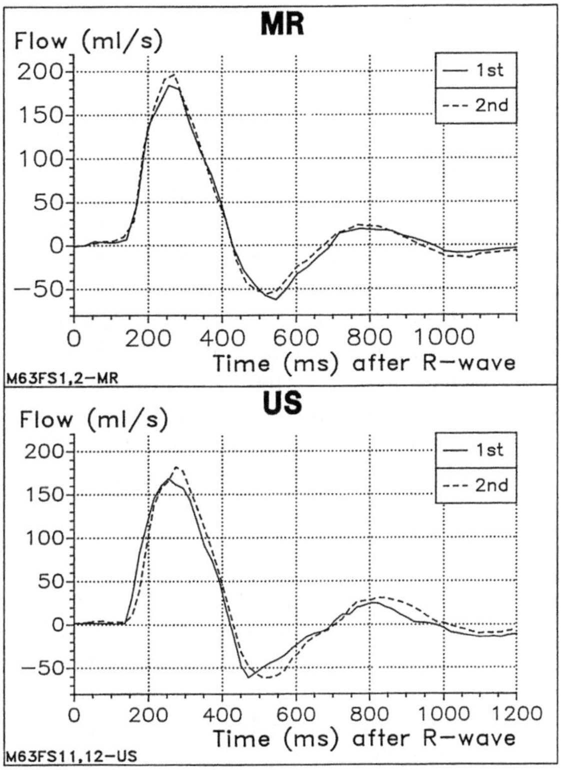

Fig. 3. Temporal flow curves of the abdominal aorta of healthy volunteer obtained with different measurements with MR (*top*) imaging and US Doppler (*bottom*). With each method, two measurements were made; the first is indicated by a *solid line* and the second by a *dotted* line. Flow was quantified at 28-ms intervals with MR and every 20 ms with US

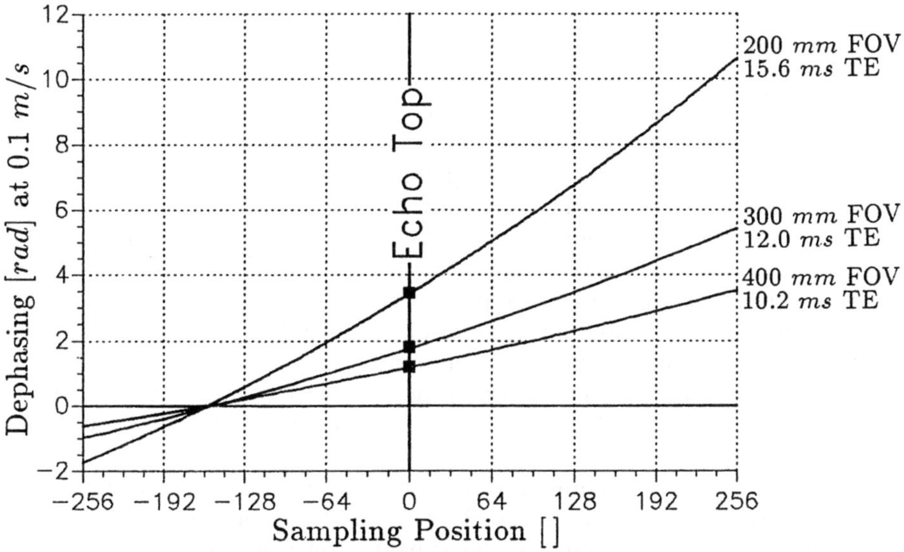

Fig. 4. Sample position-dependent dephasing for 0.1 m/s in-plane flow in the readout direction for different fields of view (*FOV*) and *TE*. Gradient strength during readout is $2.85 \cdot 10^{-3}$ T/m. Calculations are based on conventional sequences

velocity mapping due to signal noise and inhomogeneities of the main field and tissue-specific susceptibility are less severe if the full range is utilized, and the motion sensitivity should be set accordingly [16].

Phase errors of the entire phasemap, particularly in in vivo examinations, have different causes. Motion-independent phase inhomogeneities derive from local variations of the mainfield and tissue-specific susceptibility as well as eddy currents. If flow orthogonal to the selected slice occurs in a preferred direction – e.g., during systolic peak arterial flow or in a flow phantom – the top sample position and phase of the echo signal experience a slight shift. With most reconstruction software this shift is corrected by default and thus both a linear and constant phase error are introduced in the final phasemap.

For accurate velocity determination these phase errors can not be ignored, and a correction scheme has to be applied. Usually, a phasemap with different velocity encoding but almost identical phase errors is acquired and subsequently subtracted. A simpler but efficient procedure feasible with a flow phantom is the subtraction of a second phase-reference map which has been acquired with identical imaging parameters but without flow. The same approach is also possible in vivo when the measurement sites are restricted to the ascending or abdominal aorta in healthy subjects with unperturbed physiological flow, where the blood flow at the end of diastole or beginning of systole over the entire cross section of the vessel is approximately zero [17].

In a subsequent step, the phase values measured inside the vessel boundaries are extracted. To correct for constant phase deviations in each individual frame the magnitude weighted average of phase values outside the vascular lumen is considered as the zero-velocity baseline. The final result is a series of subsequent instantaneous flow-velocity profile maps, depicted in Fig. 5 as gray scale phase images representing phase values from $-\pi$ to $+\pi$. Phase aliasing is visible as an abrupt shift of brightness.

If the values near the vessel boundaries are correct, and if it can be assumed that the phase difference between two neighboring pixels never surpasses π and the phase values are not random due to noise, it is possible to correct for phase leaps in the two-dimensional phase map. Obviously, phase leaps in images that have been interpolated for display cannot be corrected. Phase correction can either be done with a sophisticated phase leap search algorithm, or if the difference between the true lowest and highest phase value within one frame is smaller than one cycle, a general phase shift δ ($-\pi < \delta < \pi$) to fit the values within a single cycle yields the same result. A selection of the same velocity profiles as in Fig. 5 after correction of the ambiguous phase values is presented in Fig. 6 as wire maps. An animation of the wire maps as a slow motion movie reveals the dynamics of the aortic blood flow far better than has any other method before.

The total instantaneous flow is computed by adding up the pixel values over the two-dimensional flow map (Fig. 3). The sum from the entire cardiac cycle gives the total volume of blood per heart beat. However, this value is not very representative as the heart rate and the size of organs to be supplied varies widely among individuals. In particular in large vessels the calculation of

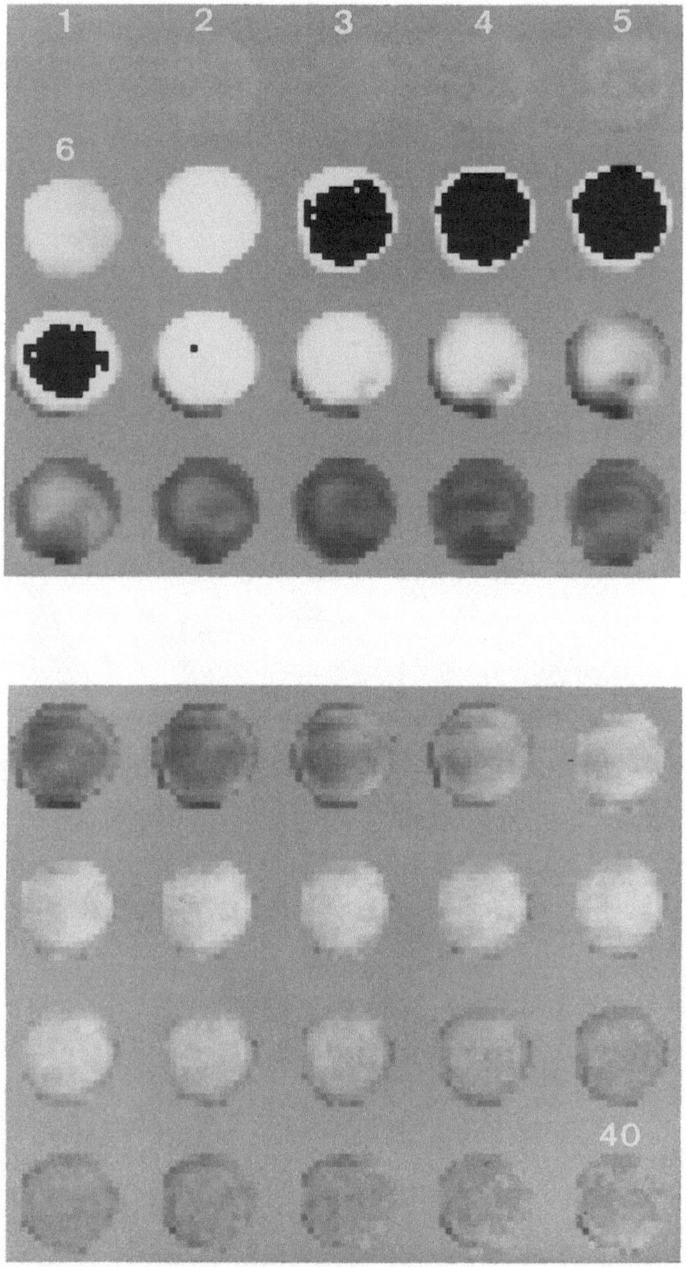

Fig. 5. Brightness representation of 40 consecutive MR phase images of abdominal aorta in a healthy volunteer. The surrounding pixel values have been masked for visualization. Phase aliasing is visible as an abrupt change of brightness in *panels 8 to 12*. (Spatial resolution of 1.2 mm led to about 250 pixel values per phase image inside the vessel area. Temporal resolution = 29 ms; velocity sensitivity in the slice selection direction with a slice thickness of 25 mm was set to 360° phase rotation per 1.4 m/s.)

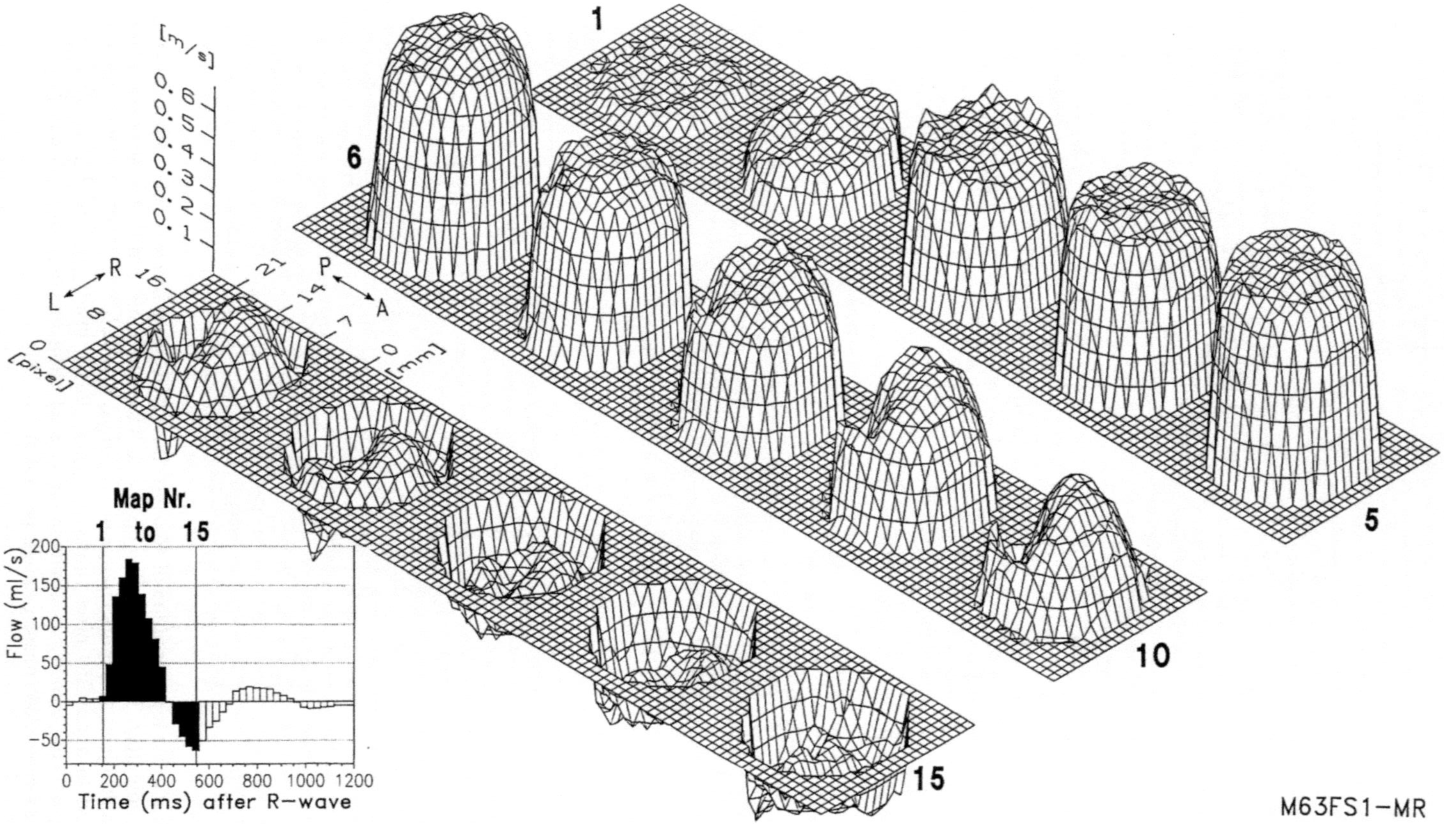

Fig. 6. Wire maps at 15 consecutive instances of the same measurement as in Fig. 5. The time interval between two consecutive maps is 29 ms. The orientation and original dimensions of the velocity profiles in millimeters and pixels are indicated in one map. The integrated flow and the temporal position in relation to the R-wave peak of the velocity patterns are shown in the *inset* (*L*, left; *R*, right; *P*, posterior; *A*, anterior)

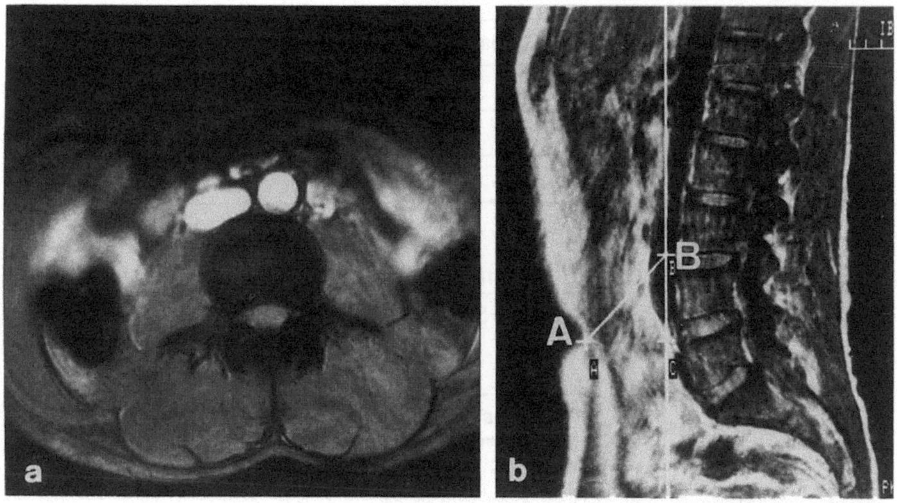

Fig. 7 a, b. **a** Transverse MR slice field of view (FOV) of 200 mm at the level of the second lumbar vertebra. **b** Sagittal MR image FOV of 400 mm of the measurement site in the abdominal aorta. *Line* between *A* and *B* indicates the course of the US beam

volume flow rate related to the body index gives a more realistic estimation of the normal blood supply.

The MR method introduced above was validated by multigated Doppler US in the abdominal aorta of healthy volunteers as represented in Fig. 7 [7, 8]. To acquire the one-dimensional velocity profiles in the anteroposterior direction across the center of the vessel a prototype Doppler US instrument with a pulse repetition rate of 10 kHz and a carrier frequency of 4 MHz was used [18]. The lateral resolution of the one-dimensional acquisition at an angle of incidence of 45° and a US beam diameter of 1.5 mm was 1.1 mm. The temporal resolution was chosen to be 20 ms. For comparison MR velocity values were linearly interpolated at the times and positions utilized by US. A selection of corresponding profiles at various instants in the cardiac cycle recorded in one volunteer is depicted in Fig. 8. A slight tendency of the diastolic US profiles to be narrower than the MR imaging profiles, particularly during early diastole, may be due to the pressure exerted on the vessel by the transducer. Unexpectedly, the profiles are skewed in the anteroposterior direction during early diastolic retrograde and sometimes during the peak forward systolic flow. As this characteristic is typical in both modalities, which are based on completely different acquisition techniques, it cannot be interpreted as a measurement artifact. Asymmetric velocity profiles were also measured in the ascending aorta (Fig. 9) and conform with US results of other groups [12, 13] and the expected behavior of blood flow pattern in a bend.

An analysis of all comparative measurements yielded a high correlation with *r* values, ranging from 0.92 to 0.97, and slopes of the regression lines between

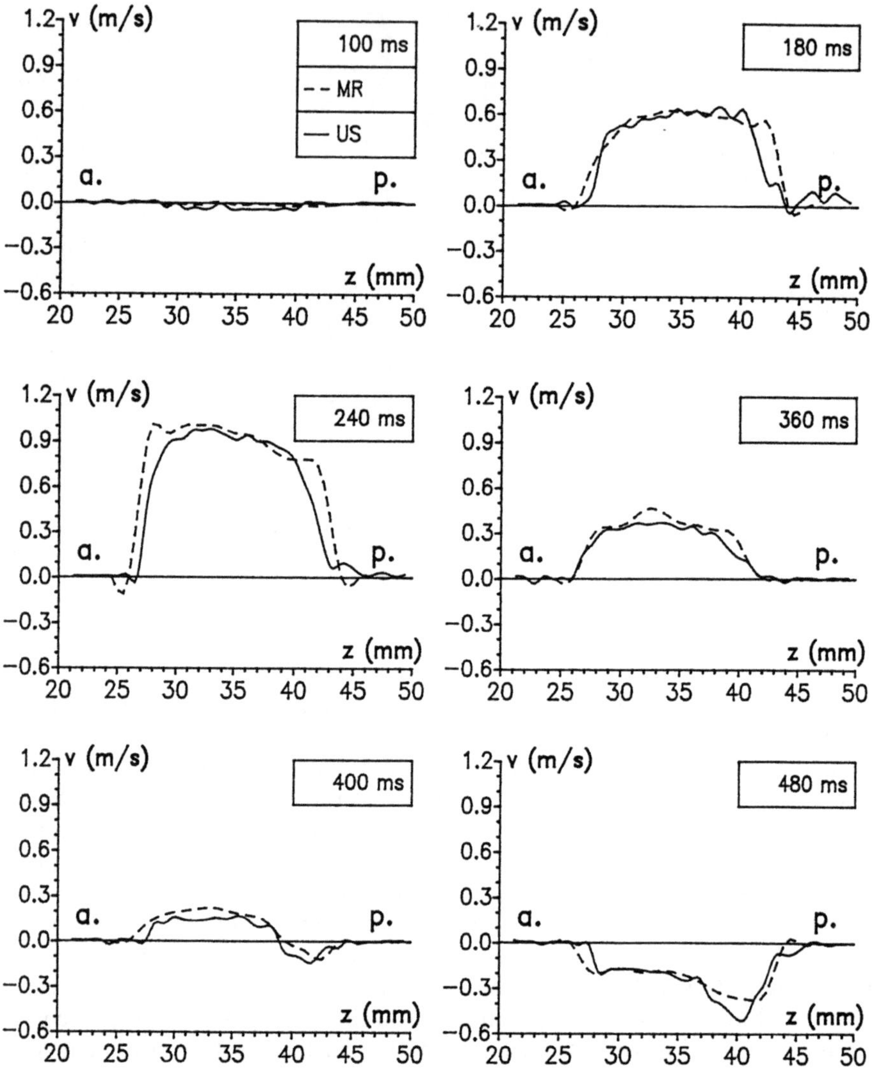

Fig. 8. One-dimensional velocity profiles of abdominal aorta of a healthy volunteer. Correspondence of velocities determined with MR imaging (*broken lines*) and Doppler US (*solid lines*) at various delays relative to the R-wave peak. The *z*-axis indicates the distance to the head of the US transducer. *a*, anterior; *p* posterior

0.86 and 1.13. Minor deviations may be due to physiological changes of the blood flow during the examination and can also be attributed to inaccuracies in the determination of the US angle of incidence.

Figure 3 depicts in separate graphs the instantaneous flow values in one healthy volunteer imaged with MR and Doppler US. Despite the symmetric complementation of the one-dimensional US velocity data, the calculated flow

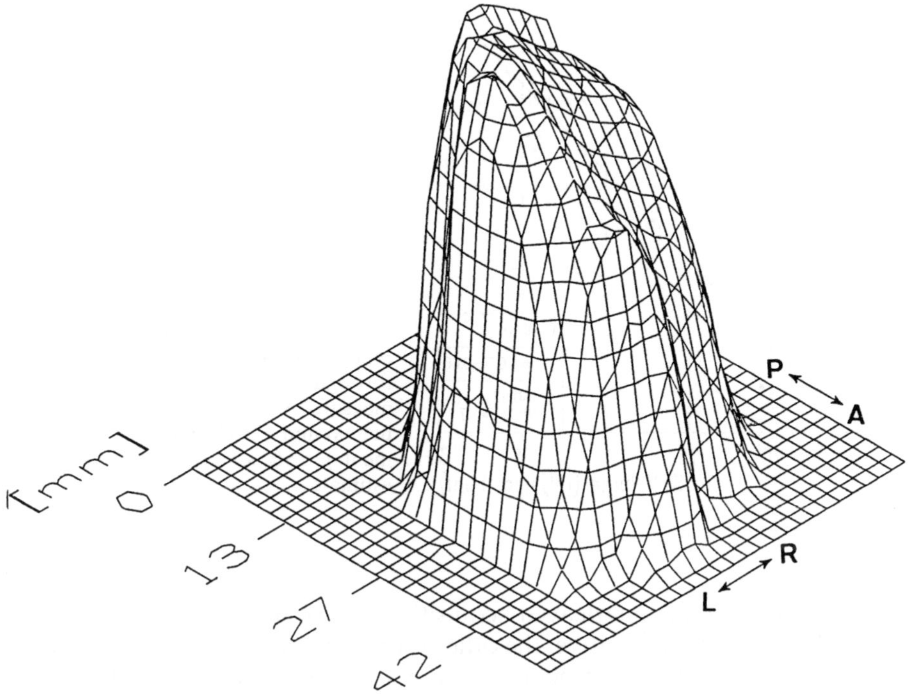

Fig. 9. Wire map depicting the velocity pattern in the ascending aorta of a healthy volunteer during systolic peak flow, 160 ms after the R wave. The distance between contour line levels is 0.1 m/s

values correlated well with the flow values derived from the velocity distributions of temporally corresponding two-dimensional MR imaging. When all estimations of instantaneous flow values (MR imaging, $n=17$; US, $n=22$) were considered in a single correlation analysis, the r value was 0.96 and the slope was 1.06. A correlation plot ($r=0.97$, slope 1.07) of the integrated blood-flow volume V per heart beat determined with the two different modalities in nine volunteers is depicted in Fig. 10.

Correction of Phase Errors Not Related to Flow

As previously mentioned, non-flow-related phase misregistrations in the MR image need to be corrected. The assumption that the errors are mainly caused by inhomogeneities of the main field and tissue-specific susceptibility, as well as eddy currents led to a first approach of simply subtracting the phase image acquired at the end of diastole or beginning of systole. This of course limits the potential measurement sites, because zero velocity over the entire cross section of the vessel is assumed. Nayler et al. [19] successfully applied a new method

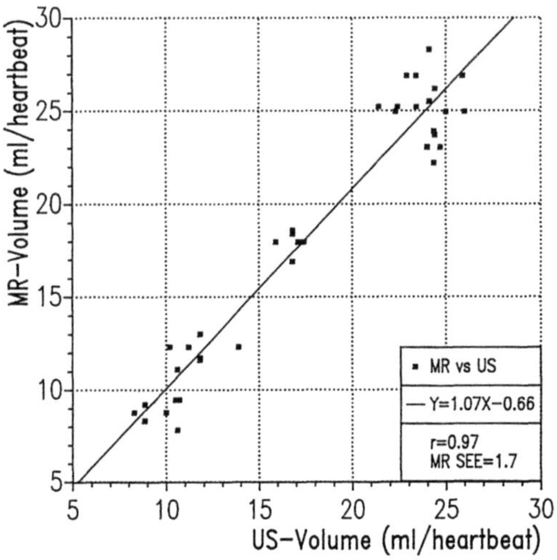

Fig. 10. Correlation analysis of the blood flow volumes *V* per heart beat through the vessels of nine volunteers measured with MR (17 measurements) and US Doppler (22 measurements). SEE, standard error of the estimate

which consisted in the interleaved acquisition of two phase maps. One of the acquisitions is motion sensitive, the other velocity insensitive due to nulling of the first-order gradient moment, referred to as gradient moment nulling (GMN). After subtraction of the two phase maps non-flow-related phase errors are greatly reduced.

However, GMN requires additional gradient lobes, leading to a prolonged TE. This also causes additional sensitivity for higher-order motion terms and introduces further eddy current-induced image artifacts. With very short TEs, on the order of a few milliseconds, achieved by separate sampling of the left and right k-space half planes (FAcE), as presented by Scheidegger et al. [20], gradient timings become very terse and GMN of the slice selection gradient is not possible without sacrificing the very short TE. In an application of this sequence for quantitative phase velocity mapping, different velocity encoding is achieved by a simple reversal of the slice selection gradient [21]. This sequence, with its inherent low sensitivity for in-plane flow, acceleration and higher-order motion, was successfully applied for flow measurements in the bifurcation of the human carotid artery where an irregular flow pattern dominates [22]. Fifteen different subtracted velocity phase maps of the common carotid artery of a healthy volunteer are shown as wiremaps in Fig. 11. The images were acquired at an interval time of 40 ms with 256 phase-encoding steps at a FOV of 200 mm and a TE of 4 ms. The imaging time, determined by electrocardiographic triggering and the acquisition of two differently velocity-encoded image series, was around 15 min. Velocity sensitivity in the slice selection direction with a slice thickness of 10 mm was set to 360° phase rotation per 0.9 m/s. The flow profile acquired with the same imaging parameters at bifurcation level of the same volunteer 100 ms after the R-wave trigger is depicted in Fig. 12.

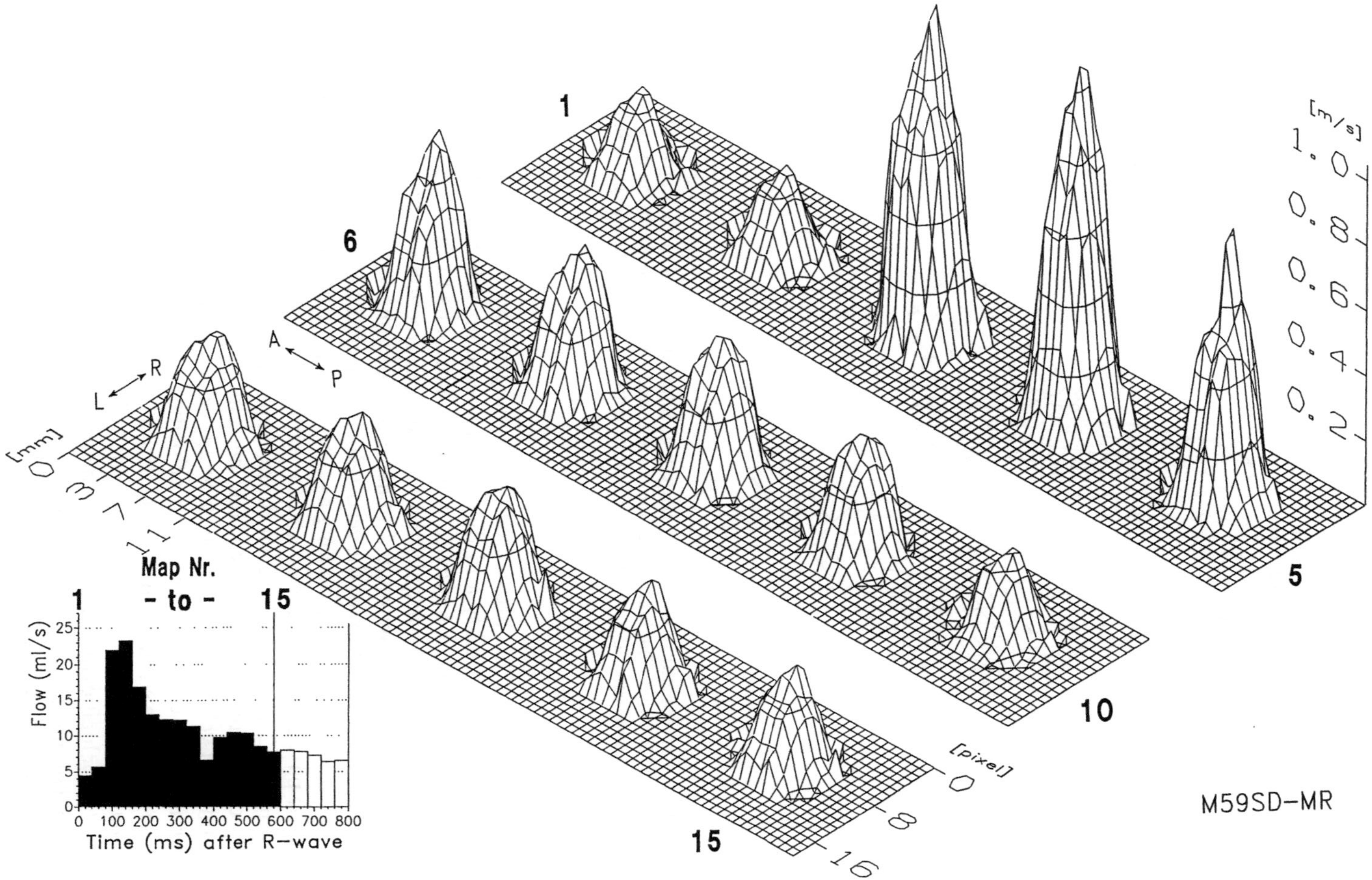

Fig. 11. Wire maps at 15 consecutive instances of right common carotid artery of healthy volunteer. The time interval between two consecutive maps is 40 ms. The orientation and original dimensions of the velocity profiles in millimeters and pixels are indicated. The integrated flow and the temporal position in relation to the R-wave peak of the velocity patterns are shown in the *inset*

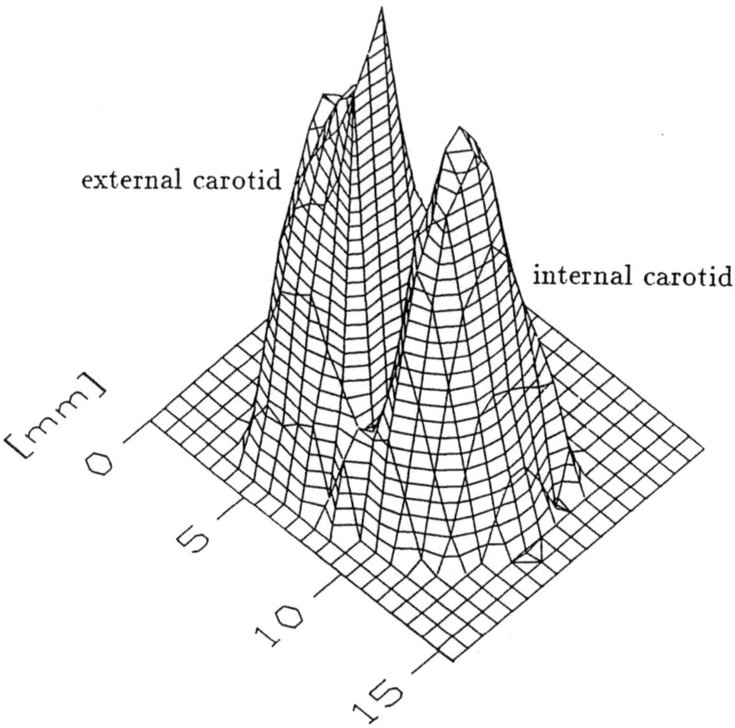

Fig. 12. Wire map depicting the velocity pattern at bifurcation level of the carotid arteries 100 ms after the R wave. The highest velocities are found close to the partition between the internal and external carotid arteries. The distance between contour line levels is 0.02 m/s

The main phase errors are caused by inhomogeneities of the main field and tissue-specific susceptibility and therefore increase linearly with the TE. The ultimate goal might be a sequence with very short TEs in the range of 1 ms. With such a sequence the acquisition of a phase correction would be unnecessary because the non-flow-related phase errors become very small.

The exact determination of the zero-velocity baseline has not received much attention in MR literature. As long as the velocity sensitivity is well adjusted to suit the particular application and the explored velocities induce dephasing in a range of 2π radians or even more, ignoring the zero-velocity baseline usually causes an error of only a few percent. However, to circumvent the ambiguities of velocity-induced phase wrap, velocity sensitivity is normally set according to the expected maximum velocities during systolic peak flow. In this case the phase shifts of the low velocities during the long diastolic interval are situated in the range of the zero-velocity baseline error. Thus the estimation of the velocities becomes highly inaccurate and strongly influences the result of volume flow rate. Buonocore et al. [23] suggest a sequence with variable velocity sensitivity matching the prevailing velocities of each cardiac phase. However, the knowledge of the expected velocity pattern is required to set the velocity sensitivity accordingly. This may be difficult as systolic acceleration and deceleration occur quite abruptly. Considering the magnitude-weighted phase average of the stationary tissue surrounding the vessel boundaries as the zero-velocity baseline usually results in sufficient correction. Of course this scheme will fail when applied at the great vessels near the base of the heart, because the neighboring tissue is only stationary for a short time during a cardiac contraction cycle. In this case the values of the zero-velocity baseline found during late diastole may serve as a less accurate correction value for all frames.

Reduction of Flow-Related Image Artifacts

Recent years have brought a better understanding of flow phenomena and new methods to avoid image artifacts caused by flow [24]. The main artifacts in two-dimensional Fourier transform imaging which appear as a blurring in phase-encoding direction are caused by view-to-view fluctuations of magnitude and phase angle of the net voxel magnetization. Cancellation of the flow-related view-to-view alterations of the signal can be achieved by upstream presaturation. But with phase velocity mapping, where preservation of the flow related signal is important, the view-to-view signal variations caused by pulsatile flow components are reduced by an electrocardiogram-synchronized succession of the phase-encoding steps. Remaining fluctuations deriving from minor irregularities of the heart action and patient motion are reduced by signal averaging. This procedure is usually sufficient for phase velocity mapping of healthy volunteers at rest.

Cardiac phases with high accelerations or decelerations are prone to image artifacts, because a small timing error of the signal acquisition causes a marked view-to-view fluctuation. Therefore special attention should be paid to an undisturbed electrocardiographic pattern and a perfect detection of the R-wave peak [25]. Minor changes of the heart rate or even a nonphasic sinus arrhythmia have less impact on the image quality than might be expected. This is because the timings of the vulnerable systolic interval are short and therefore physiologic less affected by heart rate variations [26]. In patients with irregular heart beat intervals imaging with arrhythmia rejection i.e. elimination of the data from the highly irregular heart cycle intervals is preferable. Sampling at regular intervals and retrospective gating with a simultaneously acquired electrocardiogram after the examination circumvents this disadvantage [27, 28]. At the same time contrast and average signal variations in the images as well as the incomplete sampling during the end of diastole is avoided. To ensure a fixed time position of the phase-encoding steps in relation to the R-wave peak after imaging a linear interpolation of subsequent MR-signals is necessary. However, alterations of the cardiac output in the course of the MR examination, as it occurs during exercise and pronounced breathing, will in any case lead to severe view-to-view fluctuations and, hence, image artifacts.

Other flow-related image artifacts are signal voids and distortion of the flow profile geometry. Signal voids occur if the flow and, hence, spin phase angle variation across a voxel is large. With the nonisotropic voxel dimensions common in MR multiple phase imaging this is very likely to occur. With a Taylor series of the time-dependent spin position the general expression for induced phase shift $\Delta\Phi_{tot}$ at the sampling instant t_s due to all orders of motion in the presence of the imaging gradients $\vec{G}(t) = G_x(t) + G_y(t) + G_z(t)$ is given by

$$\Delta\Phi_{tot}(t_s) = \sum_{n=0}^{\infty} \gamma \frac{\vec{r}^{(n)}}{n!} \int_{t_0}^{t_s} \vec{G}(t)(t - t_0)^n \, dt, \tag{5}$$

where $\vec{r}^{(n)}$ is the n-th derivative of the spin position at the instant of the RF pulse application. The contribution to intravoxel phase dispersion due to acceleration and higher-order motion is negligible for in vivo applications at TEs shorter than 10 ms. This is because a shortening of TE yields a cubic reduction of acceleration-induced dephasing for constant $\vec{G}$. The effect of flow in the direction of the phase-encoding gradient is only minor as the first-order moment is low and easy to compensate. The slice selection gradient is usually tailored to measure the velocities in the corresponding direction. As long as the velocities under investigation induce dephasings in a range of 2π radians – which is preferred in order to avoid phase aliasing – the spin phase angle variation across a voxel will only be a fraction of this value. In the frequency-encoding direction the motion sensitivity is extremely high and a major cause of undersired flow effects. According to Eq. (5), there is a quadratic relationship between the first-order moment and the sampling instant. As shown for a calculated example in Fig. 4 the velocity sensitivity of the readout gradient at some sample positions exceeds by far the constant sensitivity of a slice selection gradient for normal applications.

A slight misalignment introducing flow components in the frequency-encoding direction leads, apart from image artifacts, to phase misregistration. Under these circumstances correct flow rate quantification is only possible if the phase error of the echo top sample is corrected. This is either achieved by acquisition and subsequent subtraction of a correcting phase map or by the introduction of a frequency-encoding gradient with GMN. Even so, distortion of the flow profiles in the frequency-encoding direction will still be present, because a simultaneous compensation of all sampling positions is not possible by simple measures. With GMN, shifting the point of velocity insensitivity versus the echo's top position is accompanied by an inevitable trade-off in prolonged TE and increased higher-order motion sensitivity. Furthermore, the prolonged TE causes a marked increase in motion sensitivity for other sampling positions and thus an even more pronounced distortion of the velocity profiles can be expected.

In most situations of in vivo flow, quantification errors by in-plane flow in the readout direction can be avoided by a careful positioning of the imaging plane. For example, when imaging the iliac branch a readout direction orthogonal to the branching plane, i.e., in anteroposterior direction, is preferred.

Quantitative Evaluation of Velocity Phase Maps

To accurately evaluate phase maps it is best to use a high-resolution screen with overlay mode. This allows for adequate magnification of the area of interest and facilitates the manual contouring of the vessel. Other interfering vessels should be excluded for an accurate zero-velocity baseline determination based on the magnitude-weighted phase average of the neighboring tissue. Large arteries tend to change in size and to become displaced during the cardiac cycle and thus the borders must be adapted accordingly [29]. During diastole, when the flow-induced phase is rather low and no contrast between lumen and vessel wall can be distinguished, the magnitude image is useful for outlining the irregular region of interest (IROI). On the other hand, the magnitude image may be unreliable for contouring if plug flow is present, and signal voids caused by signal dispersion in zones with high velocity shear obscure the vessel's true shape.

The program for phase map evaluation should include algorithms to correct for phase aliasing. Primitive procedures such as a general phase shift in order to unwrap the phase leaps or using a mouse-supported image editor are sufficient in most cases. At the vessel boundaries the phase value of the pixels is affected by partial volume effects and velocities are underestimated. This should be considered when calculating the total flow rate by summation of the pixel values especially in vessles where the pixel size is a significant proportion of the vessel diameter [30].

The accuracy of the estimated velocity and flow rate values not only depends on the MR flow sequences but also on the evaluation scheme. Therefore further effort should be put into the development of reliable, observer-independent evaluation programs. For productive clinical use of the MR flow method, automated time-saving evaluation programs are needed.

Clinical Applications

Principally the new MR flow quantitation methods can be divided into the two main fields, velocity mapping and volumetric flow rate measurements. The latter is possible at lower imaging times by saerifycing the resolution. Quantification of flow in the great vessels to estimate the cardiac output and the regurgitant and stenotic heart valve lesions are on the way to become routine clinical methods [4]. In addition valuable physiological informations about the human peripheral circulation [31, 32] thus far not accessible to noninvasive methods will become available in the future. For example the results of angioplasty or of the noninvasive treatment of peripheral vascular disease will become easily and reproducibly quantifiable [33]. Together with magnetic resonance angiography (MRA) the MR flow quantitation should become a powerful tool to assess peripheral vascular pathology. However the present clinical experience is limited. Several authors have focused their investigations on flow measurements in the splanchnic circulation [34, 35]. The problem of an oblique course vessel can be evaded by electronic angulation, whereas the best method of encompassing displacement of the vessels by respiratory motion is real-time MR velocity mapping accomplished in a one-dimensional projection mode [36] or with echo planar imaging (EPI) [38]. Together with dynamic MR spectroscopy (MRS) real-time flow quantification and velocity mapping opens a wide spectrum of new applications, such as the study of the short lived drug effects on the circulation [37].

Until to now it has not been feasible to study cerebrospinal fluid (CSF) circulation noninvasively. Even though the velocities of the CSF are small, fluid production and pulsation in the aqueduct can be studied by MR [40, 41] and may improve our understanding of normal fluid dynamics, pathology and diagnosis of hydrocephalus [42]. Martin et al. [43] proved the feasibility of CSF shunt patency measurements with an MR phase-based method and in this way circumvented the risk of shunt infection by an invasive modality.

To detect infarcted areas in the myocardium the heart wall movement has been investigated by MR phase mapping [3]. Myocardial tagging, as presented by Zerhouni et al. [44], is a new and reliable method to assess in-plane motion of the heart wall.

Finally two-dimensional velocity mapping, which is one of the basic advantages of this method, is able to reveal detailed information about hemodynamics of atherosclerosis not accessible by other techniques. The complicated flow patterns with secondary backflow regimes and mensional phase-based method it become possible even to quantify the regional backflow in zones of high velocity shear. In addition, it is also possible to quantify the regional backflow in the carotid bulb [22, 45]. Further applications include the study of the early atherosclerotic lesions e.g. in vessel bifurcations and bends as well as the late e.g. aneurysmic stages. Due to its reproducibility the MR flow quantitation methods might in particular play an important role in the clinical follow-up of patients undergoing revascularizations. With thin sections or three-dimensional Fourier transforms at a high resolution it should be possible to explore at least in vitro even the most complex flow regimes in convoluted vessel segments. The knowledge of three-dimensional velocity distributions might become of a paramount clinical significance [46].

Perspectives

MR phase velocity mapping is a novel method allowing noninvasive quantitative assessments of regional hemodynamics in patients. The accurate definition of the cross-sectional vascular anatomy and the corresponding flow velocity fields allow detailed analysis of pulsatile flow patterns in the individual peripheral arteries as well as reliable calculations of time- or space-averaged flow velocities and volumetric flow rates to be made. Although the initial results are promising, more experience in a greater number of laboratories will be required to determine the reproducibility and clinical feasibility of this method. To explore the clinical potential of the MR phase velocity mapping, its full integration into vascular MR imaging and angiography protocols will be necessary. For clinically meaningful data interpretation systematic evaluations of regional hemodynamics in health and disease are indispensable. Although a noninvasive analysis of the coronary circulation is still a distant goal, it is already possible today to accurately analyze the peripheral regional hemodynamics noninvasively.

References

1. Lentner C (1990) Geigy scientific tables: heart and circulation (Vol 5). Ciba-Geigy, Basel
2. Axel L (1984) Blood flow effects in magnetic resonance imaging. AJR 143:1157–1166
3. van Dijk P (1984) Direct cardiac NMR imaging of heart wall and flow velocity. J Comput Assist Tomogr 8(3):429–436
4. Underwood SR, Firmin DN, Klipstein RH, Rees RSO, Longmoore DB (1987) Magnetic resonance velocity mapping: clinical application of a new technique. Br Heart J 57:404–412
5. Lipton MJ, Weikl A, Müller E, Reinhardt ER (1987) Measurement of cardiac output in man by magnetic resonance imaging. Sixth Annual Meeting of Society of Magnetic Resonance in Medicine Book of abstracts (vol 1). Berkeley, California, p 352
6. Matsuda T, Shimizu K, Sakurai T, Fujita A, Ohara H, Okamura S, Hashimoto S, Tamaki S, Kawai C (1987) Measurement of aortic blood flow with MR imaging: comparative study with Doppler US. Radiology 162:857–861
7. Maier SE, Meier D, Boesiger P, Moser UT, Vieli A (1989) Human abdominal aorta: comparative measurements of blood flow with MR imaging and multigated Doppler US. Radiology 171:487–492
8. Maier SE (1988) Vergleichende Blutflußmessungen an der Aorta abdominalis mittels Magnetresonanz und Ultraschall-Doppler. MD thesis, Universität Zürich
9. Anliker M, Casty M, Friedli P, Kubli R, Keller H (1977) Cardiovascular flow dynamics and measurements; noninvasive measurement of blood flow. University Park, Baltimore
10. Meier D, Maier S, Boesiger P (1988) Quantitative flow measurements on phantoms and on blood vessels with MR. Magn Reson Med 8(1):25–34
11. Nordell B, Ståhlberg F, Ericsson A, Ranta C (1988) A rotating phantom for the study of flow effects in MR imaging
12. Segadal L, Matre K (1987) Blood velocity distribution in the human ascending aorta. Circulation 76(1):90–100
13. Vieli A, Jenni R, Anliker M (1986) Spatial velocity distributions in the ascending aorta of healthy humans and cardiac patients. IEEE Trans Biomed Eng BME 33(1)
14. Latham RD, Westerhof N, Sipkema P, Rubal BJ, Reuderink P, Murgo JP (1985) Regional wave travel and reflections along the human aorta: a study with six simultaneous micromanometric pressures. Circulation 72:1257–1269
15. Ernst R, Anderson WA (1966) Application of Fourier transform spectroscopy to magnetic resonance. Rev Sci Instr 37(1):93–102
16. Conturo TE, Smith GD (1990) Signal-to-noise in phase angle reconstruction: dynamic range extension using phase reference offsets. Magn Reson Med 15:420–437
17. McDonald DA (1974) Blood flow in arteries (2nd edn). Edward Arnold, London
18. Vieli A, Moser U, Maier S, Meier D, Boesiger P (1989) Velocity profiles in the normal human abdominal aorta: a comparison between ultrasound and magnetic resonance data. Ultrasound Med Biol 15(2):113–119
19. Nayler GL, Firmin DN, Longmoore DB (1986) Blood flow imaging by cine magnetic resonance. J Comput Assist Tomogr 10:715–722

20. Scheidegger MB, Maier SE, Boesiger P (1991) FID acquired echoes (FAcE): A short echo time imaging method for flow artefact suppression. Magn Reson Imaging (in press)
21. Maier SE, Scheidegger M, Liu K, Boesiger P (1990) FID-acquired-echoes (FAcE) for quantitative flow measurements. 9th Annual Meeting of Society of Magnetic Resonance in Medicine, Book of abstracts (Vol 1). Berkeley, California
22. Maier SE, Scheidegger M, Liu K, Boesiger P (1990) Flow measurements in the carotid artery. 9th Annual Meeting of Society of Magnetic Resonance in Medicine, Works in progress, Berkeley, California
23. Buonocore MH, Bogren H (1990) Methodology for realistic and repeatable blood flow measurements using velocity encoded phase imaging. 9th Annual Meeting of Society of Magnetic Resonance in Medicine, Book of abstracts (vol 1). Berkeley, California
24. Ehman RL, Felmlee JP (1990) Flow artifact reduction in MRI: a review of the roles of gradient moment nulling and spatial presaturation. Magn Reson Med 14:293–307
25. Wendt RE, Rokey R, Vick W, Johnston DL (1988) Electrocardiographic gating and monitoring in NMR imaging. Magn Reson Imaging 6:88–95
26. Ashmann R, Hull E (1941) Essentials of electrocardiography (2nd edn). Macmillan, New York
27. Lenz GW, Haacke EM, White RD (1989) Retrospective cardiac gating: a review of technical aspects and future directions. Magn Res Imaging 7:445–455
28. Bohning DE, Groen JP, de Graaf RG, Whitman R, Simon HE (1990) Retrospectively gated aortic flow quantification. 9th Annual Meeting of Society of Magnetic Resonance in Medicine, Book of abstracts (vol 1). Berkeley, California
29. Lanzer P, McKibbin W, Bohning D, Pohost G (1990) Quantitation of abdominal wall dynamics in man by gradient echo NMR imaging. Magn Reson Med 13:407–415
30. Tarnawski M, Porter DA, Graves M, Smith MA (1989) Flow determination in small diameter vessels by magnetic resonance imaging. 8th Annual Meeting of Society of Magnetic Resonance in Medicine, Book of abstracts (vol 2) Berkeley, California
31. Stettler JC, Niederer P, Anliker M (1981) Theoretical analysis of arterial hemodynamics including the influence of bifurcations. Part II: critical evaluation of theoretical model and comparison with noninvasive measurements of flow patterns in normal and pathological cases. Ann Biomed Eng 9(II):165–175
32. Rieu R, Friggi A, Pelissier R (1985) Velocity distribution along an elastic model of human arterial tree. Biomechanics 18(9):703–715
33. Koch M(anola[1]), Maier SE, Baumgartner I(ris), Hagspiel KD, von Weymarn C, Boesiger P, Bollinger A, von Schulthess GK (1991) Magnetic resonance angiography and flow quantification in peripheral vessel disease before and after percutaneous transluminal angioplasty (PTA). 10th Annual Meeting of Society of Magnetic Resonance in Medicine, Book of abstracts (Vol 1), Berkeley, California
34. Edelman RR, Zhao B, Liu C, Wentz KV, Mattle HP, McArdle JPFC (1989) MR Angiography and dynamic flow evaluation of the portal venous system. AJR 153:755–760
35. Tamada T, Moriyasu F, Ono S, Shimizu K, Kajimura K, Soh Y, Kawasaki T, Kimura T, Yamashita Y, Someda H, Hamato N, Uchino H (1989) Portal blood flow: measurement with MR imaging. Radiology 173:639–644
36. Maier SE, Scheidegger MB, Tjon-A-Meeuw L, Liu K, Schneider E, Bollinger A, Boesiger P (1991) Flow measurements in the renal arteries. 10th Annual Meeting of Society of Magnetic Resonance in Medicine, Works in progress, Berkeley, California
37. Mueller E, Finelli D, Laub G (1989) Real time quantification of blood flow on a whole body MR unit using the RACE sequence. 8th Annual Meeting of Society of Magnetic Resonance in Medicine, Book of abstracts (vol 2). Berkeley, California
38. Firmin DN, Klipstein RH, Hounsfield GL, Paley MP, Longmoore DB (1989) Echo-planar high-resolution flow velocity mapping. Magn Reson Med 12:316–327
39. Pennel OJ, Underwood SR, Manzara U, Mohiaddin RH, Poole-Wilson PA, Ell P, Swanton RH, Walker JM, Longmore DB (1990) Magnetic resonance imaging of reversible myocardial ischaemia during dobutamine stress. 9th Annual Meeting of Society of Magnetic Resonance in Medicine, Book of abstracts (vol 1). Berkeley, California

40. Enzmann DR, Pelc NJ (1991) Normal flowpatterns of intracranial and spinal cerebrospinal fluid defined with phase-contrast cine MR imaging. Radiology 178:467–474
41. Ståhlberg F, Møgelvang J, Thomsen C, Nordell B, Stubgaard M, Ericsson A, Sperber G, Greitz D, Larsson H, Henniksen O, Persson B (1989) A method for MR quantification of flow velocities in blood and CSF using interleaved gradient-echo pulse sequences. Magn Reson Imaging 7:655–667
42. Schroth G, Klose U (1989) MRI of CSF flow in normal pressure hydrocephalus. Psychiatry Research 29:289–290
43. Martin A, Drake JM, Lemaire C, Henkelman RM (1989) Cerebrospinal fluid shunts: flow measurements with MR imaging. Radiology 173:243–247
44. Zerhouni EA, Parish DM, Rogers WJ, Yang A, Saphiro EP (1988) Human heart: tagging with MR imaging – a method for noninvasive assessment of myocardial motion. Radiology 169:59–63
45. Liu K, Scheidegger M, Maier SE, Boesiger P (1990) Flow measurements on a bifurcation phantom by using a very short echo time pulse sequence FAcE. 9th Annual Meeting of Society of Magnetic Resonance in Medicine, Book of abstracts (Vol 1). Berkeley, California
46. Perktold K, Resch M (1990) Numerical flow studies in human carotid artery bifurcations: basic discussion of the geometric factor in atherogenesis. J Biomed Eng 12:111–123

Subject Index

MIX
Papier aus verantwortungsvollen Quellen
Paper from responsible sources
FSC® C105338

If you have any concerns about our products,
you can contact us on
ProductSafety@springernature.com

In case Publisher is established outside the EU,
the EU authorized representative is:
Springer Nature Customer Service Center GmbH
Europaplatz 3, 69115 Heidelberg, Germany

Printed by Libri Plureos GmbH
in Hamburg, Germany